Oliver von Bohlen und Halbach and Rolf Dermietzel

Neurotransmitters and Neuromodulators

Handbook of Receptors and Biological Effects

2nd completely revised and enlarged edition

WILEY-VCH Verlag GmbH & Co. KGaA

The Authors

Dr. Oliver von Bohlen und Halbach
Department of Anatomy and Cell Biology
University of Heidelberg
Im Neuenheimer Feld 307
69120 Heidelberg
Germany

Prof. Dr. med. Rolf Dermietzel
Department of Neuroanatomy
and Molecular Brain Research
University of Bochum
Universitätsstr. 150
44780 Bochum
Germany

All books published by Wiley-VCH are carefully produced. Nevertheless, authors, editors, and publisher do not warrant the information contained in these books, including this book, to be free of errors. Readers are advised to keep in mind that statements, data, illustrations, procedural details or other items may inadvertently be inaccurate.

Library of Congress Card No.: applied for

British Library Cataloguing-in-Publication Data
A catalogue record for this book is available from the British Library.

Bibliographic information published by the Deutsche Nationalbibliothek
The Deutsche Nationalbibliothek lists this publication in the Deutsche Nationalbibliografie; detailed bibliographic data are available in the Internet at http://dnb.d-nb.de.

© 2006 WILEY-VCH Verlag GmbH & Co. KGaA, Weinheim, Germany

All rights reserved (including those of translation into other languages). No part of this book may be reproduced in any form – by photoprinting, microfilm, or any other means – nor transmitted or translated into a machine language without written permission from the publishers. Registered names, trademarks, etc. used in this book, even when not specifically marked as such, are not to be considered unprotected by law.

Printed in the Federal Republic of Germany
Printed on acid-free paper

Composition   K+V Fotosatz GmbH, Beerfelden
Printing   betz-druck GmbH, Darmstadt
Bookbinding   Litges & Dopf Buchbinderei GmbH, Heppenheim

ISBN-13: 978-3-527-31307-5
ISBN-10: 3-527-31307-9

*Oliver von Bohlen und Halbach
and Rolf Dermietzel*

**Neurotransmitters and
Neuromodulators**

Related Titles

Thiel, G. (ed.)

Transcription Factors in the Nervous System

Development, Brain Function, and Diseases

2006
ISBN 3-527-31285-4

Becker, C. G., Becker, T. (eds.)

Model Organisms in Spinal Cord Regeneration

2006
ISBN 3-527-31504-7

Bähr, M. (ed.)

Neuroprotection

Models, Mechanisms and Therapies

2004
ISBN 3-527-30816-4

Frings, S., Bradley, J. (eds.)

Transduction Channels in Sensory Cells

2004
ISBN 3-527-30836-9

Smith, C. U. M.

Elements of Molecular Neurobiology

2003
ISBN 0-471-56038-3

Contents

Preface for the Second Edition *XIII*

Preface for the First Edition *XV*

1	**Introduction** *3*	
1.1	Neuroactive Substances *3*	
1.1.1	Neurotransmitters *4*	
1.1.2	Neuromodulators *5*	
1.2	Receptors and Transporters *8*	
1.2.1	Ionotropic Receptors *9*	
1.2.2	Metabotropic Receptors *10*	
1.2.3	Receptor Regulation *13*	
1.2.4	Transporters *14*	
1.3	Distribution and Localization of Neurotransmitters and Neuromodulators *15*	
1.4	The Blood–Brain Barrier *16*	
1.5	Volume Transmission and Wiring Transmission *18*	
2	**Methods** *21*	
2.1	Bio- and Radioisotope Assays *22*	
2.2	Microdialysis and Electrochemical Detection *24*	
2.2.1	Microdialysis *24*	
2.2.2	Electrochemical Detection *24*	
2.3	Chromatography *26*	
2.3.1	Affinity Chromatography *26*	
2.3.2	High Performance Liquid Chromatography *27*	
2.3.3	Proteomics: Multidimensional Protein Identification Technology *28*	
2.4	Autoradiography *30*	
2.5	Immunohistochemical Methods *31*	
2.6	*In situ* Hybridization *33*	
2.7	Staining and Neuroanatomical Tract Tracing *34*	
2.8	Electrophysiology *36*	
2.8.1	*In vivo* Recording *36*	

2.8.2	*In vitro* Recording	37
2.9	**Behavioral Testing**	*41*
2.9.1	Classic Conditioning	41
2.9.2	Operant Conditioning	42
2.9.3	Further Behavioral Tests	42

3 Neurotransmitters 46

3.1	**Acetylcholine**	*46*
3.1.1	General Aspects and History	46
3.1.2	Localization Within the Central Nervous System	47
3.1.3	Biosynthesis and Degradation	48
3.1.4	Receptors and Signal Transduction	49
3.1.5	Biological Effects	55
3.1.6	Neurological Disorders and Neurogenerative Diseases	57
3.2	**Dopamine**	*59*
3.2.1	General Aspects and History	59
3.2.2	Differentiation and Localization of the Dopaminergic System	60
3.2.3	Biosynthesis and Degradation	62
3.2.4	Release, Re-uptake and Degradation	64
3.2.5	Receptors and Signal Transduction	65
3.2.6	Biological Effects	69
3.2.7	Neurological Disorders and Neurodegenerative Diseases	70
3.3	**γ-Amino Butyric Acid**	*75*
3.3.1	General Aspects and History	75
3.3.2	Localization Within the Central Nervous System	75
3.3.3	Biosynthesis and Degradation	76
3.3.4	GABA Transporters	77
3.3.5	Receptors and Signal Transduction	78
3.3.6	Biological Effects	84
3.3.7	Neurological Disorders and Neurogenerative Diseases	85
3.4	**Glutamate and Aspartate**	*90*
3.4.1	General Aspects and History	90
3.4.2	Localization Within the Central Nervous System	91
3.4.3	Biosynthesis and Degradation	92
3.4.4	Transporters	92
3.4.5	Receptors and Signal Transduction	93
3.4.6	Biological Effects	103
3.4.7	Neurological Disorders and Neurodegenerative Diseases	106
3.5	**Glycine**	*108*
3.5.1	General Aspects and History	108
3.5.2	Localization Within the Central Nervous System	109
3.5.3	Biosynthesis and Degradation	109
3.5.4	Receptors and Signal Transduction	110
3.5.5	Biological Effects	113
3.5.6	Neurological Disorders and Neurodegenerative Diseases	113

3.6	**Histamine** *114*	
3.6.1	General Aspects and History *114*	
3.6.2	Localization Within the Central Nervous System *115*	
3.6.3	Biosynthesis and Degradation *116*	
3.6.4	Receptors and Signal Transduction *117*	
3.6.5	Biological Effects *119*	
3.6.6	Neurological Disorders and Neurodegenerative Diseases *121*	
3.7	**Norepinephrine** *123*	
3.7.1	General Aspects and History *123*	
3.7.2	Localization Within the Central Nervous System *123*	
3.7.3	Biosynthesis and Degradation *125*	
3.7.4	Receptors and Signal Transduction *127*	
3.7.5	Biological Effects *129*	
3.7.6	Neurological Disorders and Neurodegenerative Diseases *130*	
3.8	**Serotonin (5-Hydroxytryptamine)** *132*	
3.8.1	General Aspects and History *132*	
3.8.2	Localization Within the Central Nervous System *133*	
3.8.3	Biosynthesis and Degradation *133*	
3.8.4	Receptors and Signal Transduction *135*	
3.8.5	Biological Effects *139*	
3.8.6	Neurological Disorders and Neurodegenerative Diseases *139*	
4	**Neuromodulators** *144*	
4.1	**Adrenocorticotropic Hormone** *144*	
4.1.1	General Aspects and History *144*	
4.1.2	Localization Within the Central Nervous System *145*	
4.1.3	Biosynthesis and Degradation *145*	
4.1.4	Receptors and Signal Transduction *145*	
4.1.5	Biological Effects *146*	
4.1.6	Neurological Disorders and Neurodegenerative Diseases *147*	
4.2	**Anandamide (Endocannabinoids)** *149*	
4.2.1	General Aspects and History *149*	
4.2.2	Localization Within the Central Nervous System *150*	
4.2.3	Biosynthesis and Degradation *150*	
4.2.4	Receptors and Signal Transduction *152*	
4.2.5	Biological Effects *155*	
4.2.6	Neurological Disorders and Neurodegenerative Diseases *156*	
4.3	**Angiotensin** *158*	
4.3.1	General Aspects and History *158*	
4.3.2	Localization Within the Central Nervous System *159*	
4.3.3	Biosynthesis and Degradation *159*	
4.3.4	Receptors and Signal Transduction *162*	
4.3.5	Biological Effects *165*	
4.3.6	Neurological Disorders and Neurodegenerative Diseases *166*	
4.4	**Atrial Natriuretic Factor** *169*	

4.4.1	General Aspects and History	169
4.4.2	Localization Within the Central Nervous System	169
4.4.3	Biosynthesis and Degradation	170
4.4.4	Receptors and Signal Transduction	171
4.4.5	Biological Effects	172
4.4.6	Neurological Disorders and Neurodegenerative Diseases	173
4.5	**Bombesin and Related Neuropeptides**	**175**
4.5.1	General Aspects and History	175
4.5.2	Localization Within the Central Nervous System	175
4.5.3	Biosynthesis and Degradation	176
4.5.4	Receptors and Signal Transduction	176
4.5.5	Biological Effects	177
4.6	**Calcitonin and Calcitonin Gene-related Protein**	**178**
4.6.1	General Aspects and History	178
4.6.2	Localization Within the Central Nervous System	179
4.6.3	Biosynthesis and Degradation	180
4.6.4	Receptors and Signal Transduction	180
4.6.5	Biological Effects	182
4.6.6	Neurological Disorders and Neurodegenerative Diseases	183
4.7	**Cholecystokinin**	**184**
4.7.1	General Aspects and History	184
4.7.2	Localization Within the Central Nervous System	185
4.7.3	Biosynthesis and Degradation	186
4.7.4	Receptors and Signal Transduction	187
4.7.5	Biological Effects	188
4.7.6	Neurological Disorders and Neurodegenerative Diseases	190
4.8	**Corticotropin-releasing Factor**	**192**
4.8.1	General Aspects and History	192
4.8.2	Localization Within the Central Nervous System	192
4.8.3	Biosynthesis and Degradation	194
4.8.4	Receptors and Signal Transduction	195
4.8.5	Biological Effects	196
4.8.6	Neurological Disorders and Neurodegenerative Diseases	196
4.9	**Dynorphin**	**199**
4.9.1	General Aspects and History	199
4.9.2	Localization Within the Central Nervous System	200
4.9.3	Biosynthesis and Degradation	201
4.9.4	Receptors and Signal Transduction	202
4.9.5	Biological Effects	203
4.9.6	Neurological Disorders and Neurodegenerative Diseases	203
4.10	**Eicosanoids and Arachidonic Acid**	**205**
4.10.1	General and History Aspects	205
4.10.2	Biosynthesis and Degradation	206
4.10.3	Receptors and Signal Transduction	207
4.10.4	Biological Effects	208

4.10.5	Neurological Disorders and Neurodegenerative Diseases	208
4.11	**Endorphin** *211*	
4.11.1	General Aspects and History	211
4.11.2	Localization Within the Central Nervous System	211
4.11.3	Biosynthesis and Degradation	211
4.11.4	Biological Effects	214
4.11.5	Neurological Disorders and Neurodegenerative Diseases	215
4.12	**Enkephalin** *216*	
4.12.1	General Aspects and History	216
4.12.2	Localization Within the Central Nervous System	217
4.12.3	Biosynthesis and Degradation	218
4.12.4	Receptors and Signal Transduction	219
4.12.5	Biological Effects	219
4.12.6	Neurological Disorders and Neurodegenerative Diseases	220
4.13	**Fibroblast Growth Factors** *221*	
4.13.1	General Aspects and History	221
4.13.2	Localization Within the Central Nervous System	221
4.13.3	Biosynthesis and Degradation	223
4.13.4	Receptors and Signal Transduction	224
4.13.5	Biological Effects	225
4.13.6	Neurological Disorders and Neurodegenerative Diseases	227
4.14	**Galanin** *229*	
4.14.1	General Aspects and History	229
4.14.2	Localization Within the Central Nervous System	229
4.14.3	Biosynthesis and Degradation	229
4.14.4	Receptors and Signal Transduction	231
4.14.5	Biological Effects	233
4.14.6	Neurological Disorders and Neurodegenerative Diseases	234
4.15	**Ghrelin** *235*	
4.15.1	General Aspects and History	235
4.15.2	Localization Within the Central Nervous System	236
4.15.3	Biosynthesis and Degradation	236
4.15.4	Receptors and Signal Transduction	237
4.15.5	Biological Effects	238
4.15.6	Neurological Disorders and Neurodegenerative Diseases	239
4.16	**Gonadotropin-releasing Hormone** *240*	
4.16.1	General Aspects and History	240
4.16.2	Localization Within the Central Nervous System	240
4.16.3	Biosynthesis and Degradation	241
4.16.4	Receptors and Signal Transduction	242
4.16.5	Biological Effects	242
4.16.6	Neurological Disorders and Neurodegenerative Diseases	243
4.17	**Growth Hormone-releasing Hormone** *244*	
4.17.1	General Aspects and History	244
4.17.2	Localization Within the Central Nervous System	244

4.17.3	Biosynthesis and Degradation	245
4.17.4	Receptors and Signal Transduction	246
4.17.5	Biological Effects	246
4.17.6	Neurological Disorders and Neurodegenerative Diseases	247
4.18	**Hypocretin (Orexin)**	248
4.18.1	General Aspects and History	248
4.18.2	Localization Within the Central Nervous System	249
4.18.3	Biosynthesis and Degradation	250
4.18.4	Receptors and Signal Transduction	251
4.18.5	Biological Effects	251
4.18.6	Neurological Disorders and Neurodegenerative Diseases	252
4.19	**Interleukin**	254
4.19.1	General Aspects	254
4.19.2	Localization Within the Central Nervous System	255
4.19.3	Biosynthesis and Degradation	255
4.19.4	Receptors and Signal Transduction	256
4.19.5	Biological Effects	258
4.19.6	Neurological Disorders and Neurodegenerative Diseases	259
4.20	**Melanin-concentrating Hormone**	261
4.20.1	General Aspects and History	261
4.20.2	Localization Within the Central Nervous System	262
4.20.3	Biosynthesis and Degradation	262
4.20.4	Receptors and Signal Transduction	263
4.20.5	Biological Effects	263
4.20.6	Neurological Disorders and Neurodegenerative Diseases	265
4.21	**Melanocyte-stimulating Hormone**	266
4.21.1	General Aspects	266
4.21.2	Localization Within the Central Nervous System	267
4.21.3	Biosynthesis and Degradation	267
4.21.4	Receptors and Signal Transduction	267
4.21.5	Biological Effects	269
4.21.6	Neurological Disorders and Neurodegenerative Diseases	270
4.22	**Neuropeptide Y**	272
4.22.1	General Aspects and History	272
4.22.2	Localization Within the Central Nervous System	273
4.22.3	Biosynthesis and Degradation	274
4.22.4	Receptors and Signal Transduction	275
4.22.5	Biological Effects	276
4.22.6	Neurological Disorders and Neurodegenerative Diseases	277
4.23	**Neurotensin**	279
4.23.1	General Aspects and History	279
4.23.2	Localization Within the Central Nervous System	280
4.23.3	Biosynthesis and Degradation	281
4.23.4	Receptors and Signal Transduction	282
4.23.5	Biological Effects	283

4.23.6	Neurological Disorders and Neurodegenerative Diseases	284
4.24	**Neurotrophins** 286	
4.24.1	General Aspects and History	286
4.24.2	Localization Within the Central Nervous System	287
4.24.3	Biosynthesis and Degradation	288
4.24.4	Receptors and Signal Transduction	289
4.24.5	Biological Effects	292
4.24.6	Neurological Disorders and Neurodegenerative Diseases	293
4.25	**Nitric Oxide and Carbon Monoxide** 295	
4.25.1	General Aspects and History	295
4.25.2	Localization Within the Central Nervous System	297
4.25.3	Biosynthesis and Degradation	297
4.25.4	Receptors and Signal Transduction	299
4.25.5	Biological Functions	301
4.25.6	Neurological Disorders and Neurodegenerative Diseases	303
4.26	**Nociceptin (Orphanin FQ)** 305	
4.26.1	General Aspects and History	305
4.26.2	Localization Within the Central Nervous System	306
4.26.3	Biosynthesis and Degradation	306
4.26.4	Receptors and Signal Transduction	307
4.26.5	Biological Effects	308
4.26.6	Neurological Disorders and Neurodegenerative Diseases	309
4.27	**Pituitary Adenylate Cyclase-activating Polypeptide** 311	
4.27.1	General Aspects and History	311
4.27.2	Localization Within the Central Nervous System	312
4.27.3	Biosynthesis and Degradation	312
4.27.4	Receptors and Signal Transduction	312
4.27.5	Biological Effects	313
4.27.6	Neurological Disorders and Neurodegenerative Diseases	314
4.28	**Proopiomelanocortin** 315	
4.28.1	General Aspects and History	315
4.28.2	Localization Within the Central Nervous System	316
4.28.3	Biosynthesis and Degradation	316
4.28.4	Receptors and Signal Transduction	318
4.28.5	Biological Effects	318
4.29	**Purines** 320	
4.29.1	General Aspects	320
4.29.2	Biosynthesis and Degradation	321
4.29.3	Receptors and Signal Transduction	321
4.29.4	Biological Effects	324
4.29.5	Neurological Disorders and Neurodegenerative Diseases	325
4.30	**Somatostatin** 326	
4.30.1	General Aspects and History	326
4.30.2	Localization Within the Central Nervous System	327
4.30.3	Biosynthesis and Degradation	328

4.30.4	Receptors and Signal Transduction	*329*
4.30.5	Biological Effects	*331*
4.30.6	Neurological Disorders and Neurodegenerative Diseases	*332*
4.31	**Substance P and Tachykinins**	*334*
4.31.1	General Aspects and History	*334*
4.31.2	Localization Within the Central Nervous System	*335*
4.31.3	Biosynthesis and Degradation	*335*
4.31.4	Receptors and Signal Transduction	*336*
4.31.5	Biological Effects	*338*
4.31.6	Neurological Disorders and Neurodegenerative Diseases	*339*
4.32	**Thyrotropin-releasing Hormone**	*341*
4.32.1	General Aspects and History	*341*
4.32.2	Localization Within the Central Nervous System	*342*
4.32.3	Biosynthesis and Degradation	*343*
4.32.4	Receptors and Signal Transduction	*344*
4.32.5	Biological Effects	*345*
4.32.6	Neurological Disorders and Neurodegenerative Diseases	*346*
4.33	**The Tyr-MIF-1 Family**	*347*
4.34	**Vasoactive Intestinal Polypeptide**	*349*
4.34.1	General Aspects and History	*349*
4.34.2	Localization Within the Central Nervous System	*350*
4.34.3	Biosynthesis and Degradation	*351*
4.34.4	Receptors and Signal Transduction	*351*
4.34.5	Biological Effects	*352*
4.34.6	Neurological Disorders and Neurodegenerative Diseases	*354*
4.35	**Vasopressin and Oxytocin**	*355*
4.35.1	General Aspects and History	*355*
4.35.2	Localization Within the Central Nervous System	*356*
4.35.3	Biosynthesis and Degradation	*358*
4.35.4	Receptors and Signal Transduction	*359*
4.35.5	Biological Effects	*360*
4.35.6	Neurological Disorders and Neurodegenerative Diseases	*363*
4.36	**Deorphanized Neuropeptides**	*364*
4.36.1	Apelin	*365*
4.36.2	Kisspeptin/Metastin	*366*
4.36.3	Opiod-modulating Peptides (NPFF and NPAF)	*367*
A	**Appendix**	
A1	Amino Acids	*369*
A2	Nucleotides	*372*
A3	Abbreviations for Neurotransmitters and Neuromodulators	*372*
A4	Miscellaneous Abbreviations (Enzymes and Transporters)	*374*

Subject Index *377*

Preface for the Second Edition

The overwhelming success of our handbook prompted us to embark on a second edition which keeps in line with the original concept of the book, namely to provide a comprehensive source of information on the extraordinary complex field of neurotransmitter and neuromodulator chemistry and function. Apparently the "handyness" of the format was one of the key features that made the book so popular. In this new edition we have updated the *Neurotransmitters* and *Neuromodulators* parts which now include, among others, some additional paragraphs such as D-amino acids and their relationship to the NMDA receptor, and the endocannabinoids. In addition, each chapter has been extended by a paragraph on *Neurological Disorders and Neurodegenerative Diseases*, and the list of references for further reading is renewed. Extensive progress has also been made concerning the uncovering and functional identification of growth factors and neuropeptides with neuromodulatory functions. Here we have concentrated on some factors which gained increasing attention during recent years, such as the family of fibroblast growth factors, the neurotrophins and new neuropeptides of the G-protein coupled receptors (GPCRs) like the ghrelin and the hypocretin/orexin family.

The make up of the book has also been improved by redrawing all sketches which depict the distribution of transmitters and neuropeptides. Coloring of most of the diagrammatic representations made the outfit more appealing.

We thus hope that the new edition keeps track with the speed of progress in this field and that the reader will find the requested information in a straightforward and comfortable way.

Finally, we would like to express our gratitude to Dipl. Biol. Helga Schulze who spent extraordinary efforts on the drawing of almost all the new graphical art work and the redrawing of the old sketches of neurotransmitter and neuromodulator distribution. We also would like to thank Dr. Georg Zoidl who contributed a section on proteomics in the *Methods* part.

January 2006　　　　　　　　　　　　　　PD Oliver von Bohlen und Halbach
　　　　　　　　　　　　　　　　　　　　　Prof. Dr. Rolf Dermietzel

Preface for the First Edition

A handbook on neurotransmitters and neuromodulators is necessarily a work in progress. For example, although our knowledge about neuromodulators – their numbers, molecular composition and some of their functions – has increased considerably over the past ten years, the tempo of progress in this field is likely to increase in the near future as the results of genomic cloning and proteomic research become available. Quite apart from these developments, there is now a need for a comprehensive source of information in a format, which is convenient and accessible. For this reason this handbook was conceived as a "handy" book, allowing the reader rapid access to essential information on the main classes of neurotransmitters and neuromodulators. In order to draw informative profiles, we decided to concentrate on certain basic features which characterize each class of substance, namely: *General Aspects and History, Localization, Biosynthesis and Degradation, Receptors and Signal Transduction* and *Biological Effects*. This concept is kept throughout the book. Each chapter is followed by a brief list of *Further Reading*. Although such a strict guideline places some restrictions on the flow of information and results in some overlap, we nevertheless considered it helpful in providing both a broad outline of each substance and also points of comparison between them. Necessarily, this concept is encylopaedic, but nothing more was intended. The absence of detailed discussion and justification for many statements has inevitably given the book a somewhat dogmatic tone, but the bibliographies provide further sources of information for readers wishing to pursue particular issues.

We expect this book to be a useful companion on the laboratory shelf rather than a heavyweight in a librarian's cupboard. We would appreciate comments about the work from colleagues, and from junior and senior students.

This book is dedicated to all of those who suffered from our absence during the process of data collection and writing, especially our families, our friends and coworkers.

June 2001

O. von Bohlen und Halbach
R. Dermietzel
D. Ballantyne

1
Introduction

The paramount functional property of nerve cells is their ability to receive, conduct and store information. Although diverse cell types can perform one or other of these functions most effectively, neurons are unique insofar as they integrate the ability of information transmission with network behavior, which accounts for experience-dependent mechanisms such as memory storage, learning and consciousness. In order to perform these tasks, neurons are structurally and functionally polarized. This is apparent in their tripartite structural differentiation into a cell soma, an axon and dendrites. While the soma harbors the biosynthetic machinery – the nucleus, ribosomes, the endoplasmatic reticulum and the Golgi apparatus including mitochondria for energy supply – the axon is furnished with molecular and subcellular components for the propagation of action potentials away from the cell body to distant targets. Dendrites constitute sets of branched cytoplasmic processes that extend from the cell body and result in an enlargement of the soma for signal reception.

The crucial structural links for signal transmission between neurons are specialized junctions, which are referred to as synapses. Signal transmission between chemical synapses involves the release from presynaptically located sites of molecules (neurotransmitter or neuromodulator) which bind to receptors in the membrane of the target cell, the postsynaptic membrane. Direct transmission of action potentials has also evolved in the form of electrical synapses, which are constituted by gap junctions.

With the exception of the sites of electrical synapses, nerve cells are electrically isolated from one another by an intersynaptic cleft.

A change in the electric potential of the presynaptic neuron triggers the release of a chemical substance, which diffuses across the synaptic cleft and elicits an electrical change at the postsynaptic neuron.

The release of neuroactive substances is linked to the arrival of an action potential at the presynaptic terminal, which elicits the opening of voltage-dependent Ca^{2+} channels (so-called L-type channels). The increase of Ca^{2+} in the presynaptic terminals is the key event that activates the molecular machinery for excocytosis of synaptic vesicles and ultimately triggers the release of the neuroactive molecules. Following diffusion through the intersynaptic cleft, the neurotransmitter binds to the postsynaptic receptor complex. This leads by conformational changes or allosteric mechanisms to the opening of ion channels followed by voltage change at the postsynaptic site.

Depending on the nature of the neuroactive substance and the type of receptor to which they bind, neuroactive substances produce effects that have either a rapid onset and a brief duration or a slow onset and a more prolonged duration.

Neurons are specialized to synthesize a variety of neurotransmitters, and in turn, their activity may be modulated be neurotransmitters released from other neurons. For decades, neurons were believed to constitute monofunctional units with respect to neurotransmitter production and secretion (Dale's principle). However, a large body of evidence now indicates that individual neurons are able to synthesize different neuroactive substances and process them for secretion. This evidence does not, in principle, violate Dale's idea that the neuron is a monofunctional entity, but it does lead to a modification of this paradigm, i.e. the functional phenotype of a differentiated neuron is monospecific in respcet of its neurotransmitter efficacy. The synthesis and release of more than one neuroactive substance from a single neuron substantially augments the range of variability of chemically coded signals. The full significance of this increase is far from being understood. The neuroactive messengers synthesized in an individual neuron belong to two different classes: the neurotransmitters and the neuromodulators.

The first class, the neurotransmitters, includes substances which are responsible for intersynaptic signal transmission, whereas the second, the neuromodulators, exerts a modulatory function on postsynaptic events. Thus, neurons can synthesize and release individual neurotransmitters and are able to produce and release co-transmitter in the form of the neuromodulators.

Brain tissue is composed not only of neurons, but also of supporting cells, the so-called glia. Glial cells are classified into four categories: astrocytes, ependymal cells, microglia cells and oligodendrocytes.

Astrocytes provide mechanical and metabolic support for neurons since they can synthesize and degrade neuroactive substances. They are essential for balancing ion homeostasis and may be involved in neurotransmitter-triggered signaling, thereby constituting a non-neuronal link of spatial signal transmission (Cornell-Bell and Finkbeiner 1991; Cooper 1995). Astrocytes are equipped with neurotransmitter transporters capable of taking up released neuroactive substances and so terminating signals involved by these substances. Glial neurotransmitter transporters also contribute to the synthesis of new neuroactive substances by recycling the captured and/or degraded neurotransmitter metabolites to neurons for reuse.

Ependymal cells line the internal cavities of the central nervous system and seem to play a role in stem cell generation in the central nervous system. They are also involved in controlling volume-transmitted exchanges (see below) of neuroactive substances at the cerebrospinal/interestitial fluid interphase. The primary function of oligodendrocytes is to insulate axons via myelin sheets and thereby provide the cellular substrate for saltatory action potential propagation in the central nervous system. Finally, microglia represent the brain-specific mononuclear macrophage system essential for immune response of brain tissue and repair mechanisms.

Further Reading

Bruzzone, R., Giaume, C. (1999): Connexins and information transfer in glia. *Adv. Exp. Med. Biol.* **468**: 321–337.

Byrne, J. H., Roberts, J. L. (eds) (2004): *From Molecules to Networks*. Elsevier Science, Amsterdam, and Academic Press, London.

Cooper, M. S. (1995): Intracellular signaling in neuronal–glial networks. *Biosystems* **34**: 65–85.

Cornell-Bell, A. H., Finkbeiner, S. M. (1991): Ca^{2+} waves in astrocytes. *Cell Calcium* **12**: 185–204.

Kandel, E. C., Schwartz, H. S., Jessell, T. M. (eds) (2000): *Principles of Neural Sciences*. MacGraw–Hill, New York.

Kettenmann, H., Ransom, B. R. (eds) (2003): *Neuroglia*. Oxford University Press, Oxford.

Rouach, N., Glowinski, J., Giaume, C. (2000): Activity-dependent neuronal control of gap-junction communication in astrocytes. *J. Cell Biol.* **149**: 1513–1526.

Zigmond, M. J., Bloom, F. E., Landis, S. C., Roberts, J. L., Squire, L. R. (eds) (2002) *Fundamental Neuroscience*. Elsevier Science, Amsterdam, and Academic Press, London.

1.1
Neuroactive Substances

A variety of biologically active substances, as well as metabolic intermediates, are capable of inducing neurotransmitter or neuromodulator effects. A large diversity of neuroactive substances regarding their metabolic origin exists. The molecular spectrum of neuroactive substances ranges from ordinary intermediates of amino acid metabolism, like glutamate and GABA, to highly effective peptides, proteohormones and corticoids.

Recent evidence indicates that neuronal messengers convey information in a complex sense entailing a variety of processes. These include:
- reciprocal influence on the synthesis of functionally linked neuronal messengers;
- induction of different temporal patterns in terms of short-term and long-term effects;
- shaping of network topology including synaptic plasticity during long-term potentiation.

Chemical neurotransmission is not restricted to central nervous synapses but occurs in peripheral tissues as well, including neuromuscular and neuroglandular junctions.

Neuroactive molecules target receptors with pharmacologically different profiles. The existence of multiple sets of neuronal receptors for a single neurotransmitter seems to be the rule rather than the exception. This receptor multiplicity seems to mirror at molecular level the functional diversity of neuronal networks.

Although functional overlap between neurotransmitters and neuromodulators is quite common, this classification has proven useful for practical purposes.

1.1.1
Neurotransmitters

Neurotransmitters are the most common class of chemical messengers in the nervous system. A neuroactive substance has to fulfill certain criteria before it can be classified as a neurotransmitter (R. Werman 1966).

- It must be of neuronal origin and accumulate in presynaptic terminals, from where it is released upon depolarization.
- The released neurotransmitter must induce postsynaptic effects upon its target cell, which are mediated by neurotransmitter-specific receptors.
- The substance must be metabolically inactivated or cleared from the synaptic cleft by re-uptake mechanisms.
- Experimental application of the substance to nervous tissue must produce effects comparable to those induced by the naturally occurring neurotransmitter.

A neuroactive substance has to meet all of the above criteria to justify its classification as a neurotransmitter.

Based on their chemical nature, neurotransmitters can be subdivided into two major groups: biogenic amines and small amino acids (Fig. 1.1).

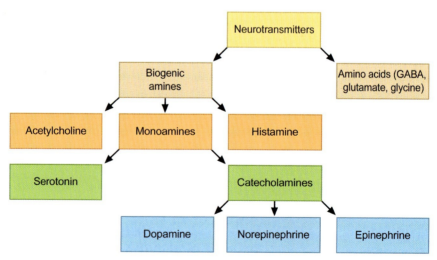

Fig. 1.1 Differentiation of neurotransmitters based on their chemical structures.

1.1.2
Neuromodulators

In contrast to neurotransmitters, neuromodulators can be divided into several subclasses. The largest subclass is composed of neuropeptides. Additional neuromodulators are provided by some neurobiologically active gaseous substances and some derivatives of fatty acid metabolism (for details see Sections 4.2, 4.10, 4.23). The history of the discovery of neuropeptides goes back to the 1960s and 1970s, when it became evident that the regulatory factors of the pituitary gland are peptidergic substances and that some enteric peptides are also synthesized in the central nervous system.

Neuropeptides are synthesized by neurons and released from their presynaptic terminals, as is the case for neurotransmitters. Like neurotransmitters, some of the neuropeptides act at postsynaptic sites, but since they do not meet all of the above criteria they are not classified as such. These neuropeptides are frequently labeled "putative neurotransmitters" (for example: endorphins).

Other neuropeptides are released by neurons, but show no effects on neuronal activity (e.g. follicle-stimulating hormone (FSH) produced in gonadotrophs of the anterior pituitary). These neuropeptides target to tissues in the periphery of the body and can therefore be classified as neurohormones. Consequently, not all neuropeptides function as neuromodulators.

The fact that peptides are synthesized in neurons and are able to induce specific effects via neuronal receptors led de Wied (1987) to formulate the neuropeptide postulate; in essence, this states that peptides which are of neuronal origin and exert effects on neuronal activities are classified as neuropeptides.

The term neuropeptide is no longer used in this restricted sense, because most neuropeptides not only influence neurons, but also non-neuronal tissues. In addition, neuropeptides seem also to link brain activities with other body functions, the most prominent of which is the neuro-immune axis. Thus, neuropeptides are involved in diverse physiological processes, including development, growth, body homeostasis, behavior and immune responses.

Neuropeptides are generated from large precursor molecules. Maturation of the precursor can lead to the formation of one or more peptides (Table 1.1). The biosynthesis requires a cascade of cellular processes to translate the genetic information into the generation of the biological active substance.

The synthesized neuropeptides are stored in vesicles and released upon an adequate stimulus. Different neuropeptides have been found to coexist in a single neuron. Equally, colocalization of neuropeptides with neurotransmitters is quite common.

In the case of colocalization with neurotransmitters, neuropeptides are capable of modulating the effects of the co-released neurotransmitter(s).

Since the formulation of the neuropeptide postulate, a large number of neuropeptides and receptors have been discovered. To date, more than 50 different neuropeptides have been described which are biologically active. We consider the list of neuromodulators and neuropeptides (shown in Table 1.1) to be far

Table 1.1 The most common neuropeptide precursor molecules which lead to one or more biologically active neuropeptides by proteolytic cleavage.

Neuropeptide family	Precursor	Active peptide
Angiotensin	Angiotensinogen	Angiotensin I
		Angiotensin II
		Angiotensin (1–7)
		Angiotensin IV
Calcitonin I gene (CALC I) products	Pro-CALC I	Calcitonin
	Pro-CGRP I	Calcitonin gene related peptide I (α-CGRP)
Calcitonin II gene (CALC II) products	Pro-CGRP II	Calcitonin gene related peptide II (β-CGRP)
Cholecystokinin (CCK)	Pro-CCK	CCK-8
		CCK-33
		CCK-58
Dynorphin gene products	Prodynorphin	Dynorphin A
		Dynorphin B
		α-Neoendorphin
		β-Neoendorphin
Enkephalin gene products	Proenkephalin	Met-enkephalin
Melanin-concentrating hormone gene products	Pro-MCH	MCH
		Neuropeptide Glu-Ile (NEI)
		Neuropeptide Gly-Glu (NGE)
Neurotensin gene products	Proneurotensin	Neurotensin (NT)
		Neuromedin N
Preprotachykinin A (PPTA) gene products	PPTA	Substance P
		Neuropeptide K
		Neurokinin A
		Neuropeptide γ
Preprotachykinin B (PPTB) gene products	PPTB	Neuromedin K
Proopiomelanocortin (POMC)	POMC	α-Melanocyte-stimulating hormone (α-MSH)
		β-Melanocyte-stimulating hormone (β-MSH)
		ACTH
		β-Endorphin
		α-Endorphin
		γ-endorphin
		β-Lipoprotein (β-LPH)
		Corticotropin-like intermediate peptide (CLIP)

Table 1.1 (continued)

Neuropeptide family	Precursor	Active peptide
Somastostatin	Prosomatostatin	Somatostatin-12
		Somatostatin-14
		Somatostatin-28
Vasoactive intestinal peptide gene products	Pro-VIP	Vasoactive intestinal peptide (VIP)
		PHM-27/PHI-27
Vasopressin gene products	Provasopressin	Vasopressin (VP)
		Neurophysin II (NP II)

from being complete and expect an increase in number in the near future, including the precursor molecules. Some derivatives of fatty acids are also known to elicit biological effects in neurons. These lipid neuromodulators have effects which are functionally similar to that of neuropeptides. Like neuropeptides, the neuroactive fatty acids are generated from precursors by a sequence of diverse enzymatic steps.

Similar to neurotransmitters and neuropeptides, the neuroactive derivatives of fatty acids bind to membranous receptors, which leads to downstream signal transduction. Some diffusable gases, like nitric oxide and carbon monoxide, can also act as chemical messengers. The gaseous messengers are generated intracellularly by the activity of specific sets of enzymes. Because of their short half-life and diffusion-dependent radius of activity, they operate in restricted areas, mostly in the proximity of their synthesizing neurons.

In contrast to neurotransmitters and most of the common neuromodulators, nitric oxide does not bind to its own high-specific membraneous receptors because its lipophilic character allows it to pass membranes freely (however, it could bind to guanylyl-cyclase).

Further Reading

Byrne, J. H., Roberts, J. L. (eds) (2004): *From Molecules to Networks*. Elsevier Science, Amsterdam, and Academic Press, London.

Brogden, K. A., Guthmiller, J. M., Salzer, M., Zasloff, M. (2005): The nervous system and innate immunity: the neuropeptide connection. *Nat. Immunol.* **6**: 558–564.

de Wied, D. (1987): The neuropeptide concept. *Prog. Brain Res.* **72**: 93–108.

de Wied, D., Diamant, M., Fodor, M. (1993): Central nervous system effects of the neurohypophyseal hormones and related peptides. *Neuroendocrinology* **14**: 251–302.

Hökfelt, T. (1991): Neuropeptides in perspective: the last ten years. *Neuron* **7**: 867–879.

Hökfelt, T. (2003): Neuropeptides: opportunities for drug discovery. *Lancet Neurol.* **2**: 463–472.

Sossin, W. S., Sweet, C. A., Scheller, R. H. (1990): Dale's hypothesis revisited: different neuropeptides derived from a common prohormone are targetted to different processes. *Proc. Natl Acad. Sci. USA* **87**: 4845–4848.

Werman, R. (1966): Criteria for identification of a central nervous transmitter. *Comp. Biochem. Physiol.* **18**: 745–766.

1.2
Receptors and Transporters

Signal transduction at the cellular level is defined as the transmission of a signal from the outside of a cell into the cell interior. The initial step in signal transduction is performed by the "first messengers", which overcome the barrier of the plasma membrane by interacting with a receptor or a ligand-gated channel. In nervous tissue, "first messengers" are neurotransmitters and neuromodulators, which signal the postsynaptic cell to modify its electrical behavior.

Most common receptors are membrane-bound oligomeric proteins which:
- recognize a ligand with high affinity and selectivity;
- convert the process of recognition into a signal that results in secondary cellular events.

The molecular make-up of a membrane-bound receptor consists of three distinct structural and functional regions: the extracellular domain, the transmembrane-spanning domain and the intracellular domain.

Receptors are characterized by their affinity to the ligand, their selectivity, their number, their saturability and their binding reversibility. So-called isoreceptors form families of structurally and functionally related receptors which interact with the same neuroactive substance. They can be distinguished by their response to pharmacological agonists or antagonists.

Isoforms of receptors can occur in a tissue-restricted manner, but expression of different isoreceptors in the same tissue is also found.

The binding of a ligand to a receptor induces a modification in the topology of the receptor by changing its conformation; and this allows either an ion current to flow, so-called ionotropic receptors, or elicits a cascade of intracellular biochemical events, the metabotropic receptor. Mediators of the intracellular events often consist of "second messengers", like cAMP or cGMP (see below).

The design of intramembraneous receptors is quite variable. Some receptors consist of single polypeptides exhibiting three domains: an intracellular and an extracellular domain linked by a transmembrane segment. Other receptors are also monomeric, but folded in the cell membrane and thus form variable intra- and extracellular as well as transmembrane segments. A large group of receptors consists of polymeric structures with complex tertiary topology.

Receptor categories are completed by cytosolic and nuclear receptors, though these are functionally less relevant so far as neuronal signal transmission is concerned. After the binding of a ligand to a cytosolic receptor, a complex is formed which consists of the receptor and the ligand. This complex is translocated to the cell nucleus and can influence gene transcription. The most common cytosolic receptors are corticoid receptors, which are the targets of the membrane-permeable steroids. Neurotransmitter receptors are commonly located on both the presynaptic and the postsynaptic site and can be ionotropic or metabotropic.

A special class of receptors is referred to as autoreceptors. These are located presynaptically and they bind neuroactive substances released by the presynaptic cell. Through the activation of autoreceptors, the turnover of released neuroactive substances can be modulated by feedback mechanisms. Thus, autoreceptors seem to be involved in limiting transmitter release and thereby balancing the amount of neurotransmitter concentration in the synaptic cleft.

Another class of receptors comprises the so-called heteroreceptors. Heteroreceptors are also located at presynaptic sites. In contrast to autoreceptors, they can bind neuroactive substances different from the neurotransmitter released by the "host" neuron. Heteroreceptors are therefore able to influence the release of neuroactive substances from their own "host" cell after stimulation by a neurotransmitter from a different source.

The primary signal, which is elicited by the binding of the ligand to the receptor, must be amplified to generate the transmembrane signal. Two distinct amplification mechanisms are of major importance:

- A change in membrane potential, resulting from an agonist-induced change in ion flux. The change in ion flux can arise either directly via ligand-gated ion channel receptors or, indirectly, via an effector-mediated change (for example via G proteins) in channel conductance.
- Activation of a phosphorylation–dephosphorylation cascade. Such a cascade can be initiated either directly, via a receptor that possesses intrinsic ligand-modulated tyrosine kinase activity or indirectly via effector-generated signal mediators, such as cAMP, diacylglycerol (DAG), or elevated intracellular calcium, that on their own can activate protein kinases.

On the basis of their signal transduction pathways, receptors can be classified into two major groups. One group consists of receptors belonging to the gated ion channels, the ionotropic receptors.

The other consists of receptors which act through activation of several enzymes and thus need a cascade of enzymatic events for their signal transduction: this is the group of the metabotropic receptors.

1.2.1
Ionotropic Receptors

Signal transduction of membrane-bound receptors can entail a single step consisting of the activation of gated ion channels. In this case, the receptor forms an ion channel. Binding of the ligand to the receptor opens the ion channel and ions can cross the membrane, driven by an electrical gradient. This current produces a net inward flow of positive or negative charge with the consequence of a change in conductance followed by a postsynaptic potential response. The receptor allows the signal to flow (in the form of ions) from outside the cell into the cell or vice versa.

The activation of ionotropic receptors results in fast signal propagation, but the induced electrical effects are mainly short-lasting without metabolic consequences.

1.2.2
Metabotropic Receptors

More complex forms of signal transduction of membrane-bound receptors involve a coupling of ligand and receptor followed by the activation of different intracellular signal transduction pathways.

After binding of the ligand, the signal is transferred into the cell and leads to the activation of "second messengers", like cyclic nucleotides (cAMP, cGMP), calcium (Ca^{2+}), inositol-trisphosphate (IP3), diacylglycerol (DAG), as well as to the phosphorylation of proteins. Phosphorylation involves in many cases the activity of cAMP-dependent protein kinase (PKA) and DAG-activated protein kinases.

The downstream response modifies intracellular processes, for example the release of neuroactive substances, an altered activity of ion channels or changes in enzymatic activity, particularly kinase cascades. The modifications are induced slowly and the resulting effects can be long-lasting.

G protein-coupled receptors

The most prominent group of metabotropic receptors consists of G protein-coupled receptors (GPCRs). These reveal a uniform molecular composition. The extracellular domain is formed by the N-terminal sequence of the receptor, which has potential glycosylation sites. Since this domain is exposed to the extracellular space, it constitutes the ligand-binding site. The tertiary structure of G protein-coupled receptors exhibits a conserved motif of seven membrane-spanning segments, which are connected by alternating extracytoplasmic and cytoplasmic loops. The C-terminal segment is in the cytoplasm (see Fig. 1.2).

The signal arising form the ligand–receptor interaction is forwarded to membrane-bound GDP/GTP-binding proteins inside the cell. These proteins are termed G proteins.

Each G protein is a heterotrimer consisting of a GTP-binding subunit (a-subunit) and two further subunits (β- and γ-subunits). Subclasses of the a-subunit form the different G protein subtypes G_s, $G_{i/o}$ and G_q. In response to receptor activation, the G_s proteins stimulate adenylate cyclase. This stimulation leads to catalytic formation of cAMP from cytosolic ATP.

The most profound effect of cAMP is the activation of cAMP-dependent, calcium-independent protein kinases through binding to the regulatory subunits of the latter. By allosteric mechanisms the kinases alter their conformation into the active enzymatic form. In this activated form they are able to catalyze the transfer of the γ phosphate from ATP to specific amino acid residues, such as serine and threonine for PKA, and tyrosine for PKC, resulting in phosphorylation of the protein and so, as a consequence, resulting in changes in energy equilibrium.

G_i proteins inhibit adenylate cyclase in response to receptor activation. In turn, they can activate K^+ channels and, additionally, they decrease the influx of calcium through voltage-gated Ca^{2+} channels. G_q proteins activate phospholipase C in response to receptor activation.

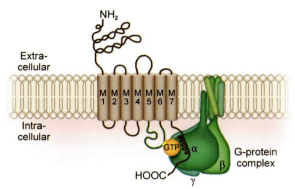

Fig. 1.2 Prototype of a G protein-coupled receptor (GPCR). The domain topology of the single subunits consists of seven transmembrane domains (TM1–TM7) with the amino terminus on the extracellular site and the carboxyl terminus on the cytoplasmic site. GPCRs are homologous to rhodopsin for which detailed structural data are available. The receptor can actively isomerize between the inactive and active state. A receptor domain of the third intracytoplasmic loop significantly affects the specificity of the G protein coupling and contains the main site for receptor coupling to G proteins. Each G protein is a heterotrimer consisting of a GTP-binding subunit (α subunit) and two further subunits (β and γ subunits). For further details see text and Fig. 3.15.

Phospholipase C hydrolyzes the membrane-bound lipid phosphatidyl-4,5-bisphosphate (PIP2) to form diacylglycerol (DAG) and inositol-trisphosphate (IP3). Subsequently, the cytosolic IP3 binds to specific receptors on the endoplasmic reticulum (ER) where it induces the opening of ER-bound Ca^{2+} channels and the release of Ca^{2+} from this store into the cytoplasm. Intracellular elevation of calcium elicits multiple effects on cellular metabolism. These include the activation of Ca^{2+}-dependent enzymes, e.g. calmodulin and its related kinases, activation of motor ATPases for vesicle trafficking, a mechanism that underlies exocytosis of neurotransmitter and receptor dynamics at postsynaptic sites.

Cytosolic DAG binds to cAMP-independent protein kinase C (PKC) and leads to its activation. In addition, DAG induces the opening of plasmalemmal Ca^{2+} channels and thus augments the effects of IP3.

Not all metabotropic receptors target to the G protein system for signal transduction. Other signal transduction systems take advantage of pathways involving guanylate cyclase or tyrosine kinase.

Deorphanizing GPCRs

A total of about 800 GPCRs are found in the human genome, which seems to represent the largest of all gene families. GPCRs can bind a variety of ligands which can be classified into chemosensory receptors, which bind olfactory-gustatory ligands, chemokines and chemoattractants, and the neuromodulator receptors. Some 367 receptors have been found in the human genome on the basis of their sequence homology. Most of them were "orphans" because the ligands were unknown. "Deorphanizing" the GPCRs by exogenous expression in adaequate expression systems has been used to identify a large scale of new potential neuromodulators and neurotransmitters. The first to be found were the serotonin receptor 5-HT$_{1A}$ and the dopamine D2 receptors. By the mid-1990s, about 150 GPCRs had been paired to 75 known transmitters. Since the number of potential transmitters was about 15 whereas ~200 GPCRs remained orphans, there was a need to identify further new ligands. The use of tissue extracts as a source of new transmitters instead of potential transmitters was rendered successful with the discovery of a new neuropeptide, nociceptin/orphanin FQ (see Section 4.24), as the neuroactive ligand of the GPCR ORL-1.

Deorphanization of GPCRs is still an expanding field of pharmacological reseach and the estimated number of deorphanized GPCRs are ~7–8 per year. Some of the most prominent ligands paired by this approach are: hypocretins/orexins, prolactin-releasing peptide, apelin, ghrelin, metastin, neuropeptide B, neuropeptide W and neuropeptide S.

Guanylate cyclase-coupled receptors
Cyclic guanosine monophosphate (cGMP) resembles cAMP in its chemical composition, with a substitution of guanosine for adenosine as nucleotide.

In contrast to the membrane-bound adenylate cyclase, guanylate cyclase occurs in both membrane-bound and cytosolic forms.

The two forms of guanosine cyclases have different functions and follow different routes of activation. The soluble form is activated by nitric oxide (NO) and by free radicals, whereas the membrane-bound form is a part of a transmembranous receptor. Cyclic GMP activates a specific cGMP-dependent protein kinase (PKG) and leads to a downstream activation of a 23-kDa protein (known as G substrate).

It is thought that the cGMP-induced signaling pathway involves the inhibition of phosphatases through the G substrate, so prolonging the effects of phosphorylation, catalyzed by other signal transduction cascades.

Tyrosine kinase-coupled receptors
The molecular backbone of tyrosine kinase-coupled receptors is a single membrane-spanning polypeptide which separates the N-terminal segment from the cytoplasmatic C-terminal domain.

The binding of a ligand to the extracellular domain is followed by receptor activation. Tyrosine kinases initiate a cascade of intracellular phosphorylation steps.

Prominent members of the tyrosine kinase-coupled receptor family are the insulin receptor and receptors for diverse growth factors, e.g. the group of neurotrophic growth factors.

Cytokine receptors
Cytokine receptors are classified into four families (type I, II, III and IV) but only type I receptors seem to be expressed in the central nervous system. The prolactin receptor, some interleukin receptors and growth hormone (GH) recep-

tors belong to the type I cytokine receptors. The exact signal transduction pathways of cytokine receptors are largely unknown and are the subject of intensive investigation.

1.2.3
Receptor Regulation

The affinity of receptors to their appropriate ligands can be decreased by various mechanisms, one of which involves allosteric changes in the receptor molecule, which can result in decreased affinity. Binding of the ligand to its receptor is reversible. The dissociation of the ligand from the receptor results in a decay of the evoked effect after its removal or inactivation.

Receptors comprise highly dynamic moieties, in terms of lateral mobility, membrane insertion and turnover. The concept of a receptor as a "mobile" or "floating" intramembranous constituent has evolved along with the understanding of the general properties of cell surface proteins. Receptors, like other cell surface constituents, can perform complex protein–protein interactions, resulting in the internalization and intracellular redistribution of both, the receptor and the ligand.

Neuromodulators as well as neurotransmitters can influence the number and sensitivity of receptors. Under normal physiological conditions, the number of receptors is finite and the receptors are saturable. Depending on the concentration of neuroactive substances, the number can increase (up-regulation) or decrease (down-regulation).

For instance, the chronic presence of receptor antagonist can lead to an increase in receptor number, which is often accompanied by an increase in sensitivity to the specific agonist.

In contrast, receptor down-regulation occurs in response to continuous stimulation (for example due to the chronic administration of an agonist) and is frequently accompanied by desensitization.

The regulation of receptor sensitivity is orchestrated by several feedback mechanisms, which may involve the receptor itself (i.e. by allosteric mechanisms at the level of the receptor itself), or subsequent steps in the signal transduction pathways, which result in a reduction in receptor number, a loss of coupling to G proteins or phosphorylation of the receptor by protein kinases. The latter mechanism induces a conformational change in the receptor, which modifies its affinity to the ligand.

Desensitization can also occur after binding of the ligand to the receptor and subsequent internalization of the ligand–receptor complex by endocytosis. This internalization rapidly limits the duration of the signal of the "first messenger".

Once the endosome carrying the ligand–receptor complex is transported intracellularly, it becomes acidified by an ATP-dependent mechanism. This results in dissociation of the ligand from the receptor. The receptors are further targeted to the Golgi apparatus, where sorting takes place, with either partial degradation or recycling of the receptors to the cell membrane.

1.2.4
Transporters

In order to fuel cells with essential hydrophilic metabolites, transporter systems are required which overcome the phospholipid barrier of the plasma membrane. At nerve terminals, transporters have evolved for high-affinity uptake of neurotransmitters and some neuromodulators. These transporters control the temporal and spatial concentrations of extracellular transmitters via a rapid re-uptake into the nerve terminals. Equally, astrocytes participate in transporter-mediated uptake of neurotransmitters, serving a kind of servo-function in extracellular neurotransmitter cleansing. It is assumed that as much as 80% of released GABA is recaptured by GABA-transporters, indicating that transporter systems play an essential role in controlling the concentration of released neurotransmitters.

Termination of the activity of neuroactive substances is an essential prerequisite for controlled neuronal excitation. For instance, overexcitation of neurons induced by some excitatory neurotransmitter, which are not recaptured adequately, can exert severe damage to neurons (so-called "excitotoxity"). Glutamate, for example, is the most widespread excitatory neurotransmitter in the central nervous system and influences numerous neuronal networks. To limit receptor activation during signaling and to prevent overstimulation of glutamate receptors that would trigger excitotoxic mechanisms and cell death, extracellular concentration of glutamate is strictly controlled by transport systems from both neuronal and glial sites.

Further Reading

Barry, P.H., Lynch, J.W. (2005): Ligand-gated channels. *IEEE Trans Nanobiosci.* **4**: 70–80.
Bolander, F.F. (1989): *Molecular Endocrinology*. Academic Press, San Diego.
Civelli, O. (2005): GPCR deorpanization: the novel, the known and the unexpected neurotransmitter. *Trends Pharmacol. Sci.* **26**: 15–19.
Kennedy, M.B. (1992): Second messengers and neuronal function. In: *Effects and Mechanisms of Action*, vol. 3, ed. Negro-Vilar, A., Conn, P.M., CRC Press, Boca Raton, pp 120–142.
Keramidas, A., Moorehouse, A.J., Schofield, P.R., Barry, P.H. (2004): Ligand-gated ion channels: mechanisms underlying ion selectivity. *Prog. Biophys. Mol. Biol.* **86**: 161–204.
Kubo, Y., Tateyama, M. (2005): Towards a view of functioning dimeric metabotropic receptors. *Curr. Opin. Neurobiol.* **15**: 289–295.
LeBeau, F.F., El Manira, A., Griller, S. (2005): Tuning the network: modulation of neuronal microcircuits in the spinal cord and the hippocampus. *Trends Neurosci.* **28**: 552–561.
Mons, N., Cooper, D.M.F. (1995): Adenylate cyclases: critical foci in neuronal signaling. *Trends Neurosci.* **18**: 536–542.
Simon, M.I., Strathmann, M.P., Gautam, N. (1991): Diversity of G proteins in signal transduction. *Science* **252**: 802–808.
Sonders, M.S., Quick, M., Javitch, J.A. (2005): How did the neurotransmitter cross the bilayer? A closer view. *Curr. Opin. Neurobiol.* **15**: 296–304.
Vizi, E.S. (2000): Role of high-affinity receptors and membrane transporters in nonsynaptic communication and drug action in the central nervous system. *Pharmacol. Rev.* **52**: 63–89.

1.3
Distribution and Localization of Neurotransmitters and Neuromodulators

Neurotransmitters and neuromodulators reveal a spatially organized distribution in the central nervous system. For instance, glycine is the most abundant inhibitory neurotransmitter in the spinal cord, norepinephrine is focally synthesized in the locus coeruleus of the brain stem and neuropeptides are most abundant in hypothalamic nuclei. When describing neurotransmitter distribution, one has to take into account that axonal transport can carry neuroactive substances anterogradually to reach distant targets. Consequently, high concentrations of neuroactive substances can be found in nerve terminals, well separated from the source of their biosynthesis. Thus, in constructing a descriptive map of neurotransmitter distribution, it is important to differentiate between soma-bound and terminal localization.

Additional techniques, e.g. *in situ* hybridization, provide further insights into a descriptive neurotransmitter-based map of the brain. This technique has the limitation that it provides signals mainly from the cell soma where the highest concentration of cRNA occurs and gives no information about terminal localizations.

The coexistence of neurotransmitter and neuropeptides as originally described by Hökfelt adds an additional level of complexity to a descriptive approach of local maps of neuroactive substances.

The fact that neurotransmitter and neuropeptide coexist in single neurons necessitates the presence of pre- and postsynaptic receptors for both classes of neuroactive substances to induce their effects. In describing the functional relationships of both neuroactive substances, Lundberg and Hökfelt (1985) summarized possible effects which arise from their coexistence:
- A neurotransmitter acts on one type of postsynaptic receptor.
- A neurotransmitter acts on multiple types of postsynaptic receptors.
- A neurotransmitter also acts on presynaptic receptors to affect its own release.

Nerve terminals carry different neuroactive substances, which are stored in small vesicles (neurotransmitters) and in large dense-cored vesicles (neurotransmitter and neuromodulator). If these vesicles are co-released, they can interact in the following ways:
- inhibition of the release of the neuromodulator by the neurotransmitter via presynaptic action;
- inhibition or enhancement of the release of the neurotransmitter by the neuromodulator via presynaptic action;
- modulation of the activity of the postsynaptic neuron via postsynaptic receptors by the neuromodulator.

In some cases, coexpressing neurons exhibit specific projection patterns: this is the case in the basal ganglia.

Gibbins (1989) summarized some principles in neurotransmitter architecture which result from the coexistence of variable neuroactive substances:
- A target cell can be innervated by different types of neurons, which express the same neuroactive substances.
- Functionally closely related neurons can express different neuropeptides.
- Neurons can express biochemically related neuroactive substances, since different neuromodulators can be synthesized from a common precursor.
- Although some different members of a neuromodulator family are generated by differential processing of a common precursor, not all active forms coexist in a single neuron.

Further Reading

Fuxe, K., Grillner, S., Hökfelt, T., Ölson, L., Agnati, L. F. (eds) (1998): *Towards an Understanding of Integrative Brain Functions.* Elsevier, Amsterdam.

Gibbins, I. L. (1989): Co-existence and co-function. In: *The Comparative Physiology of Regulatory Peptides*, ed. Holgren, S., Chapman & Hall, New York, pp 308–343.

Lundberg, J. M., Höckfelt, T. (1985): Coexistence of peptides and classical neurotransmitters. In: *Neurotransmitters in Action*, ed. Bousfield, D., Elsevier, New York, pp 104–118.

1.4
The Blood–Brain Barrier

Neuroactive substances are not exclusively expressed in the central nervous system, but are also common in peripheral tissues. One prominent example is represented by the gastro-intestinal peptides which possess a chimeric function as enteric hormones and neuromodulators. This dual function requires mechanisms that restrict or facilitate the entry of neuroactive substances from the periphery into the central nervous system. The highly regulated system for controlled exchange of neurobiologically relevant substances resides at the border between the general blood circulation and the brain parenchyma in the form of the blood–brain barrier.

The existence of a functional barrier between the blood and the brain was first demonstrated by Paul Ehrlich at the beginning of the 20th century. He observed that, when injected into the circulation, dyes like methylene blue stained the parenchyma of most organs of the body but not the brain. Injection of dyes into the cerebrospinal fluid, however, led to a staining of the brain, but not the body. These experiments were the first demonstration that a barrier between the blood and the brain exists and that this barrier blocks all free transport, regardless of the direction from which the barrier is approached by the substance.

While the capillaries of most non-neural tissues are permeable to molecules smaller than 30 kDa, capillaries in the mammalian central nervous system show a high degree of selectivity concerning permeation coefficients. The permeability of substances into brain is governed by specific chemical properties of the molecules. Although the term blood–brain barrier implies a general impermeability, it is best considered as selectively permeable. The blood–brain barrier does not provide a passive barrier between the brain and the body, but constitutes a kind of active filter that regulates the flow of substances between both compartments by structural and metabolic elements. The most effective structure that restricts the free access of dissolved hyrophilic molecules into the brain is the high-resistance tight junction that seals the interendothelial cleft. The active components, which regulate exchange in both directions, are specific transporters and enzymes which, for example, include amino acid transporters, glucose transporters and transporters for essential elements like the transferrin transporter for iron.

Among the neuroactive substance that are transported through the barrier are some neuromodulators and their analogs. Lipid-soluble substances like alcohol and steroids can penetrate the endothelial barrier freely.

A second barrier exists at the circumventricular organs (CVOs). The CVOs are located close to the ventricles of the brain and they include the *chorioid plexuses,* the median eminence, the *organum vasculosum* of the *lamina terminalis,* the subfornical organ, the subcommissural organ, the area postrema, the neurohypophysis and the pineal gland.

In contrast to the common brain parenchyma, these structures are equipped with leaky fenestrated capillaries, which allow the transfer of substances through the endothelium, and a tight junction between the covering ependymal cells blocks free passage into the cerebro-spinal fluid.

In general, substances can cross the blood–brain barrier by four different pathways:
- penetration via pores and trancytosis;
- transmembrane diffusion;
- carrier-mediated mechanisms and transporters;
- retrograde neuronal transport, so by-passing the blood–brain barrier.

Some chemical properties enable substances to cross the blood–brain barrier. The properties which affect permeability include lipid solubility, molecular weight and the ability to form electro-neutral complexes. Some neuroactive substances, like a-MSH, are known to cross the blood–brain barrier by this mechanism.

The most convenient route for molecules to cross the blood–brain barrier is by making use of specific receptor-mediated mechanisms or by transporters. In order to do this, the substance has first to bind to a receptor on the endothelium; and second, the formed ligand–receptor complex has to be internalized and transferred via endosomes into the endothelial cytoplasm. Finally, the ligand–receptor complex has to be degraded and the ligand can then be released by exocytosis on the opposite side of the barrier.

Some peptides, like MSH, can cross the barrier primarily by transmembrane diffusion, a non-saturable mechanism largely dependent on the lipid solubility of the peptide. Other neuroactive substances utilize highly specific transporters for crossing the blood–brain barrier, which can operate unidirectionally or bidirectionally. The unidirectional transport can be from the blood to brain parenchyma, as is the case for Leu-enkephalin, or in opposite direction, as is the case for the neuropeptides Tyr-MIF-1 and Met-enkephalin. This latter seems to depend on the peptide transport system-1 (PTS-1), which carries small peptides with an N-terminal tyrosine from the brain to the blood. This system preferentially transports two peptides, namely Tyr-MIF-1 and Met-enkephalin.

Bi-directional transport has been described, for example, for the gonadotropin-releasing hormone GnRH.

The transport systems are highly specific and each system carries its own complement of substrates. Another route of entry to the brain parenchyma utilized by some neurotoxins and viruses is to by-pass the barrier by retrograde transport from peripheral nerve endings.

Specific transport systems, which carry peptides through the blood–brain barrier, have important neuro-regulatory functions. Pharmacological manipulation of the blood–brain barrier can therefore provide therapeutic strategies for the efficient delivery of drugs which are impermeable under normal conditions.

Further Reading

Banks, W. A., Kastin, A. J. (1985): Permeability of the blood–brain barrier to neuropeptides: the case for penetration. *Psychoneuroendocrinology* **10**: 385–399.

Banks, W. A., Kastin, A. J., Fischman, A. J., Coy, D. H., Strauss, S. L. (1986): Carrier-mediated transport of enkephalins and N-Tyr-MIF-1 across blood–brain barrier. *Am. J. Physiol.* **251**: E477–E482.

Begley, D. J. (1994): Peptides and the blood–brain barrier: the status of understanding. *Ann. N.Y. Acad. Sci.* **739**: 89–100.

Dermietzel, R., Krause, D. (1992): Molecular anatomy of the blood–brain barrier as defined by immunocytochemistry. *Int. Rev. Cytol.* **127**: 57–109.

Dermietzel, R., Spray, D. C., Nedergaard, M. (eds) (2006): *Blood–Brain Barrier: From Ontogeny to Artificial Membranes.* Wiley–VHC, Weinheim.

Greenwood, J., Begley, D. J., Segal, M. B. (1995): *New Concepts of a Blood–Brain Barrier.* Plenum Press, New York.

Jaspan, J. B., Bank, W. A., Kastin, A. J. (1994): Study of passage of peptides across the blood–brain barrier: biological effects of cyclo (His-Pro) after intravenous and oral administration. *Ann. N.Y. Acad. Sci.* **739**: 101–107.

Nag, S. (ed) (2003): *The Blood–Brain Barrier.* Humana Press, Totowa, N.J.

Pardridge, W. M. (2001): *Brain Drug Targetting. The Future of Brain Blood Development.* Cambridge University Press, Cambridge.

Pardridge, W. M. (2000): Receptor-mediated peptide transport through the blood–brain barrier. *Endocrinol. Rev.* **7**: 314–330.

1.5
Volume Transmission and Wiring Transmission

Intercellular communication in the brain can be grouped in two broad classes, as proposed by Agnati and coworkers (1986). Based on some general features of signal transmission, the authors differentiate between wiring transmission (WT) and volume transmission (VT). Transmission by WT is defined as a mode for intercellular communication, which is mediated via a relatively constrained cellular chain (wire), while transmission by VT consists of a three-dimensional diffusion of signals in the extracellular fluid for distances larger than the synaptic cleft. Transmission by WT thus exhibits (like classic synaptic transmission) a one-to-one ratio with respect to the number of signal source structures and the number of signal targets, whereas transmission by VT shows a one-to-many ratio. With respect to the neuroactive substances and their function in brain tissues, as described above, neurotransmitters as well as neuromodulators convey both modes of function. In fact, it is now generally believed that there exists in the brain some kind of non-synaptic, hormone-like, modulatory transmission besides synaptic transmission. This concept is supported by recent findings of the neurotrophic effects of some neuropeptides and neurotrophins. Furthermore, data on gaseous transmitters like NO have given strong support to this view. For instance NO, once released, can affect the electro-metabolic state of numerous neurons not in synaptic contact with the neuron source of the signal. In addition to the classic mode of wiring transmission in the form of synaptic communication as described above, we will give here a brief account of the basic elements of transmission by VT, as summarized recently by Zoti et al. (1998).

The general features of VT are:
- a cell source of the VT signal, neuronal or non-neuronal, from which a signal molecule can be released into the extracellular cerebral fluid (ECF);
- a VT signal diffusing in the ECF for a distance larger than the synaptic cleft;
- communication trails in the extracellular space in form of preferred diffusion pathways;
- a cell target of the VT signal, that is, a cell possessing molecules capable of detecting and decoding the message.

Model systems for VT are the highly divergent monoaminergic pathways of the brain, e.g. the dopaminergic mesostriatal system. Both morphological and functional evidence indicates that dopamine acts as a VT signal in the striatum.

The existence of a functionally coupled syncytium provided by astrocytes and postnatal neurons has led to an extension of the concept of VT transmission (Dermietzel 1998). If the route of VT is generally regarded as through extracellular cerebral fluid, primarily via diffusion, then gap junctions, which represents the structural correlate of electrical synapses, may provide a second highly regulated route of VT which can be defined as intercellular as opposed to the extracellular VT. This intercellular VT could serve as a route parallel to the extracellular VT, allowing coordinated responses of functionally coupled neurons or glial compartments. In fact, recent evidence indicates that neurotransmitter coupling via gap junctions exists. For instance, the inhibitory neurotransmitter glycine can be provided by glycinergic amacrine cells to cone bipolar cells in the retina (Vaney et al. 1998). Although evidence for intercellular volume transmission of neurotransmitters is still circumstantial, one has to expect new concepts in transmitter trafficking and function in the near future.

Further Reading

Agnati, L.F., Fuxe, K., Zoli, M., Zini, I., Toffano, G., Ferraguti, A. (1986): A correlation analysis of the regional distribution of central enkephalin and beta-endorphin immunoreactive terminals and of opiate receptors in adult and old mice rats. Evidence for the existence of two main types of communication in the central nervous system; the volume transmission and the wiring transmission. *Acta Physiol. Scand.* **128**: 201–207.

Alexander, D.B., Goldberg, G.S. (2003): Transfer of biologically important molecules between cells through gap junction channels. *Curr. Med. Chem.* **10**: 2045–2058.

Dermietzel, R. (1998): Gap junction wiring: a new principle in cell-to-cell communication in the nervous system. *Brain Res. Rev.* **26**: 176–183.

Vaney, D.I., Nelson, J.C., Pow, D.V. (1998): Neurotransmitter coupling through gap junctions in the retina. *J. Neurosci.* **18**: 10594–10602.

Zoti, M., Torri, C., Ferrari, R., Jansson, A., Zini, I., Fuxe, K., Agnati, L.F. (1998): The emergence of the volume transmission concept. *Brain Res. Rev.* **26**: 136–147.

2
Methods

A broad spectrum of methods is available to analyze the nature and effects of neuroactive substances. For isolation and purification, common preparative and analytical methods are used, e.g. differential centrifugation, PAGE-SDS electrophoresis, 2-D electrophoresis, affinity chromatography, etc. This arsenal of biochemical techniques makes it possible to isolate and purify receptor proteins in high concentrations so that crystallography and structural analysis can be applied. A classic example of the effectiveness of this strategy is the molecular analysis of the acetylcholine receptor, which was enriched and purified from electroplaques of the electric eel. By subjecting crystallized samples to X-ray diffraction and low-dose electron microscopy, the quaternary structure of this receptor was successfully elucidated.

Molecular biological techniques and molecular genetics provide a further methodological facet for the analysis of genes and their encoded substrates, e.g. neurotransmitters, precursor molecules and receptor complements. In particular, the unraveling of the enormous multiplicity of neurotransmitter receptors which are presently known would have not been possible without molecular biological approaches.

A further field of intensive study is the search for specific binding sites of neuroactive substances and analysis of their topographical distribution and pharmacological profiles. This research has been furthered by autoradiography and immunohistochemistry.

By combining immunohistochemistry with tracing methods, it has become possible to map both the neurotransmitter and neuroreceptor characteristics of brain nuclei and their projection patterns.

Electrophysiological methods are frequently employed to study the effects of neurotransmitters and drugs on neuronal activities. Further strategies to look for behavioral effects include batteries of cognitive and psychomotor tests to which laboratory animals can be subjected.

By creating genetically engineered animals – transgenic and knockout mice – neurotransmitters, neuromodulators and their specific receptor complexes can now be studied against different genetic backgrounds.

Transgenic animals carry artificial genes which are integrated into their genome. This makes it possible to study gain of function or loss of function effects of genes which code for a neuroactive substance, their precursors or receptors.

Various techniques are available to create transgenic animals. One technique is to integrate a new gene into the embryo by transfecting it with retroviruses. Another is to injection a gene into the pronucleus of an intact oocyte. A further technique is the transfection of embryonic stem cells and their injection into embryonic blastocystes, followed by implantation into a foster mother.

The insertion of a new gene by homologue recombination allows the creation of animals which express the new gene under the control of the recipient's genome. The same strategy can be applied to inactivate or "knock out" single genes. Homolog recombination can cause the normal (active) gene to be replaced by the inactivated gene. When the new gene or the deleted gene is integrated into the germ line, the litter-mates will carry the "manipulated" gene in a Mendelian pattern.

Analysis of the effects of genetic manipulation requires the entire spectrum of modern analytical techniques, including physiological, morphological and behavioral studies. In the following, we will describe some basic methods commonly used in the neurosciences to explore the function, expression pattern and behavioral effects of neuroactive substances.

Further Reading

Champtiaux, N., Changeux, J.P. (2004): Knockout and knockin mice to investigate the role of nicotininc receptors in the central nevous system. *Prog. Brain Res.* **145**: 235–251.

Shepard, G.M. (1994): *Neurobiology, 3rd ed.* Oxford University Press, New York.

Unwin, N. (2003): Structure and action of the nicotinic acetylcholine receptor explored by electron microscopy. *FEBS Lett.* **555**: 91–95.

2.1
Bio- and Radioisotope Assays

Bioassays

Assays are frequently used to characterize the function of bioactive substances. The principle behind assays is to remove the endogenous source of a substance and apply the substance to the deprived organism or cell culture systems. The removal of the source can elicit a variety of metabolic and physiological dysfunctions in the organism or culture system. If application of the particular substance compensates for these dysfunctions, this is taken as an indication that the substance is involved in the normal regulation of function. Since most biologically active substances are not monofunctional, different assays have to be performed to analyze their effects.

Radioisotope assays

Radioisotope assays are used to detect low amounts of substances or to determine the affinity of a ligand towards a receptor or receptor population.

A sensitive assay to detect minute amounts of substances with antigen properties is the radioimmunoassay (RIA), which was developed by S. Berson and R. Yalow in 1960. As compared with bioassays, the radioimmunoassay is more sensitive and permits the detection of physiological and even subphysiological concentrations of biologically active substances in the femtomolar range.

The physical principal of RIA is based on an antigen–antibody reaction. The probe containing the antigen to be measured is incubated with an identical radioactively labeled antigen and its corresponding antibody. By competition between the labeled and the unlabeled antigen, measurement of the radioactivity of the bound antigen–antibody complex gives a measure of the relative amount of the unlabeled antigen when compared with standardized competition curves.

A further radioisotope assay is the so-called radioreceptorassay (RRA), also known as a Scatchard diagram. This method is applied to determine the dissociation constant K_D. The K_D value indicates the affinity of a ligand to its receptor. The Scatchard diagram also allows the determination of receptor densities or number of receptors (B_{max}) in a given tissue. It can also be applied to determine the binding kinetics of bioactive substances in radioligand studies. In this case, the specific binding of a labeled ligand to a receptor is studied at different concentrations. The data obtained by measuring the radioactivity is expressed in femtomoles (fmol) per milligram (mg) substance. When the amount of labeled binding varies linearly with the ligand concentration, then the existence of a monospecific receptor of the ligand can be assumed. If the relation differs from linearity, then multireceptor binding is most likely.

The radioreceptorassay is superior to RIA with respect to receptor studies, since it allows measurement of the binding of a ligand to its biologically relevant receptor. A limitation of this method is that it does not distinguish between ligand effects, e.g. antagonistic or agonistic properties.

Further Reading

Davenport, A.P., Hill, R.G., Hughes, J. (1989): Quantitative analysis of autoradiograms. *Experientia Suppl.* **56**: 137–153.

Deodhar, S.D., Genuth, S.M. (1970): Radioimmunoassay of hormones and a critical review of its application to angiotensin assay. *Crit. Rev. Clin. Lab. Sci.* **1**: 119–134.

Goldsmith, S.J. (1975): Radioimmunoassay: review of basic principles. *Semin. Nucleic Med.* **5**: 125–152.

Mousah, H., Jacqmin, P., Lesne, M. (1987): The quantification of gamma-aminobutyric acid in the cerebrospinal fluid by a radioreceptorassay. *Clin. Chim. Acta* **170**: 151–159.

Quirion, R. (1982): Bioassays in modern peptide research. *Peptides* **3**: 223–230.

Schmidt, K.C., Smith, C.B. (2005): Resolution, sensitivity and precision with autoradiography and small animal positron emission tomography: implications for functional brain imaging in animal research. *Nucleic Med. Biol.* **32**: 719–725.

2.2
Microdialysis and Electrochemical Detection

2.2.1
Microdialysis

Microdialysis is used to introduce or withdraw a constant amount of a biologically active substance into or from the extracellular cerebro-spinal fluid in living animals. The microdialysis probe is designed to mimic a "capillary" system in which the substrate has to cross a semi-permeable membrane. The direction of flow depends primarily on concentration gradients. The gradient built up by the substance in question is not exclusively dependent on the differences in concentration between sample and extracellular fluid, but depends also on the velocity of flow inside the microdialysis chamber. For instance, when a physiological salt solution is dialyzed from inside the chamber, the solution equilibrates with the extracellular fluid, i.e. solutes diffuse from the cerebro-spinal fluid across the membrane into the probe. After a period of time, the chamber contains the substance (and other solutes) in an amount representative of that dissolved in the extracellular cerebral fluid. On the basis of *in vitro* calibration data, the chamber is estimated to recover 10–20% of the actual extracellular moities. Thus, microdialysis can also be used to collect molecules from the cerebral fluids. The rate of diffusion is a function of the area of the membrane, the flow rate and the diffusion coefficient of the substances. Since the microdialysate is applied continuously, the amount of substrate which is exchanged also depends on the duration of microdialysis. By using microdialysis as a diffusion trap, femtomolar concentrations of molecules in the extracellular fluid can be captured.

Microdialysis is applicable to living animals. The major disadvantages are that only small amounts of substance(s) can be collected and that the surgical insertion of the chamber causes lesioning or irritation of brain tissue; and this may change the physiological composition of the extracellular fluid.

2.2.2
Electrochemical Detection

The term "electrochemical detection" refers to a number of detection techniques which involve the application of an electric oxidation–reduction potential via appropriate electrodes to a sample solution containing oxidizable or reducible solutes. The resulting current is measured as a function of time. Electrochemical detection is used where high sensitivity or selectivity is required.

The principles underlying electrochemical techniques require the employment of a three-electrode system, consisting of a recording (indicator), an auxiliary (counter) and a reference electrode.

Electrochemical detection can only be used for molecules which are electrochemically active. To overcome this problem, electrochemically inactive molecules must be converted to active molecules, for example, by an enzymatic reaction.

Voltammetry

Voltammetry allows the direct monitoring of molecules within extracellular cerebral fluid. Voltammetry makes use of the fact that certain compounds are readily oxidizable. Typically, the method employs a working electrode, a reference electrode and an auxiliary electrode. These electrodes are positioned in electrical continuity with one another, the working electrode being positioned in the brain structure which is of interest. A controlled potential difference is then applied between the working and reference electrodes; and the resultant current that flows from the working electrode provides a measure of the amount of electroactive material in the solution. The current which results from the reaction of the analyzed species at the working electrode is measured.

Voltammetry makes use of the fact that several neurotransmitters, especially catecholamines, are readily oxidizable at potentials at which most other substances in the extracellular fluid do not oxidize.

However, even in the case of catecholamines, additional molecules can oxidize, so complicating their detection. One way to apply volammetry to molecules with similar oxidation potentials is to separate them by HPLC before analysis.

Amperometry and coulometry

The current produced for an electrochemical reaction can be measured using an electrochemical sensor either in the dialysates obtained from microdialysis or in the eluent from HPLC. The application of an electrochemical detector results in a highly selective and sensitive detection tool.

Amperometry is a technique in which a fixed potential is applied to a working electrode with respect to a reference electrode. During amperometric detection, the mobile phase is passed onto the electrode and only substances which contact the electrode surface directly are oxidized and measured. The substance to be detected undergoes a reaction if the applied potential has appropriate polarity and magnitude.

When the reaction is incomplete, i.e. only a fraction of the total analyte reacts, the detection mode is termed "amperometry". When the working electrode has larger surface area and the reaction is complete, the mode is called "coulometry". The former method is easier to maintain, but it is somewhat less sensitive than the latter.

Coulometric electrochemical detection is similar to the amperometric method. As mentioned above with the amperometric electrode, only a fraction of the substance comes into direct contact with the electrode when the mobile phase flows across it. Consequently, a relatively small proportion of the compound is oxidized. This is in contrast to coulometric detection, where all of the analyte is oxidized or reduced. The difference is the result of perfusing the mobile phase through a coulometric electrode (at pressure) rather than across it. The electrode used in colormetric detection consists of a carbon rod with many small pores which greatly increase the surface area exposed to the mobile phase.

Further Reading

Humpel, C., Ebendal, T., Olson, L. (1996): Microdialysis: a way to study *in vivo* release of neurotrophic bioactivity: a critical summary. *J. Mol. Med.* **74**: 523–526.

Landolt, H., Langemann, H. (1996): Cerebral microdialysis as a diagnostic tool in acute brain injury. *Eur. J. Anaesthesiol.* **13**: 269–278.

Parsons, L. H., Justice, J. B. Jr (1994): Quantitative approaches to *in vivo* brain microdialysis. *Crit. Rev. Neurobiol.* **8**: 189–220.

Sharp, T., Hjorth, S. (1990): Application of brain microdialysis to study the pharmacology of the 5-HT1A autoreceptor. *J. Neurosci. Methods* **34**: 83–90.

2.3
Chromatography

Chromatography allows substances to be purified on the basis of their size, charge or affinity to an antibody. In order to isolate and purify a substance, the tissue must first be collected, homogenized under protease inhibition and then centrifuged to separate the various cell components, i.e. nuclei, membranes and cytosol. These fractions can then be subjected to detergent treatment or treatment with chelatotropic substances (e.g. urea). Further separation is achieved by differential centrifugation, for example on a sucrose gradient.

The fractions which contain the enriched and solubilized substance are collected and subjected to further purification by chromatography. In the case of microdialysis, samples the dialysate can be used directly in chromatographic analysis.

The separation of compounds not only provides a means of purifying them for further molecular analysis, but is also used in studying changes in brain metabolism, i.e. by collecting and analyzing samples before and after treatment with a drug.

Purification is the process of separating or extracting the target substance from other (possibly structurally related) compounds. The substance of interest should separate when subjected to chromatography. This depends usually on the velocity of migration through a column, which is a function of substance size or charge. Separation can successfully be achieved when the substance differs significantly in its migration velocity from other (contaminating) substances.

2.3.1
Affinity Chromatography

Affinity chromatography provides a method by which specific antibodies are coupled to a matrix (the most commonly used matrix is bromocyane-activated Sepharose). Samples containing the substance to be purified are loaded onto the column. If the substance displays antigenic properties to the antibody, it binds with high affinity. Contaminating substances are not recognized by the antibody and so run freely through the column. The column is then eluted with a low pH glycine buffer (pH 2–3) for acid hydrolysis of the antigen–antibody complex; and the efflux containing the substance is collected. Affinity chromatography also allows pharmacologically active substances to be bound to the column, as for example in the isolation of neurotransmitters or their receptor complex. A frequently used approach for the isolation of $GABA_A$ receptors, for instance, is the binding of benzodiazepin to a matrix.

2.3.2
High Performance Liquid Chromatography

High performance liquid chromatography (HPLC) has become a powerful tool in analytical and preparative neurochemistry. Modern HPLC methods can be used for several purposes, including separation, purification and quantitative assessment of neurochemical compounds. It is important to understand the theoretical background to HPLC if it is to be used effectively.

In brief, HPLC consists of two main components:
- A separation column or stationary phase. Compounds that contain functional groups capable of strong hydrogen bonding adhere more tightly than less polar compounds to the stationary phase. Thus, less polar compounds elute from the column faster than compounds that are highly polar.
- An aqueous mobile phase. The mobile phase in HPLC refers to the solvent which is continuously applied to the column, or stationary phase. The mobile phase is a buffered water/acetone solvent and exhibits polar properties.

The mobile phase is pumped through the column under high pressure (between 30 and 200 bar; or 3–20 MPa). When a sample is introduced into the system via an injector, it is dissolved in the mobile phase. The molecules in the sample separate according to their relative affinity to the non-polar matrix and the polar mobile phase. Molecules which are neutral or of low charge (more apolar) exhibit a greater affinity for the stationary phase and thus become trapped by the column, with the result that their elution is delayed. Molecules which are charged (and therefore more polar) possess a greater affinity to the mobile phase and elute faster. Increasing the length of the column or decreasing the particle size of the packing material improves separation, but also results in an increase in the pressure required to pump the mobile phase. When the column length is increased, the dilution of samples is also increased.

Preparative HPLC refers to process of isolation and quantitative purification of compounds. Important requirements for a successful preparation are the degree of solute purity and the "throughput", i.e. the amount of compound produced per time. In contrast in analytical HPLC, the focus is mainly on obtaining information about the composition of the sample. Recording the efflux of compounds from the HPLC column is a crucial part of any HPLC approach. In order to monitor the efflux, a UV detector must be selected and adjusted to optimal detection settings. Setting can be checked by standard separation assays and adjusted so that, in the standard samples, there is a sharp (detection) peak in the range expected from the substance being studied.

Quantification of compounds by HPLC is the process of determining unknown concentrations of a substrate. It involves the injection of a series of known concentrations of a standard compound into the HPLC column. The chromatogram of the known concentrations provides a series of peaks which correlate with the concentration of the injected compound. Normograms can be used to estimate unknown concentrations.

2.3.3
Proteomics: Multidimensional Protein Identification Technology

During the past decade, the necessity for large-scale (shotgun) identification of complex protein mixtures using peptide mass mapping has fostered technological developments in proteomics research with a dramatic impact on the biology of complex systems. One of the technical developments in shotgun proteomics – termed multidimensional protein identification technology (MudPIT) – couples two-dimensional (2D) chromatography of peptides in mass spectrometry-compatible solutions directly to electrospray ionization tandem mass spectrometry (ESI-MS/MS), allowing the identification of proteins from highly complex mixtures with high confidence. Since the initial description of MudPIT in 1999 (Link et al. 1999), this approach has been implemented in the analysis of whole proteomes (Durr et al. 2004; Cagney et al. 2005; Jessani et al. 2005), organelles (Skop et al. 2004) and protein complexes (de Bruin et al. 2004; Sato et al. 2004). Key aspects of many of the reported analyses focus on the validation of MudPIT datasets with alternate strategies and the integration of MudPIT datasets with other biochemical, cell biology or molecular biology approaches. In these studies, MudPIT provides high resolution, automation, high throughput and the ability to analyze complex mixtures of proteins in a single run.

Technically, MudPIT incorporates multidimensional high-pressure liquid chromatography (LC/LC), tandem mass spectrometry (MS/MS) and advanced database-searching algorithms. Protein mixtures can derive from a wide variety of sources, including tissues, biofluids or subcellular protein fractions. It is particular noteworthy that MudPIT, but not classic 2D protein gel electrophoresis separation techniques, allows analysis of both membrane and soluble proteins from complex membrane-containing samples. The separation problems commonly associated with the isolectric focussing of hydrophobic, detergent-solubilized proteins during classic 2D gel-based approaches do not occur.

Protein lysates are enzyme-digested by a site-specific protease and loaded onto a biphasic capillary column integrating a strong cation exchange resin (SCX) capillary column and then are gradually released from a reversed phase resin (RP). Peptides elute off the SCX phase by increasing pI; and elution off the SCX material is evenly distributed during an analytical run. Eluted peptides are on-line directed to a microflow electrospray interface using quadrupole TOF/TOF instrumentation. Next, protein and peptide sequences are identified by comparison of theoretical and actual MS/MS spectra via SEQUEST software. Wolters et al. (2001) described the chromatographic benchmarks of MudPIT:

two analyses were reproducible within 0.5% and a dynamic range of 10 000 to 1 between the most abundant and least abundant proteins/peptides in a complex peptide mixture measured. SEQUEST is a powerful program that takes uninterpreted tandem mass spectra as input and produces peptide identifications as output. The general principle is that, for each of the tandem mass spectra produced in a typical analysis, the program searches through entire protein databases for fragments that are around the right mass. The top 500 sequences are then used to produce theoretical tandem mass spectra, with fragment ions produced depending on the amino acid sequence of the peptide. The experimental spectrum is then compared to the theoretical spectra produced using cross-correlation analysis, and the best match is reported as the identification of the peptide. Finally, extensive filtering of SEQUEST results has to be performed to exclude false-positive results. This can be done by programs like DTAselect (Tabb et al. 2002). However, it is important to remember that peptide identification and protein identification are two different things and any result obtained by MudPIT needs reconfirmation by other methods such as Western blotting.

In summary, MudPIT compared to the current one- or two-dimensional electrophoresis/MS method has the advantage of higher identification capacity, higher sensitivity, higher throughput and a higher degree of automation. Thus, MudPIT is an essential tool for proteomic analysis and its combination with improved sample preparation techniques will aid in the overall analysis of proteomes by identifying proteins of all functional and physical classes.

Further Reading

Cagney, G., Park, S., Chung, C., Tong, B., O'Dushlaine, C., Shields, D.C., Emili, A. (2005): Human tissue profiling with multidimensional protein identification technology. *J. Proteome Res.* **4**: 1757–1767.

de Bruin, R.A., McDonald, W.H., Kalashnikova, T.I., Yates, J. 3rd, Wittenberg, C. (2004): Cln3 activates G1-specific transcription via phosphorylation of the SBF bound repressor Whi5. *Cell* **117**: 887–898.

Durr, E., Yu, J., Krasinska, K.M., Carver, L.A., Yates, J.R., Testa, J.E., Oh, P., Schnitzer, J.E. (2004): Direct proteomic mapping of the lung microvascular endothelial cell surface *in vivo* and in cell culture. *Nat. Biotechnol.* **22**:985–992.

Jessani, N., Niessen, S., Wei, B.Q., Nicolau, M., Humphrey, M., Ji, Y., Han, W., Noh, D.Y., Yates, J.R. 3rd, Jeffrey, S.S., Cravatt, B.F. (2005): A streamlined platform for high-content functional proteomics of primary human specimens. *Nat. Methods* **2**: 691–697.

Kischka, U., Wallesch, C.-W., Wolf, G. (eds) (1997): *Methoden der Hirnforschung: eine Einführung*. Spektrum-Akademischer Verlag. Heidelberg.

Link, A.J., Eng, J., Schieltz, D.M., Carmack, E., Mize, G.J., Morris, D.R., Garvik, B.M., Yates, J.R. III. (1999): Direct analysis of protein complexes using mass spectrometry. *Nat. Biotechnol.* **17**: 676–682.

Sato, S., Tomomori-Sato, C., Parmely, T.J., Florens, L., Zybailov, B., Swanson, S.K., Banks, C.A., Jin, J., Cai, Y., Washburn, M.P., Conaway, J.W., Conaway, R.C. (2004): A set of consensus mammalian mediator subunits identified by multidimensional protein identification technology. *Mol. Cell* **14**: 685–691.

Simpson, R.C., Brown, P.R. (1986): High-performance liquid chromatographic profiling of nucleic acid components in physiological samples. *J. Chromatogr.* **379**: 269–311.

Skop, A.R., Liu, H., Yates, J. 3rd, Meyer, B.J., Heald, R. (2004): Dissection of the mammalian midbody proteome reveals conserved cytokinesis mechanisms. *Science* **305**: 61–66.

Tabb, D. L., McDonald, W. H., Yates, J. R. 3rd.(2002): DTASelect and Contrast: tools for assembling and comparing protein identifications from shotgun proteomics. *J. Proteome Res.* **1**: 21–26.

Wisniewski, R. (1992): Principles of the design and operational considerations of large scale high performance liquid chromatography (HPLC) systems for proteins and peptides purification. *Bioseparation* **3**: 77–143.

Wolters, D. A., Washburn, M. P., Yates, J. R. 3rd. (2001): An automated multidimensional protein identification technology for shotgun proteomics. *Anal. Chem.* **73**: 5683–5690.

Wu, C. C., MacCoss, M. J., Howell, K. E., Yates, J. R. 3rd. (2003): A method for the comprehensive proteomic analysis of membrane proteins. *Nat. Biotechnol.* **21**: 532–538.

2.4
Autoradiography

Autoradiography is used to detect the distribution and to measure the amount of a radioisotope deposited in a specimen. Conventional autoradiography utilizes direct contact of the radiolabeled samples with a photographic film or emulsion. Receptor autoradiography is a special form of autoradiography in neurobiology, which is widely used to localize neurotransmitter and neuromodulator binding sites. For visualizing binding sites the neuroactive substance must be labeled with a radioactive isotope. This labeling is normally made with $[^{3}H]$, $[^{13}C]$ or $[^{125}J]$.

Control experiments make use of specific antagonists which bind to the receptor sites with higher affinity than the substance being tested. Coapplication of both substances, the labeled substance and the unlabeled antagonist, should result in extinction of labeling. When the ligand is capable of binding to more than one type of receptor, occupancy of one class of receptor by a specific antagonist allows the labeled ligand to bind to the remaining other binding sites. In this way the distribution of the non-occupied receptors can be monitored. A typical protocol for autoradiography is as follows. Tissue samples or whole animals are exposed to the radioactively labeled compound for receptor–ligand binding. Following receptor occupancy, the tissue is fixed and embedded in paraffin. Alternatively, tissue samples can be frozen and stored at –70 °C for further cryostat sectioning. Slices are collected on slides and thoroughly washed in a cooled buffer to remove unbound ligand, thereby leaving specific receptor–ligand complexes on the section. Finally, the slides are removed from the wash and dried.

After drying, the slides can be coated with a liquid photo-emulsion or covered by X-ray-sensitive films and stored in light-tight cassettes at 4 °C for periods ranging from a few hours to months. Coating with liquid photo-emulsion usually gives higher resolution than coating with X-ray films. After sufficient exposure, the films are developed in photographic developer in order to visualize the silver grains.

Manipulation of the binding of a ligand to its receptor is done for a variety of reasons. For example, by adding unlabeled ligand (or a pharmacological analog) in excess to the incubation sample pharmacological issues such as specificity and pharmacokinetics can be addressed (K_D, B_{max}).

The validity of the autoradiographic method has been verified by data derived from other techniques, as e.g. immunohistochemistry and *in situ* hybridization. Autoradiography has the advantage over immunocytochemistry that it combines histological resolution (including electron microscopical resolution) with the information about affinity and density of receptor-binding sites.

A limitation of autoradiography is that receptor proteins are commonly transported along dendritic processes; and this makes it difficult to distinguish between positive signals from perikarya and their dendrites. In addition, the best binding conditions for an endogenous ligand may be different from those of a pharmacological analog.

It should be also kept in mind that autoradiography is an indirect method based on ligand labeling rather than on direct labeling of the receptor.

Further Reading
Palacios, J.M., Dietl, M.M. (1989): Regulatory peptide receptors: visualization by autoradiography. *Experientia Suppl.* **56**: 70–97.
Stumpf, W.E. (1970): Localization of hormones by autoradiography and other histochemical techniques. A critical review. *J. Histochem. Cytochem.* **18**: 21–29.
Unnerstall, J.R., Niehoff, D.L., Kuhar, M.J., Palacios P. (1982): Quantitative receptor autoradiography using (^{3}H) Ultrofilm: application to multiple benzodiazepine receptors. *J. Neurosci. Methods* **6**: 59–73.
von Bohlen und Halbach, O., Dermietzel, R. (1999): *Methoden der Neurohistologie*. Spektrum Akademischer Verlag. Heidelberg.

2.5
Immunohistochemical Methods

Immunohistochemistry is a method for studying proteins in an anatomical context. The principle of immunohistochemistry is the formation of an antigen–antibody complex and the visualization of the binding sites through labeled antibodies.

The first antibody (so-called primary antibody) applied to tissue sections can be labeled with a fluorophore or a different reporter molecule which can be visualized by a secondary histochemical reaction. Antibodies are raised towards proteins or peptides with antigenetic properties. Since antigenetic determinants are independent of the sites determining the biological activity of a protein, antibodies can recognize inactive forms or even metabolic fragments of a protein. Therefore immunoreactive sites are frequently referred to in terms of substance-like reactivity (e.g. angiotensin-like immunoreactivity). A common protocol for immunohistochemical staining works as follows. The tissue is fixed in order to keep the antigen in place. Normally, mild fixatives are preferred to maintain antigenicity of the substrate, e.g. paraformaldehyde or acetone fixation. Thereafter, the fixed specimen is rinsed thoroughly in a buffered solution [phosphate-buffered saline (PBS) is the most common rinsing buffer], followed by embedding in paraffin after dehydration or slam-freezing for cryostat sectioning.

The antibody is then applied to a section in an appropriate dilution. Incubation time is in the order of 60 min at room temperature. In the case of the so-called direct immunofluorescence, the primary antibody is labeled with a fluorophore and can be visualized directly after rinsing. When the more common indirect immunofluorescence is used, a second antibody is used to sandwich the first antibody. In this case, the second antibody is labeled with the fluorophore. The secondary antibody normally recognizes IgGs or other immunoglobulin moieties. Common target proteins in neuroimmunohistochemistry are neurotransmitters or specific enzymes involved in neurotransmitter metabolism, precursors and receptor complexes.

The direct technique has the disadvantage that it necessitates coupling of each primary antibody with a fluorophore; and sensitivity is lower than the indirect method, which allows signal enhancement by applying multiple labeling steps. A common strategy for signal amplification is the coupling of an enzyme [e.g. horseradish peroxidase (HRP) to the secondary antibody].

This allows either the enzyme HRP to react with its substrate, hydrogen peroxide, in the presence of a chromogen (e.g. 3'-3'-diaminobenzidine) which results in a colored precipitate, or the application of a further anti-HRP antibody for amplification of the signal.

Immunohistochemistry is a powerful method for analyzing the pattern of neurotransmitter and neuroreceptor distribution and has even been used to localize amino acid neurotransmitters by raising antibodies to multimeric complexes of single amino acids, e.g. polyglycine or polyGABA.

A limitation of the immunohistochemical technique, however, is the difficulty of being sure that the labeling reaction is specific, particularly in view of the potential for cross-reactivity. The interpretation of immunohistochemical data can be problematic in the presence of different homologous isoforms of neuroreceptors or where amino acid neurotransmitters should be detected which are also common in ordinary protein metabolism.

Further Reading

Bullock, G.R., Petrusz, P. (eds) (1982): *Techniques in Immunocytochemistry.* Academic Press, New York.

Javois, L.C. (ed) (1999): *Immunocytochemical Methods and Protocols.* Humana Press, Totowa, N.J.

Sternberger, L.A. (1979): *Immunohistochemistry, 2nd edn.* Wiley, New York.

Swanson, P.E. (1988): Foundations of immunohistochemistry. A practical review. *Am.J. Clin. Pathol.* **90**: 333–339.

von Bohlen und Halbach, O., Dermietzel, R. (1999): *Methoden der Neurohistologie.* Spektrum Akademischer Verlag, Heidelberg.

2.6
In situ Hybridization

In situ hybridization (ISH) allows the detection of cRNA in cells or sections. In a sense, ISH is similar to receptor autoradiography and immunohistochemistry, since all three techniques require the binding of a probe to a target molecule to form a stable complex.

ISH can be used to visualize the intracellular distribution of cRNA of neuroactive substances (mainly their precursors) or receptor proteins. A prerequisite is that the cDNA or cRNA sequence of the molecule in question is known.

The availability of standardized kits has made ISH a widely used method in neuroscience research.

For *in situ* hybridization, one needs a considerable amount of complementary DNA (cDNA) or cRNA for hybridization. Sufficient amounts can be achieved by polymerase chain reaction (PCR).

The basic step in ISH is the process of "hybridization". Hybridization can be accomplished by incubating single-stranded cRNA or cDNA of the protein to appropriate cDNA or cRNA targets in the tissue. Hybridization is achieved through specific hydrogen bonding between the complementary nucleotide residues. In order to visualize the hybrid complex, nucleotides of the probe have to be labeled either by introducing radioactivity bound to adenosine-triphosphates (^{35}S-dATP) or by binding an antigenic target molecule such as dioxygenine to uridine triphosphates (dUTP). Dioxygenin can be visualized in a secondary step by immunhistochemistry.

A common protocol for *in situ* hybridization is as follows. The first step to achieve RNA–RNA hybrids requires the denaturation of the target cRNA in the tissue. After denaturation, labeled RNA (riboprobes) are added under renaturation conditions. The labeled probe (which is applied in large excess) can now form stable hybrids with endogenous cRNA. It is the hybrid which provides the signal for intracellular cRNA localization. Following hybridization, the tissue is incubated in a RNAse solution. The purpose of this is to digest the remaining single-stranded RNA and the non-hybridized RNA probe so as to eliminate background labeling. After the RNase step, the sections are washed in low-salt buffers at 45–65 °C. Unbound RNA is washed off and complementary binding sites can be detected by their label (see above).

An alternative method is to use short cDNA probes for hybridization with cRNA. These oligonucleotide probes have the advantage over riboprobes that, because of their small size, they penetrate the tissue more readily. They are also easier to handle than cRNA probes because of their stability and resistance to RNAses; but also because of their small size they cannot be labeled to the same extent. This deficit in sensitivity can be overcome by using mixtures of non-overlapping oligonucleotide probes.

In spite of the limitation of ISH, the technique provides a useful tool – in particular if it is combined with immunocytochemistry – to map the distribution of cRNA of neuroactive substances in an anatomical context.

In terms of resolution, *in situ* hybridization is inferior to immunohistochemistry since cRNA localization can not be resolved in very small processes, e.g. dendrites. This is a particular disadvantage when translational sites of receptor protein are considered.

Further Reading

Callea, F., Sergi, C., Medicina, D., Pizzorni S., Brisigotti, M., Fabbretti. G., Bonino, F. (1992): From immunohistochemistry to *in situ* hybridization. *Liver* **12**: 290–295.

Coghlan, J. P., Aldred, P., Haralambidis, J., Niall, H. D., Penschow, J. D., Tregear, G. W. (1985): Hybridization histochemistry. *Anal. Biochem.* **149**: 1–28.

Kawata, M., Yuri, K., Sano, Y. (1991): Localization and regulation of mRNAs in the nervous tissue as revealed by *in situ* hybridization. *Comp. Biochem. Physiol. C* **98**: 41–50.

Knippers, R., Phillipsen, P., Schäfer, K. P., Fanning, E. (1990): *Molekulare Genetik, 5th edn.* Thieme Verlag, Stuttgart.

Lewin, B. (1988): *Gene.* VCH, Weinheim.

2.7
Staining and Neuroanatomical Tract Tracing

Staining provides a means of selectively "coloring" structures in an otherwise transparent tissue. The common Nissl-staining, for example, is useful for visualizing the distribution pattern of neuronal somata in brain tissue. Silver-staining allows the impregnation of neuronal processes and somata and in the form of the Golgi technique, for example, the classification of neuronal cell types by their specific topology is possible. Various methods have been developed in order to obtain structural details of connectivity.

One classic approach to determine projection patterns is to selectively damage nerve cells in defined nuclei or brain regions. Since the soma of a neuron is the metabolic source of the entire cell, destruction of the soma causes the axons to die. By specific silver-staining techniques, for example the Fink-Heimer or Nauta-Gygax methods, degenerating axons can be stained. By this means, projection patterns can be traced throughout the nervous system.

The disadvantage of the degeneration method is that only a fraction of degenerating populations of neurons is stained, so that tracing their projection pattern may give rise to spurious results. In addition, damaging neurons by toxins or mechanical lesioning does not allow single cells to be traced.

Considerable progress has been made in the development of highly sensitive dyes for neuronal tracing during the past decade.

Many modern neuroanatomical tracing techniques require the injection of compounds or dyes that are endocytosed and subsequently transported in an anterograde or retrograde direction (or both). Frequently used anterograde tracers, for example, are PHA-L and biocytin, which may be coupled to dextranamines, while for retrograde tracing fluoro-gold and the cholera toxin subunit B coupled to HRP is commonly used.

Most compounds can be visualized by fluorescence microscopy or through the application of standard histochemical or immunohistochemical techniques.

Afferent or efferent projections can usually be discriminated by placing injections within a given brain region. By using multiple tracers with different transport characteristics, it is possible to address issues such as axonal collateralization or network topology.

Intracellular injection of a tracer allows single neurons with their specific axonal and dendritic patterns to be visualized with a Golgi-like accuracy. The advantage over the Golgi technique, however, lies in its selectivity for physiologically or functionally characterized neurons.

Some tracer molecules (lucifer yellow or calcein) are small enough to allow transfer to neighboring neurons via gap junctions. Depending on the rational of study, this can be advantageous or disadvantageous.

Two categories of tracer molecules are frequently used:
1. Fluorescent dyes or a molecule coupled to a fluorophore (which allows the direct identification by means of a fluorescence microscope).
2. Tracers coupled with an enzyme (the most commonly used enzyme is horseradish peroxidase).

Since the enzyme-coupled tracers lack a color signal, they have to be further processed by histochemical methods (mainly by the use of 3'-3'-diaminobenzidine as a chromogene). The latter technique is time-consuming but has the advantage that the reaction product is durable and light-stable and does not require a fluorescence microscope.

Various types of tracers with different biological and physicochemical properties are available. Before choosing a particular tracer for an experimental application, the individual characteristics of the tracer must be taken into consideration, e.g. toxicity, diffusability and predominant transport direction and velocity.

Some tracers can also be used with electron microscopy to define synaptic connectivity. A straightforward approach for this purpose is the combination of light microscopical preselection with subsequent electron microscopical examination by a fluorescent dye, which is photoconverted into an electron dense precipitate by UV irradiation. Most tracer techniques are compatible with immunohistochemistry or *in situ* hybridization, making simultaneous detection of projections and transmitter or receptor labeling possible.

Further Reading
Honig, M.G., Hume, R.I. (1989): DiI and DiO: versatile fluorescent dyes for neuronal labelling and pathway tracing. *Trends Neurosci.* **12**: 333–341.
Horikawa, K., Armstrong, W.E. (1988): A versatile means of intracellular labeling: injection of biocytin and its detection with avidin conjugates. *J. Neurosci. Methods* **25**: 1–11.
McDonald, A.J. (1992): Neuroanatomical labeling with biocytin: a review. *NeuroReport* **3**: 821–827.
Nance, D.M., Burns, J. (1990): Fluorescent dextrans as sensitive anterograde neuroanatomical tracers: application and pitfalls. *Brain Res. Bull.* **25**: 139–145.
Richmond, F.J.R., Gladdy, R., Creasy, J.L., Kitamura, S., Smith, E., Thomson, D.B. (1994): Efficacy of seven retrograde tracers, compared in multi-labelling studies of feline motoneurons. *J. Neurosci. Methods* **53**: 35–46.

von Bohlen und Halbach, O., Dermietzel, R. (1999): *Methoden der Neurohistologie*. Spektrum Akademischer Verlag, Heidelberg.

Zimmermann, R.P. (1986): Specific neuronal staining by *in vitro* uptake of Lucifer yellow. *Brain Res.* **383**: 287–298.

2.8
Electrophysiology

2.8.1
In vivo Recording

Electrophysiological techniques can be applied to *in vivo* or *in vitro* recording of neurotransmitter and neuromodulator effects. *In vivo* experiments can monitor the electrical activity of a single cell (single unit recording) or many cells (multi-unit recording). Multi-unit recordings are more difficult to interpret because neurons recorded in a multi-cell array can be physiologically heterogeneous. For example, in multi-unit recording, opposing changes in different cells can appear as no electrical change, leading to misinterpretations. It is also often difficult to ensure stability of the recorded signal over time with multiple-cell activity.

In vivo recording from the brain is performed in the following way. One goal of *in vivo* recording is to explore electrical activity in intact brain matter. For this purpose, the animal is anesthetized or decerebrated and placed in a stereotaxic apparatus. Microelectrodes are positioned at the desired coordinates, which can be determined by reference to a stereotaxic atlas.

It is mandatory that the electrodes are placed accurately and this requires that their precise localization has to be verified after each experiment by histological examination.

In the case of single unit recordings, the impulse activity of neurons is typically recorded extracellularly and the tip of a microelectrode is positioned close to the neuron of choice. When the microelectrode is in proximity to the neuron, current fields generated by action potentials of the recorded neuron are detected through the microelectrode as small voltage deflections.

In the case of neurotransmitter studies, *in vivo* electrophysiology is often combined with iontophoresis and stimulation recording.

Neurotransmitter iontophoresis is a technique that allows charged neuroactive substances to leave a micropipette by electrical current flow of the same polarity as the net charge of the applied substance. By passing current through a glass micropipette, one can apply neuroactive substances close to a neuron which is being recorded simultaneously by a second adjacent electrode. Multiple substances can be applied by using multibarrel glass micropipettes. These pipettes are manufactured in such a way that up to seven micropipettes are packed in close together. A single recording micropipette is used to monitor extracellular neuronal activity (the recording pipette may be one capillary of the multibarrel pipette assembly, or it may be an adjacent pipette fixed to the iontophoretic multibarrel electrode). When stable electrical activity is recorded from a neuron, a current of appropriate polarity is applied to the barrel containing the substance and the resulting effects on neuronal activity can be monitored.

Stimulation can be used to measure the postsynaptic electrical effects of an afferent input to a neuron. In this case, electrical pulses are applied to a stimulating electrode to activate neurons (or axons) that project to the area where a target cell is recorded. Typically, extracellular recording is used to measure the effects of afferent input. Data are plotted in the form of histograms where the effects of several experiments are accumulated. By accumulating activity in histograms, even weak responses can be detected as a result of summation effects and averaging-out "noise". This type of analysis allows response magnitude, duration and onset latency to be quantified and compared with the response measured after experimental manipulation, e.g. drug or neurotransmitter application.

In vivo electrophysiology possesses some advantages over *in vitro* techniques. A major advantage is that the tissue is in a more intact condition than during *in vitro* studies. One disadvantage is that it may be difficult to gain access to the desired structure; and there may be side-effects of the anesthetics which may influence the responses to the applied neuroactive substance. In addition, some neuroactive substances can modulate blood pressure and this in turn may change neuronal activity physiologically or artefactually because of mechanical instability.

Neuropsychopharmacological experiments require intact and non-anesthetized animals. Thus, behavioral studies constitute the primary object of *in vivo* recordings. The most common types of microelectrodes used for recording from neurons in behaving animals are tungsten or platinum–iridium wires. Typically, multi-electrode recording is combined with simultaneous monitoring of other parameters, including EEG and EMG. Behavioral electrophysiology allows neural activity to be correlated with the behavioral patterns of an animal. Confounding effects of anesthetics are also avoided since recording takes place in the awake animal.

To minimize variability in external effects or movement artefacts which may impair electrical activity in the behaving animal, the animal has to be conditioned and trained over long periods before *in vivo* recording can be used.

2.8.2
In vitro Recording

In vitro recording is applied to cell cultures, to acute slice preparations, or to the isolated superfused brainstem. A variant of acute slice preparation is *in vitro* recording from acutely isolated neurons. The latter are prepared by mild enzymatic digestion of brain slices, where they appear to retain many of the electrophysiological properties of neurons in slice preparations.

Chronically cultured primary neurons and differentiated neuroblastoma cell lines have also been used neurophysiological experiments.

Acutely isolated or cultured neurons have the advantage of direct microscopic visualization and manipulation. A disadvantage is that these preparations lack the connectivity present in the *in vivo* situation. In order to study the network behavior of neurons, slice preparations are currently the method of choice.

Acute slices of brain tissue can be obtained by rapidly removing of the intact brain from the skull. The brain is then placed in ice-cold saline or artificial cerebrospinal fluid (ACSF) saturated with carbogen (95% O_2, 5% CO_2).

Slices can be obtained by vibratome sectioning of the brain (thickness of the sections is about 300–400 µm). Slices can be stored in an oxygenated chamber containing ACSF. From here, they are transferred to a recording chamber, where they are superfused with an oxygenated artificial cerebrospinal fluid at 35 °C. Under these conditions, slices remain "healthy" over a period of several hours.

Recording is usually performed after 1–2 h of incubation to allow the slices to recover from the acute insult of dissection.

An advantage of slice preparations is that deep-brain areas can easily be accessed and electrodes can be positioned under visual control. In addition, slice preparations are independent of blood circulation and stable recordings can be obtained without the need of anesthetics or immobilization.

A further advantage of the slice preparation is that it allows repeated application of defined concentrations of drugs to the tissue. In particular, if receptor studies are performed, information on pre- and postsynaptic events can simultaneously be obtained.

A disadvantage is that the slice isolates the neurons from more distant projections and thereby creates an artificially restricted area devoid of long-range connectivities.

Generally, one can distinguish two different types of *in vitro* recordings:
- extracellular recordings
- intracellular recordings.

For extracellular recording, the electrode is placed close to the neuron. In this position, the current flows from the nerve cell into the surrounding extracellular fluid, from which it can be recorded. For intracellular recording, the neuron is impaled by the electrode and recordings are obtained from inside the cell.

Extracellular recording
Nerve cells generate extracellular current flow by fluctuations of the membrane potential of dendrites and cell bodies. Postsynaptic potentials elicit an outflow of negative (excitatory) or positive (inhibitory) ionic charges into the extracellular fluid. The bioelectrical activity of neurons can be registered by recording the sum of different potentials (so-called field potentials). Under these conditions, the influence of physiological and pharmacological manipulation can be measured as a change in the amplitude of field potentials. The underlying physical phenomenon is the flow of current through the extracellular space, which is induced by the transmembranous current of the activated neurons. Measurements do not record these currents directly, instead the difference of current between changes of potentials in the activated neurons and the recording electrode is measured.

Intracellular recording

Measurements of membrane potentials with an intracellular or cell-attached electrode are collectively called intracellular recordings. For impaling the neuron, special microelectrodes are used (so-called "sharp" microelectrodes). These are commonly applied in traditional intracellular recordings, which are performed either under current-clamp or voltage-clamp conditions.

Current-clamp is a method involving measurement of the voltage difference across the membrane while injecting constant positive or negative currents into a cell. Current-clamp recording is usually performed by inserting a single sharp micropipette into a neuron and recording the voltage and injected currents through the same pipette. By applying a neuroactive substance to the cell, a change in the size of the voltage response to the current pulse indicates a change in ionic conductance (the conductance is the reciprocal value of resistance ($g = 1/R$) with the unit Siemens [1 Siemens (S) = 1 volt per ampere]. By incrementally varying the amplitudes of the current steps over an appropriate range, a set of voltage responses can be obtained which can be plotted as a current–voltage (I/V) curve. The curve indicates the "macroscopic" currents passing through the membrane at different membrane potentials. Treatment with neurotransmitters or drugs that alter ionic conductance thus influences the slope and shape of the I/V curve.

Under the voltage-clamp recording, a neuron is held at a constant voltage (membrane potential) and the current required to hold this voltage is measured, allowing a direct measurement of ionic currents. By applying voltage-command jumps, the changes in current flow (including their kinetics) can be measured. Thus, this method allows voltage-dependent and time-dependent ionic conductances to be directly recorded under experimental conditions.

A further intracellular recording method is the patch-clamp technique. This technique constitutes a modification of the common intracellular recording method with sharp electrodes. In the case of the patch-clamp technique, a small negative pressure is applied to the recording electrode (i.e. the patch pipette), mostly by suction with a tube-fixed syringe. If the tip of the pipette is placed on the surface of the cell, the membrane attaches to the tip of the polished micropipette and a tight mechanical and electrical high-resistance (giga-ohm) seal results.

The patch-clamp method can be applied in at least four configurations:
1. the cell-attached mode
2. the inside-out mode
3. the outside-out patch mode
4. the whole cell patch mode.

1. In the cell-attached configuration, recording of single channels is possible without disrupting the cell membrane.
2. In the inside-out and in the outside-out configurations, the membrane patch is detached from the neuron after a giga-ohm seal is formed. This mode enables one to record single-channel activity in isolation from the cell. In the inside-out configuration, the patch of membrane is pulled apart from the neu-

ron, leaving the patch attached to the pipette with its cytoplasmic surface exposed to the bathing solution.
3. In the outside-out patch, a membrane patch is formed in which the extracellular membrane surface is exposed to the bathing solution.
4. When whole-cell patches are applied, the cell membrane is ruptured by the tip of the electrode, after the giga-ohm seal has established, and recording from inside the cell can be obtained. A modification of this technique (perforated patch) is to include a substance (e.g. Nystatin) in the patch pipette, which perforates the membrane sealed to the pipette, and so provides electrical continuity between cell interior and pipette.

The patch-clamp technique offers a valuable method for the study of biophysical properties of single channels. Single-channel, cell-attached and inside-out patch-clamp modes are particularly useful in studying second-messenger systems. These modes are, however, difficult to use with slice preparations. For this purpose, a whole-cell configuration is better suited. In addition, the whole-cell patch mode allows the ionic content of a single neuron to be modified merely by changing the buffer and salt concentrations in the pipette solution. Second messengers and drugs affecting channel activity can be introduced into the pipette solution for diffusion into the neuron. Unfortunately, the whole-cell configuration does not allow the exact ionic or second-messenger concentration in the cell interior to be determined; and there a constant risk that essential metabolites, e.g. cAMP, can diffuse from the cell into the pipette.

Further Reading

Bezanilla, F., Vergata, J., Taylor, R. E. (1972): Voltage clamping of excitable membranes. In: *Methods of Expertimental Physics, vol. 20*. ed. Ehrenstein, G., Lecar, H., Academic Press, New York.

Dingledine, R. (ed.) (1983): *Brain Slices*. Plenum Press, New York.

Finkel, A. S., Redman, S. J. (1984): Theory and operation of a single microelectrode voltage clamp. *J. Neurosci. Methods* **11**: 101–127.

Kandell, E. R., Schwartz, J. H., Jessell, T. M. (2000): *Principles of Neural Science, 3rd edn*. Elsevier, New York.

Kettenmann, H., Grantyn G. (1993): *Practical Electrophysiological Methods*. Wiley Liss and Sons, New York.

Kischka, U., Wallesch, C.-W., Wolf, G. (eds.) (1997): *Methoden der Hirnforschung: eine Einführung*. Spektrum Akademischer Verlag, Heidelberg.

Llinás, R., Nicholson, C. (1974): Analysis of field potentials in the central nervous system. In: *Handbook of Electroencephalography and Clinical Neurophysiology, vol. 2*. ed. Stevens, C. F., Elsevier, Amsterdam.

Numberger, M., Draguhn, A. (1996): *Patch-clamp Technik*. Spektrum Akademischer Verlag, Heidelberg.

Sakman, B., Neher, E. (eds.) (1995): *Single-channel Recording, 2nd edn*. Plenum Press, New York.

Shepard, G. M. (1994): *Neurobiology, 3rd edn*. Oxford University Press. New York.

Sherman-Gold, R. (ed) (1993): *The Axon Guide*. Axon Instruments, Forster City.

Stuart, G. J., Dodt, H.-U., Sakmann, B. (1993): Patch-clamp recordings from the soma and dendrites of neurons in brain slices using infrared video microscopy. *Pflügers Arch.* **423**: 511–518.

2.9
Behavioral Testing

To understand the functional effects of neuroactive substances, it is necessary to study their behavioral effects in addition to their electrophysiological properties. In principle, animal behavior can be studied in natural settings or under laboratory conditions. Most behavioral studies are performed in laboratory settings, because these conditions allow strict control of parameters which could influence behavior. Behavioral studies in laboratory settings focus mainly on the following subjects: learning tasks, behavioral development, psychopharmacology.

Behavioral testing of animals allows information to be obtained about the influence of neuroactive substances in perception, learning and memory, emotional, sexual and social behavior.

Behavior can be defined as an animal response to a stimulus. A stimulus can be an endogenous signal (e.g. action of a neuroactive substance) or an environmental signal (light, sound). The behavior of animals can be modified by conditioning. Two types of conditioning paradigms are differentiated:
- classic conditioning, as introduced by the Russian physiologist, I. Pavlov (1849–1936);
- operant conditioning, which was developed by the American psychologist, B. F. Skinner (1904–1990).

2.9.1
Classic Conditioning

In classic conditioning, an animal can be trained to respond to an external stimulus in an inappropriate manner. For example, a dog can be conditioned to salivate when a bell rings – a response that is naturally irrelevant to the animal behavior. When undergoing classic conditioning, the animal is repeatedly offered two different stimuli in timed sequences. The first stimulus, called the neutral or conditioned stimulus (CS), does not usually cause the animal to respond in the desired way, since ringing of a bell is a neutral stimulus for a dog in terms of salivation.

The second stimulus, called the unconditioned stimulus (UCS), causes the desired behavior. In this example presentation of food constitutes the unconditioned stimulus because it causes the dog to salivate. In classic conditioning, the CS is followed by the UCS. When conditioning is successful, the dog responds with the CS to the bell even though no food is presented.

Fear conditioning is a special form of Pavlovian (classic) conditioning. In fear conditioning, the extent of a fear response is measured when a conditioned stimulus is paired with an aversive unconditioned stimulus.

A further test used in investigating fear-related behavior is the so-called fear-potentiated startle. This test measures an animal's (usually a rat's) startle behavior in response to a loud tone. For conditioning the tone is preceded by a light, which has previously been paired with footshock. As a result of this pairing the light increases (potentiates) the startle response of the rat to the tone.

2.9.2
Operant Conditioning

In contrast to classic conditioning, the animal is no longer a passive participant, because it has to learn a task or to solve a problem. In operant conditioning, an animal is exposed to some type of reward or punishment whenever it behaves in a certain way, e.g. pushing a lever, or moving from one place to another. The reward or punishment, also called reinforcement, follows the action. Food or water may be used as reward and electric shocks as punishment. Rewarding the animal usually increases the probability that it will repeat the action, while punishment decreases the probability. In the simplest case, the reward is delivered every time the animal produces the behavior desired by the experimenter. This is termed a continuous reinforcement (CRF) schedule.

The strength of the response depends on the amount of reward or punishment. Especially in the case of instrumental conditioning, the strength of the response can be used as a measurement of the animal's "drive" to escape the punishment or to obtain a reward. Thus, these experiments could provide a basis for studying motivation.

2.9.3
Further Behavioral Tests

In addition to classic and operant conditioning, other behavioral tests are also applied in psychopharmacological research. A test which is widely used in examining the effects of neuroactive substances on behavior is the active avoidance reaction.

In active avoidance, the animal starts a trial in an environment where it is exposed to a constant aversive stimulus, e.g. electric shock application, while an adjacent environment is safe. If the animal does not move between both areas in a fixed time, the shock is delivered. The animal is then placed back in the dangerous compartment for the start of the next trial. Under this experimental condition, learning consists in a decrease in the latency to leave the dangerous compartment, initially resulting in learned escape and then in avoidance.

Besides tests that explore emotional or conditioned behavior, maze tests can be used to record spatial learning (a type of learning, which is related to the hippocampal formation).

The simplest type of maze is the T-maze in which the animal has only two possible choices to go: a correct and an incorrect way.

The next level of complexity is represented by spatial spontaneous alternation. In this procedure, the animal is placed in the stem of a T-maze and allowed to explore freely. The first choice of arm (left or right) is recorded and the procedure is repeated. On the second trial, about 75% of normal animals typically choose the arm opposite to the one chosen on the first trial.

In a radial-arm maze task, the rat is placed in the center of a maze in the shape of a star burst. The simplest form of this task consists in the arms being supplied with a piece of food at the beginning of any one trial. The most efficient performance for the rat is to visit each arm only once. This seems a very simple behavioral pattern insofar as these results can be achieved by a rule such as "turn sharp left as you come out of each arm". In practice, rats do not solve the task in this way, but tend to choose arms roughly opposite to the one which they have visited before, a behavior that indicates a considerable amount of working memory information (possibly spatial information).

The most complex form of maze task is the Morris water maze. This consists of a circular featureless swimming pool, which contains a submerged and hence invisible platform, which is permanently located in the same spatial position. The animal is placed into the pool at different positions. Normal rats learn quickly to find the platform and to swim almost directly to it. The invisibility of the platform also allows the use of transfer tests to determine what the animal has learned. In transfer tests, the animal is placed in the water as before, but the platform has been removed from the pool after a cycle of trials. Control animals show that they are aware of the precise position of the platform. They swim to this position and start circling around the no longer present platform.

These tests, as well as others, are useful in determining the effects of neuroactive substances on behavioral pattern. Test batteries have to be performed in two groups: a non-treated group which serves as control and a group which is treated with a neuroactive substance (by injection of the substance or by inactivation of the substance). Comparative statistical analysis of the behavioral patterns of the two groups provides information about any effects of the neuroactive substance on behavior.

Since behavior requires complex mental processing, it is necessary to apply different tasks in order to obtain valid information.

Behavioral studies are time-consuming and require considerable practice on the part of the experimenter.

Further Reading

Davis, M. (1992): The role of the amygdala in fear-potentiated startle: implications for animal models of anxiety. *Trends Pharmacol.* **13**: 35–41.
Gray, J. A. (1987): *The Psychology of Fear and Stress.* Cambridge University Press, London.
Kandell, E. R., Schwartz, J. H., Jessell, T. M. (1991): *Principles of Neural Science, 3rd edn.* Elsevier, New York.
McFarland, D. (1998): *Animal Behaviour. Psychobiology Ethology and Evolution.* Prentice Hall, Englewood Cliffs, N.J.
Mackintosh, N. J. (1983): *Conditioning and Associative Learning.* Clarendon, Oxford.
Mischel, W. (1993): Behavioral Conceptions. In: *Introduction to Personality.* W. Mischel, Harcourt Brace, New York.
Morris, R. G. M. (1984): Development of a water-maze procedure for the study of spatial learning in the rat. *J. Neurosci. Methods* **11**: 47–60.
Pavlov, I. P. (1927): *Conditioned Reflexes.* Oxford University Press, London.
Shepard, G. M. (1994): *Neurobiology, 3rd edn.* Oxford University Press, New York.
Sinz, R. (1981): *Lernen und Gedächtnis, 3rd edn.* Fischer Verlag, Stuttgart.
Viana, M. B., Tomaz, C., Graeff, F. G. (1994): The elevated T-maze: a new animal model of anxiety and memory. *Pharmacol. Biochem. Behav.* **49**: 549–554.

3
Neurotransmitters

The classic neurotransmitters represent amino acids and biogenic amines, which are able to initiate synaptic transmission in the nervous system. A collection of criteria have been established to differentiate neurotransmitters from neuromodulators (see also under Introduction):

- The classic neurotransmitters are produced and stored within neurons and are released upon an adequate electrical signal.
- Neurotransmitters are localized in presynaptic terminals and are released into the synaptic cleft in the form of quantal portions for mediating postsynaptic excitatory (EPSP) or inhibitory (IPSP) events.
- Neurotransmitters are selectively released upon nerve stimulation in a calcium-dependent manner.
- Neurotransmitters react with receptors on the postsynaptic or presynaptic sites. The effects can be prevented by specific antagonists and facilitated by specific agonists, which mimic the action of the transmitter.
- Neurotransmitters are inactivated rapidly after release. This inactivation is mediated by specific enzymes or by re-uptake mechanisms.
- Experimental application of a neurotransmitter at postsynaptic sites elicits effects identical to the endogenous substrate.

The structure of the neurotransmitter is conserved and consequently its biological activity does not vary due to a strict structural–functional relationship.

Neurotransmitters are either fast-acting substances, which open ligand-gated ion channels and elicit an immediate flow of current through the activated channel or they behave as slow-acting agents which induce long-lasting changes at the postsynaptic site. The slow-acting neurotransmitters affect the membrane permeability indirectly through second-messenger systems. All neurotransmitters, with the exception of histamine, are recaptured by highly specific transport systems. The transporters play a significant role in the rapid inactivation of the released neurotransmitter, limiting its temporal and spatial action. Some neuropeptides act like neurotransmitters. However, it has not been shown in detail whether they meet all the criteria noted above. Neuropeptides with classic neurotransmitter effects are named putative neurotransmitters or co-transmitters.

Additional classes of substances with neurotransmitter properties are gaseous molecules like nitric oxide (NO) or carbon monoxide (CO). These gaseous sub-

Neurotransmitters and Neuromodulators. Handbook of Receptors and Biological Effects. 2nd Ed.
Oliver von Bohlen und Halbach and Rolf Dermietzel
Copyright © 2006 WILEY-VCH Verlag GmbH & Co. KGaA, Weinheim
ISBN: 3-527-31307-9

3 Neurotransmitters

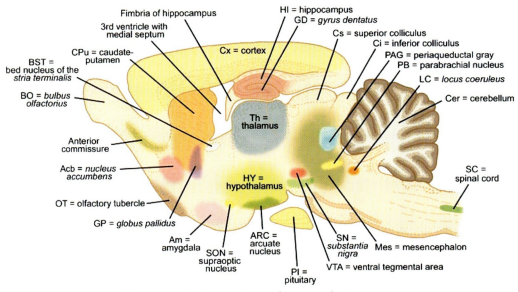

Fig. 3.1 Master map for the orientation of important brain structures with the nomenclature and captions used throughout the handbook.

stances meet some criteria of neurotransmitters. Some essential neurotransmitter criteria, like receptor interaction and specific re-uptake and degradation mechanisms, are not fulfilled by this group and thus they do not qualify as true neurotransmitters.

The putative neurotransmitters as well as the gaseous oxides will therefore be described under the section "Neuromodulators".

The following chapter will give an account on general aspects, localization function and disorders of the most important neurotransmitters and neuromodulators. As a guide for better neuroanatomical orientation we will first provide a master map which includes the major neuroanatomical structures and captions mentioned throughout the handbook (Fig. 3.1).

3.1 Acetylcholine

3.1.1 General Aspects and History

Acetylcholine (ACh) was the first neurotransmitter discovered. ACh plays a significant role in synaptic transmission in the central and peripheral nervous system. In 1907, Hunt and, in 1914, Sir Henry Dale were able to demonstrate that esters of choline produce physiological effects. Additionally, Dale distinguished two main classes of ACh receptors: the muscarinic and the nicotinic receptors. In 1921, Loewi discovered that stimulation of the vagus nerve results in the release of a chemical substance: the so-called vagal substance. The chemical substrate was identified later (Loewi and Navratil 1926) to be acetylcholine.

In 1933, Chang and Gaddum demonstrated that acetylcholine is present in the brain of mammals as well. The discovery of specific receptors capable of binding acetylcholine and the detection of enzymes which catalyze its biosynthesis and degradation have allowed demonstration of both the general distribution of acetylcholine within the nervous system and its role as a neurotransmitter.

3.1.2
Localization Within the Central Nervous System

The cholinergic system of brain tissue can be divided into three different subsystems:

- *Cholinergic motoneurons in the spinal cord:* The collaterals of these neurons activate small interneurons in the ventral horn of the spinal cord (Renshaw cells), which express nicotinic receptors.
- *Interneurons and local projection neurons*: The most representative neurons of this type are interneurons in the striatum. These interneurons interact with the dopaminergic terminals of neurons which project from the substantia nigra into the striatum. In addition, sparsely distributed cholinergic interneurons are located in the cortex, the hippocampus and in the olfactory bulb.
- *Projection neurons*: Different groups of cholinergic projection neurons can be distuinguished according to a nomenclature which was established in 1983 by Mesulam and coworkers. Group Ch1 and Ch2 correspond with cholinergic neurons in the region of the medial septal nucleus and with neurons in the diagonal band of Broca. These neurons project to the hippocampus. Group Ch3 is located in the horizontal band of Broca. Neurons of this group innervate the olfactory bulb. Members of group Ch4 are represented by neurons of the magnocellular region of the preoptic nucleus, the magnocellular region of the nucleus basalis of Meynert and in the substantia innominata. These neurons project to the cerebral cortex and to the amygdala. Members of groups Ch5 and Ch6 are located in tegmental areas of the brain. They possess ascending projections to the thalamus and to the hypothalamus as well as descending projections. The descending projections approach the pons, the nucleus vestibularis, the locus coeruleus and various raphe nuclei. The neurons of group Ch7 occur in the habenula. They project to the interpeduncular nucleus. Finally, neurons of group Ch8 are located in the parabigeminal nucleus and send projections into the superior colliculus (Fig. 3.2).

Acetylcholine has been shown to occur in colocalization with GABA, though the physiological relevances of this coexpression is not known.
Terminals of these neurons can form excitatory asymmetric synapses with dendrites and symmetric, inhibitory synapses with additional dendrites. This ar-

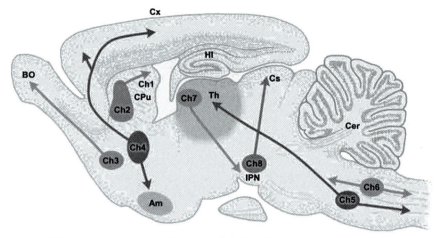

Fig. 3.2 Schematic representation of the central cholinergic projections in the brain of rats. *Abbreviations*: Am=amygdala; BO=*bulbus olfactorius*; Cer=cerebellum; Cs=*superior colliculus*; CPu=caudate-putamen; HI=hippocampus; IPN=interpeduncular nucleus; Th=thalamus.

borization allows a single neuron to exert different effects via functionally differentiated synapses.

3.1.3
Biosynthesis and Degradation

The synthesis of acetylcholine is straightforward and follows a simple scheme. ACh is synthesized from acetylCoA and choline by an enzyme known as choline acetyltransferase (ChAT). The choline acetyltransferase is located almost exclusively in the cytoplasm of cholinergic fibers. Cholinergic neurons contain high concentrations of ChAT, indicating that this enzyme is not rate-limiting in the synthesis of acetylcholine. The acetyl-CoA which serves as a donor derives from pyruvate generated by glucose metabolism in the mitochondria. Choline derives from dietary sources and from phosphatidylcholine. Furthermore, choline can be supplied by ACh hydrolysis. The source of the hydrolized acetylcholine is the extracellular acetylcholine, which has been released into the synaptic cleft and reimported into the cholinergic neurons by re-uptake. This choline transporter system which serves for the re-uptake is known as sodium-dependent high-affinity choline uptake (SDHACU). This system is localized exclusively at cholinergic nerve terminals. Its function is to transport choline from the extracellular space into the neuron. The SDHACU system is selectively inhibited by hemicholinium. The production of acetylcholine takes place in the cytoplasm of cholinergic neurons. Acetylcholine is stored in vesicular form at the nerve terminals of cholinergic fibers. In response to an action potential, it is released by exocytosis into the synaptic cleft, from which it diffuses to the postsynaptic site for interaction with appropriate receptors (or postjunctional site outside the central nervous system), which results in specific effects according to the receptor type (nicotinic or muscarinic).

Botulinus toxin, as well as tetanus toxin, inhibits the release of acetylcholine, while the toxin of the black widow (a spider) enhances the release of ACh.

Acetylcholine is metabolized by esterases. Several species of specific esterases exist. Two of them are essential for acetylcholine metabolism: acetylcholine esterase (AChE, EC 3.1.1.7) and butyrylcholine esterase (pseudocholinesterase).

AChE is expressed in neurons, mucscle cells and certain hematopoetic cells. In excitable tissue, AChE is localized primarily in the synaptic complex, extracellular to nerve and muscle cells. AChE is also present in moderate abundance in non-cholinergic neurons that receive cholinergic input (so-called cholinoceptive neurons). Acetylcholine esterase has a molecular mass of about 80 kDa and can occur in different forms: the globular monomeric form (G1), the dimeric (G2) and tetrameric (G4) forms as well as the collagen-tailed asymmetric forms A4, A8 and A12.

The G1 form is preferentially located in the cytoplasm, while the G4 form is found on plasma membranes. The asymmetric form A12 seems to occur at synaptic sites of ganglion cells. In the central nervous system, the globular G4 form predominates.

The different forms of acetylcholine esterases are transported by active neuronal transport mechanisms in retrograde and anterograde directions.

Since inactivation of the acetylcholine esterase prolongs the half-life of acetylcholine in the synaptic cleft, agents that inhibit AChE activity are used for therapeutic or toxicological purposes.

Three types of AChE blockers have become famous: the carbamyl esters, which are employed therapeutically as adjuncts to anaesthetics, the organophosphates which are widely used insecticides and anticholinesterase like sarin which have been used as "nerve gas". These substances bind in competition with acetylcholine to the acetylcholine esterase. Since acetylcholine has a half-life of about 50 s, the carbamyl esters effectively compete with the binding site due to their half-life of about 30 min. Organophosphates bind irreversibly because they are not degradable by the acetylcholine esterase.

3.1.4
Receptors and Signal Transduction

The ACh receptors consist of two major groups: the muscarinic and the nicotinic receptors. They can be distinguished by their selectivity to the alkaloids nicotine and muscarine. Dale originally introduced this classification in 1914 and it is still valid, in spite of the fact that several subtypes of nicotinic and muscarinic receptors have been described meanwhile.

The nicotinic receptors belong to the ligand-gated ion channel superfamily. This comprises the nicotinic receptor (neuronal and muscle type), 5-HT, GABA

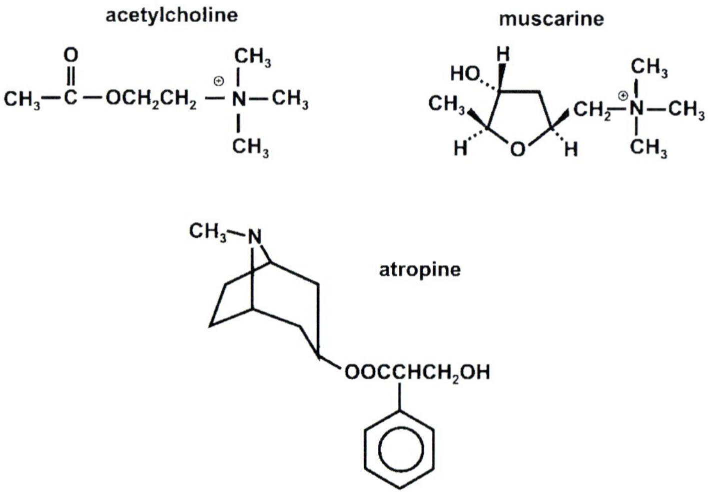

Fig. 3.3 Structure of acetylcholine, muscarine and atropine.

and glycine receptors. It is believed, on the basis of sequence similarity, that all members of this family have a similar tertiary structure.

Muscarinic receptors are activated by the alkaloid muscarine and are blocked by atropine (see Fig. 3.3) and scopolamine. The nicotinic receptors are activated by nicotine and inhibited by curare.

Further agonists are substances like butyrylcholine, carbachol, DMPP and tetramethylammonium, while α-bungarotoxin, dihydro-β-erythroidine, hexamethionine, mecamylamine and Procain elicit antagonistic effects.

The nicotinic receptors

The nicotinic receptors are activated by low concentrations of nicotine, whereas high concentrations block the receptors.

The nicotinic receptors are classified into two subclasses: a neuronal type and a muscle type. However, both receptor subtypes share a high amino acid homology, indicative of a long evolutionary heritage that has primarily conserved the receptor sequence.

The nicotinic receptor was the first neurotransmitter receptor to be isolated (1970), purified and chemically defined (1973) by classic biochemical techniques. This was due to its enrichment in the electroplaques of the electric eel *Torpedo mamorata*, which serves as a rich source for the isolation of nicotinergic membranes. The simplest form of the nicotinic receptor is a glycoprotein with a molecular mass of about 300 kDa. The receptor constitutes of a heterooligomeric complex of four distinct protein subunits, which were classified in order of their molecular size ($a=40$ kDA, $\beta=50$ kDA, $\delta=60$ kDA, $\gamma=65$ kDA). The probable quaternary arrangement of the subunits consists of a pentameric complexes of the $a2, \beta, \delta$ and γ, which encompass the ion channel (Fig. 3.4).

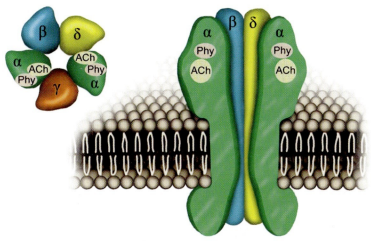

Fig. 3.4 Schematic drawing of a nicotinic acetylcholine receptor, nAChR, which is regarded as a prototype of an ionotropic ligand-gated receptor. The nAChR forms an ion channel, which is composed of five different subunits (a, β, δ, a, γ). Each of these subunits consists of three domains: extracellular, transmembrane-spanning and intracellular. Extracellular domains of the a-subunits form the acetylcholine (ACh)-binding sites. Two ACh molecules bind to the nAChR for receptor activation which directly translates into channel opening. Phy = physiostigmine acts as a non-competitive agonist.

Several putative combinations of subunits are possible to build up functional receptors, either by expression of a single type of subunit or by expression of a single type of subunit together with the β-subunit. In nerve cells, up to three different β-subunits have been described ($\beta2$, $\beta3$ and $\beta4$). An additional eight subunits can contribute to a high diversity of pentamers, which possess distinct pharmacological properties. Although the subunit composition of some native neuronal nicotinic receptors is still unknown, differences in subunit compositions are believed to account for the variance in physiological and pharmacological properties of the receptors. Autoradiographic labeling with [^{3}H] nicotine reveals differences in the binding pattern in the brain as compared to labeling with [^{125}I]bungarotoxin (bungarotoxin is an irreversible ligand and a specific antagonist of the nicotinic receptor, see above), indicative of regional heterogeneity among naturally occurring nicotinic receptors.

Binding of two molecules of acetylcholine to the a-subunit induces a change in the conformation of the pentamer and thereby induces an opening of the ligand-gated ion channel, which subsequently allows the influx of ions. This influx results in a depolarization of the neuron.

In contrast to the muscarinic acetylcholine receptors, no second messengers are involved in the signal transduction. Classic excitatory cholinergic transmission was primarily studied through the muscle-type AChR containing $a1$-, $\beta1$-, δ- and γ-subunits embryologically or ε-subunits in the adult and ganglionic

AChR consisting of $\alpha 3$-, $\beta 4$- and $\alpha 5$-subunits. These are the best studied AChR; and they mediate depolarizing, inward Na^+ currents involved in classic excitatory neurotransmission at the neuro-muscle junction and through autonomic nerve ganglia. There is ample evidence for actions of AChR in the mediation of excitatory transmission at some sites in the CNS, but nicotinergic signaling is not nearly as prominent in the central nervous system as muscarinergic signaling. Some systems such as the limbic system seem to make use exclusively of muscarinergic receptors.

Nicotinergic receptors are present in the hippocampus, the cerebral cortex, the thalamus, the hypothalamus, the superior colliculus as well as in some cholinergic nuclei of the forebrain and brain stem.

Nicotinic receptors have also been demonstrated at presynaptic endings of cortical and mesostriatal neurons.

AChR can play roles other than mediating excitatory neurotransmission. For example, AChR containing $\alpha 7$-subunits mediate very short-lived, nicotine-gated responses of high Ca^{2+} permeability; and other CNS acetylcholine receptors also reveal significant Ca^{2+} permeability. Consistent with their presynaptic location, the permeability of the receptor channel to Ca^{2+} speaks in favor of a control of ACh release through presynaptic ACh autoreceptors. Activation of these receptors in the brain results in transmitter release or facilitates the release of other neurotransmitters. Via nicotinic heteroreceptors, acetylcholine increases the level of glutamate, serotonin, GABA or dopamine and, in addition, acetylcholine increases acetylcholine levels via nicotinic autoreceptors. Recent evidence indicates that acetylcholine can coactivate the release of dopamine, serotonin and glutamate through nicotinic receptors. The identification of presynaptic AChR at sites where other subsets of neurotransmitters occur, i.e. in the hippocampus (norepinephrine, GABA, 5-HT), the striatum and the nucleus accumbens (dopamine), was interpreted as the ability of AChR to modulate the release of other neurotransmitters, perhaps without requiring action potential propagation from the cell body.

The muscarinic receptors

The muscarinic receptors share little similarity in their tertiary structure and physiological function with the nicotinic receptors.

The muscarinic receptors are metabotropic receptors, which means that they are coupled to G proteins and thus belong to the superfamily of G protein-coupled receptors.

The muscarinic receptors are monomers of 440–540 amino acids with seven membrane-spanning domains, the N-terminus residing on the extracellular side and the C-terminus on the intracellular side (Fig. 3.5).

Selective agonists of the muscarinic receptors are substances like arecholine, betanechol, carbachol, etacholine, oxotrimorin and pilocarpin. Selective antagonists are atropine, gallamine, pirenzepin, scopolamine, telenzepin and 4-DAMP.

Based on their pharmacological properties, the muscarinic receptors have been divided into two classes, which are designated M1 and M2. This classifica-

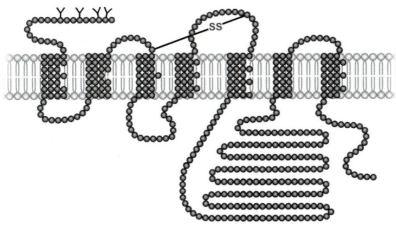

Fig. 3.5 Schematic drawing of a muscarinic acetylcholine receptor (mAChR). In contrast to the nicotinic receptor, this receptor does not form an ion channel. The receptor consists of seven transmembrane-spanning domains and belongs to the G protein-coupled receptor superfamily (GPCRs).

tion corresponds to their different selectivity to the antagonists pirenzepine (M1) and gallamine (M2). Because of the discovery of a competitive cardioselective antagonist, AF-DX 116, the M2 receptors have been subdivided into two further subclasses: M2α and M2β (or M3) receptors. The different subtypes are also present in the central nervous system of mammals.

The genes of five different subtypes of muscarinic receptors have been identified, cloned and sequenced. The five subtypes are labeled as m1, m2, m3, m4 and m5 (Table 3.1). They differ in their distribution and their signal transduction pathways. The amino acid sequence of the membrane-spanning regions is highly conserved among the five subtypes.

The actylcholine-binding side of the muscarinic receptor consists of a pocket-shaped indentation which is formed by the transmembrane domains and which exhibits a site for allosteric regulation by several compounds.

The binding of acetylcholine to muscarinic receptors activates different signal transduction pathways, depending on the type of muscarinic receptor. The two main pathways are:
- inhibition of adenylate cyclase activity with subsequent reduction of intracellular cAMP levels;
- activation of phosphatidyl inositol to form diacylglycerol and inositol trisphosphate.

The M1 receptor group (m1, m3, m5) is coupled to a pertussis toxin-insensitive G protein of the G_q family, which activates phospholipase C. Activation of phospholipase C leads to the generation of the second messengers diacylglycerol (DAG) and inositol trisphosphate (IP_3) from phosphatidylinositol. Inositol tri-

Table 3.1 Essential properties and antagonists of the different muscarinergic receptor subtypes.

Properties	Subtype				
	m1	m2	m3	m4	m5
Molecular mass (kDa)	51240	51715	66127	53058	60186
Amino acids	460	466	590	479	532
G protein coupling	G_q, G11	G_i, G_o	G_q, G11	G_i, G_o	G_q, G11
Effectors	IP3/DAG	cAMP(–)	IP3/DAG	cAMP(–)	IP3/DAG
Tissue	Brain, autonomic ganglia, vas deferens, secretory glands, sympatic ganglia	Brain, heart, sympatic ganglia, lung, Uterus, smooth muscles	Brain, secretory glands, smooth muscles	Brain, lung	?
Brain areas	Cerebral cortex, hippocampus, amygdala, striatum, bulbus olfactorius, olfactory tubercle, nucleus accumbens	Basal forebrain, bulbus olfactorius, thalamus, cerebellum, brain stem	Cerebral cortex, piriform cortex, bulbus olfactorius, thalamus, striatum, brain stem	Cortex, hippocampus, olfactory tubercle, thalamus, striatum	a)
Subtype-selective antagonists	Pirenzepine, telenzepine	Gallamine, himbacine	4-DAMP	Tropicamide, hexocyclium	?

a) Only small amounts of the m5 subtype have as yet been discovered and its distribution the CNS is not fully understood.

sphosphate initiates the release of calcium from intracellular stores, while DAG activates protein kinase C.

Muscarinic receptors of type M2 (m2, m4) are coupled to G_i proteins. Activation of the M2 group inhibits the activity of adenylate cyclase.

This classification has been extended by recent studies, indicating that additional pathways are involved in signal transduction:
- stimulation of the phospholipase A2 by receptors of type M1 (m1, m3, m5);
- stimulation of the phospholipase D (m1 and m3);
- stimulation of adenylate cyclase mediated by the β-subunit of G proteins;
- inhibition of phosphodiesterases.

Activation of the muscarinic receptors initiates a number of depolarizing and hyperpolarizing currents through direct or indirect mechanisms. These effects include:
- stimulation of the inwardly rectifying potassium conductance by muscarinic receptors of type M2 (m2 and m4);

- inhibition of calcium conductance by muscarinic receptors of type M2 (an effect mediated either directly via G proteins or indirectly via a reduction in cAMP levels);
- activation of calcium-dependent potassium, chloride and cation conductance by muscarinic receptors of type M1 (m1, m3, m5);
- inhibition of voltage and time-dependent potassium conductance (M-current) by m1 and m3.

The muscarinic receptors of type M1 are preferentially expressed in the cortex, the hippocampus (including the dentate gyrus), the nucleus accumbens, the striatum and the amygdala.

Type M2 receptors occur in cholinergic nuclei of the thalamus as well as in the superior colliculus, olfactory bulb and in the brain stem. Muscarinic receptors of type M3 (or the subtype M2β) are found in the hippocampus and in the cerebral cortex.

In situ hybridization studies revealed that mRNA of m1 is formed in cortical and striatal brain areas and in the amygdala, the hippocampus, the nucleus accumbens, the olfactory bulb and the olfactory tubercle. By the same technique, mRNA of the m2 subtype has been demonstrated in cholinergic neurons of the cortex and some subcortical areas. These data are in accordance with the distribution of muscarinic receptors of type M2.

Messenger RNA of m3 has been found in the cerebral cortex, the piriform cortex, the hippocampus, the thalamus and in the caudate-putamen, while mRNA of m4 was demonstrated in high concentrations in the cortex, the hippocampus, the thalamus, the caudate-putamen complex and the olfactory tubercle. Messenger RNA of m5 is widely distributed throughout the central nervous system, though the expression of the corresponding protein is rather low (Table 3.1).

3.1.5
Biological Effects

Several toxins can impair the functions of the acetylcholine system. These toxins can be used for pharmacological purposes because of their agonistic or antagonistic properties (Table 3.2).

In addition, further pharmacological substances have been developed, which selectively act upon muscarinic or nicotinic receptors (Table 3.3).

Effects which are mediated through nicotinic receptor are rare in the central nervous system. However, some effects have been demonstrated in Renshaw cells of the spinal cord as well as in neurons of some thalamic areas and the prefrontal cortex. These effects are characterized by a fast excitation since the neuronal nicotinic receptors are permeable for calcium ions. The nicotinic receptors are considered to be involved in synaptic plasticity, which uses calcium as a second messenger.

Muscarinic receptors are present in autonomic ganglia (as are nicotinic receptors) and in organs innervated by postganglionic parasympathetic fibers. The sig-

Table 3.2 Some biologically active substances which possess cholinergic agonistic or antagonistic properties.

	Substance	Derivation	Effect
Agonists	Nicotine	Alkaloid from the tobacco plant	Activates nicotinic receptors
	Muscarine	Alkaloid from the mushroom *Amanita muscaria*	Activates muscarinic receptors
	α-Latrotoxin	A toxin of the spider black widow	Massive release of acetylcholine
Antagonists	Atropine and scopolamine	Alkaloid from the plant *Atropa belladonna*	Blocks muscarinic receptors
	Botulinus toxin	A toxin from *Clostridium botulinum*	Inhibits the release of acetylcholine
	Bungarotoxin	A toxin of the snake *Bungarus*	Inhibits acetylcholine receptors
	d-Tubocurarin	A compound of curare	Blocks acetylcholine receptors of the motor end plate

Table 3.3 A list of some principal agonists and antagonists of nicotinic and muscarinic receptors. *Abbreviations*: 4-DAMP = 4-diphenyl-acetoxy-N-methylpiperidine; DMPP = dimethyl-4-phenylpiperazinium.

	Nicotinic receptors	Muscarinic receptors
Agonist	Nicotine Butyrylcholine Tetramethylammonium DMPP Carbachol	Muscarine Pilocarpine Arecholine Oxoremorine Carbachol
Antagonist	Hexamethonium Dihydro-β-erythroidine Mecamylamine Bungarotoxin	Atropine Scopolamine Pirenzepine (M1) Telenzepine (M1) 4-DAMP (M2) AF-DX 116 (M2)

nificance meaning of this peripheral distribution is that the muscarinic acetylcholine receptors participate in parasympathic effects, which include such principal physiological features as a decrease in heart rate, smooth muscle contraction and blood vessel dilation.

The function of muscarinic receptor activation is best studied in the hippocampus and in the cerebral cortex. Some of the effects of receptor activation have also been demonstrated in some other brain areas. In all cases, acetylcholine possesses an excitatory effect which becomes most obvious in the enhancement of neuronal firing rates.

Muscarinic effects have been described in several brain areas: these include the cerebral cortex, the locus coeruleus and some thalamic nuclei. The effects are preferentially of excitatory nature, but in some cases postsynaptic inhibition has also been described, for instance in the reticular nucleus of the thalamus and in the parabrachial nucleus.

In the central nervous system, acetylcholine is involved in the control of certain motor activities and in processes coupled to learning and memory. Lesioning of cholinergic regions in the septo-hippocampal area results in a dysfunction in memorizing and in spatial memory. Injection of cholinergic antagonists induces comparable effects. In contrast, injection of a cholinergic agonist (scopolamine) in the same region seems to have positive effects on learning and memory.

3.1.6
Neurological Disorders and Neurogenerative Diseases

A dysfunction of the cholinergic system occurs in some degenerative diseases of the brain, like Alzheimer's disease. The progression of Alzheimer's disease is accompanied by:
- a strong reduction in the activity of acetylcholine esterase in several cerebral structures, especially in the cortex, the hippocampus and the amygdala;
- a reduction in the biosynthesis of acetylcholine;
- a reduction in the high-affinity uptake of choline;
- a loss of cholinergic neurons in the nucleus basalis of *Meynert*;
- a loss of nicotinic receptors in the cortex and in the hippocampus, in contrast to the muscarinic receptors which show no obvious reduction.

The neurotoxic effect of neurofibrillary tangles and beta-amyloid plaques are hallmarks of Alzheimer's disease. Disease-modifying approaches which might lead to neuroprotection and enhanced survival of neurons are at the focus of therapeutic strategies.

Recent data suggest a role for cholinergic stimulation in counteracting beta-amyloid toxicity. Especially, the *a*7 nicotinic acetylcholine receptor is considered to be a strategic target for inducing neuroprotective effects. Galantamine, which is a modest acetylcholinesterase inhibitor in addition to being an allosteric modulator of nicotinic acetylcholine receptors, has therefore been applied in preclinical studies with promising effects; and new disease-modyfing agents capable of stimulating the *a*7 nicotinic receptor system are of further interest to discern potential neuropretection in this degenerative disease.

Imbalance of the cholinergic system also seems to be involved in Parkinson's disease. This is apparent from the hyperactivity of cholinergic interneurons in the striatum, following the reduction of the dopaminergic influence. Furthermore, the cholinergic system is impaired in Huntington's chorea, since an essential feature of this inherited disease is a characteristic loss of cholinergic interneurons in the striatum.

In the periphery, the most prominent disease which involves the cholinergic transmission is *Myastenia gravis*, an autoimmune disease which manifests itself at the motor endplate. Autoantibodies directed to the nicotinic receptors of motor endplates produce a masking of the receptors, which finally leads to their degradation. The functional consequence is a reduction of cholinergic transmission at skeletal muscles which is causal for the most prominent clinical sign of severe muscle weakness in these patients. The inhibition of acetylcholine esterases in order to prolong the action of endogenous acetylcholine is one of the therapeutical regimes used to treat this disease.

Further Reading

Albuquerque, E.X., Pereira, E.F.R., Castor, N.G., Alkondon, M., Reinhardt, S., Schröder, G., Maelicke, A. (1995): Nicotin receptor function in the mammalian central nervous system. *Ann. N.Y. Acad. Sci.* **757**: 47–72.

Aoki, C., Joh, T.H., Pickel, V.M. (1987): Ultrastructural localisation of beta-adrenergic receptor-like immunreactivity in the cortex and neostriatum of rat brain. *Brain Res.* **437**: 264–282.

Barbeau, A. (1978): Emerging treatments: replacement therapy with cholin or lecthin in neurological diseases. *Can. J. Neurol. Sci.* **5**: 157–160.

Barrantes, F.J. (ed) (1998): *The Nicotinic Acetylcholine Receptor*. Springer, Berlin, Heidelberg, New York.

Bartus, R.T., Dean, R.L., Beer, B., Lippa, A.S. (1982): The cholinergic hypothesis of geriatric dysfunction. *Science* **217**: 408–414.

Breining, S. (2004): Recent development in the synthesis of nicotinic acetylcholine receptor ligands. *Curr. Top. Med. Chem.* **4**: 609–629.

Brenner, H.R., Witzemann, V., Sakmann, B. (1990): Imprinting of acetylcholine receptor messenger RNA accumulation in mammalian neuromuscular synapses. *Nature* **344**: 544–547.

Caffé, A.R., Hawkins, R.K., De Zeeuw, C.I. (1996): Coexistance of cholin acethyltransferase and GABA in axon terminals in the dorsal cap of the rat inferior olive. *Brain Res.* **724**: 136–140.

Dajas-Bailador, F., Wonnacott, S. (2004): Nicotinic acetylcholine receptors and the regulation of neuronal signalling. *Trends Pharmacol.* **25**: 317–324.

Ding, Y.-S., Fowler, J. (2005): New generation radiotracers for nAChR and NET. *Nucleic Med. Biol.* **32**: 707–718.

Duclert, A., Changeux, J.P. (1995): Acetylcholine receptor gene expression at the developing neuromuscular junction. *Physiol. Rev.* **75**: 339–368.

Dutar, P., Bassant, M.H., Senut, M.C., Lamour, Y. (1995): The septohippocampal pathway: structure and function of a central cholinergic pathway. *Physiol. Rev.* **75**: 393–425.

Geerts, H. (2005): Indicators of neuroprotection with galantamine. *Brain Res. Bull.* **64**: 519–524.

Giacobini, E. (2004): Cholinesterase inhibitors: new roles and therapeutic alternatives. *Pharamacol. Res.* **50**: 433–440.

Giniatullin, R., Nistri, A., Yaskel, J.L. (2005): Desensitization of nicotininc Ach receptors: shaping cholinergic signaling. *Trends Neurosci.* **28**: 371–378.

Gotti, C., Clementi, F. (2004): Neuronal nicotinic receptors: from structure to pathology. *Prog. Neurobiol.* **74**: 363–396.

Grutter, T., Novere, N., Changeux, J.P. (2004): Rational understanding of nicotinic receptors drug binding. *Curr. Top. Med. Chem.* **4**: 645–650.

Hulme, E.C., Birdsall, N.J.M., Buckley, N.J. (1990): Muscarinic receptor subtypes. *Annu. Rev. Pharmacol. Toxicol.* **30**: 633–673.

Jones, S.V.P. (1993): Muscarinic receptor subtypes: modulation of ion channels. *Life Sci.* **52**: 457–464.

Kosaka, T., Tauchi, M., Dahl, J.L. (1988): Cholinergic neurons, containing GABA-like and/or glutamic acid decarboxylase-like immunreactivities in various brain regions of the rat. *Exp. Brain Res.* **70**: 605–617.

Krnjevic, K. (2004): Synaptic mechanisms modulated by acetylcholine in cerebral cortex. *Prog. Brain Res.* **145**: 81–93.

Kuhar, M.J., Taylor, N., Wamsley, J.K., Hulme, E.C., Birdsall, N.J.M. (1981): Muscarinic cholinergic receptor localisation in brain by electron microscopic autoradiography. *Brain Res.* **216**: 1–9.

Li, Y., Champs, S., Taylor, P. (1993): Tissue-specific expression and alternative mRNA processing of the mammalian acetylcholinesterase gene. *J. Biol. Chem.* **268**: 5790–5797.

Loewi, O., Navratil, E. (1926): Über humorale Übertragbarkeit der Herznervenwirkung. *Pflügers Arch.* **214**: 689–696.

Mesulam, M.M., Mufson, E.J., Wainer, B.H., Levey, A.I. (1983): Central cholinergic pathways in the rat: an overview based on an alternative nomenclature (Ch1–Ch6). *Neuroscience* **10**: 1185–1201.

Metherathe, R. (2004): Nicotininc receptors in sensory cortex. *Learn Mem.* **11**: 50–59.

Michelson, M.J., Zeimal, E.V. (1973): *Acetylcholine.* Pergamon Press, New York.

Sacco, K.A., Bannon, K.L., George, T.P. (2004): Nicotininc receptor mechanisms and cognition in normal states and neuropsychiatric disorders. *Psychopharmacology* **18**: 457–474.

Sarter, M., Parikh, V. (2005): Choline transporters, cholinergic transmission and cognition. *Nat. Rev. Neurosci.* **6**: 48–56.

Shor, R.G.L., Lefkowitz, R.J., Caron, M.G. (1981): Purification of the β-adrenergic receptor. *J. Biol. Chem.* **256**: 5820–5826.

Steriade, M., Biesold, D. (1990): *Brain Cholinergic System.* Oxford University Press, Oxford.

Unwin, N. (1995): Acetylcholine receptor channel imaged in the open state. *Nature* **373**: 37–43.

Venter, J.C. (1983): Muscarinic cholinergic receptor structure. *J. Biol. Chem.* **258**: 4842–4848.

Vizi, E.S. (2000): Role of high-affinity receptors and membrane transporters in nonsynaptic communication and drug action in the central nervous system. *Pharmacol. Rev.* **52**: 63–89.

Woolf, N.J. (1991): Cholinergic systems in mammalian brain and spinal chord. *Prog. Neurobiol.* **37**: 475–524.

3.2
Dopamine

3.2.1
General Aspects and History

The first successful synthesis of dopamine (3,4-dihydroxyphenethylamine, or 3-hydroxytryptamine) was achieved in 1910, but in contrast to the other members of the catecholamine family (epinephrine, norepinephrine), little attention was paid to this monoamine. It was thought for a long time that dopamine was simply an intermediate product in the synthesis of norepinephrine. However, since then it has been shown that dopamine is a prominent neurotransmitter in the brain with several potential functions and a distinct distribution. Dopamine has been found to be enriched, for example, in the substantia nigra and in the striatum, whereas norepinephrine is absent from these brain regions. The differential distribution of dopamine is suggestive of a specific function for this neurotransmitter in neuroregulative processes.

With respect to the peripheral nervous system, dopamine was long considered to constitute the precursor of the other catecholamines. However, dopaminergic neurons were found by Bell (1989) to occur in peripheral nerves. Consequently, dopamine can also be regarded as a transmitter in peripheral nervous tissues.

In 1962, Falck and Hillarp developed a fluorescent method (glyoxcyl acid method) for the visualization of monoamines in tissue sections. Dahlström and Fuxe (1964) applied the Falck–Hillarp method to the central nervous system and were able to demonstrate monoaminergic neurons, including neurons which expressed dopamine, norepinephrine and serotonin. This method, together with immunohistochemical studies, has provided important insights into the distribution and function of dopamine in the central nervous system.

The essential biochemical difference between noradrenergic neurons and dopaminergic neurons is the presence of the enzyme dopamine-β-hydroxylase. This enzyme can be used as a marker for the differentiation of noradrenergic from dopaminergic neurons. Within dopaminergic neurons, the dopamine-β-hydroxylase catalyzes the formation of norepinephrine from dopamine.

From a clinical point of view, dopamine attracted considerable interest since it became evident that this monoamine is involved in several major brain disorders, like Alzheimer's, Parkinsonism and Schizophrenia.

3.2.2
Differentiation and Localization of the Dopaminergic System

The dopaminergic system comprises three classes of neurons based on the length of their projections:

The class with ultra-short projections: This includes amacrine-like neurons in the retina and the periglomerular cells of the olfactory bulb and is characterized by very short intralaminar dendritic extensions.

The class with short projections: This class comprises three different subsystems:
- The system of dopaminergic cells which are located in the arcuate nucleus of the hypothalamus (area A12). These project to the pituitary and to the median eminence.
- The system of dopaminergic neurons of the posterior hypothalamus (area A11), the zona incerta (area A13) and dopaminergic neurons close to the paraventricular nucleus of the hypothalamus (area A14). Together, these projections represent the so-called intradiencephalic dopaminergic neurons (Fig. 3.6A).
- The system of dopaminergic neurons located in the nucleus of the tractus solitarius and in the periaqueductal gray.

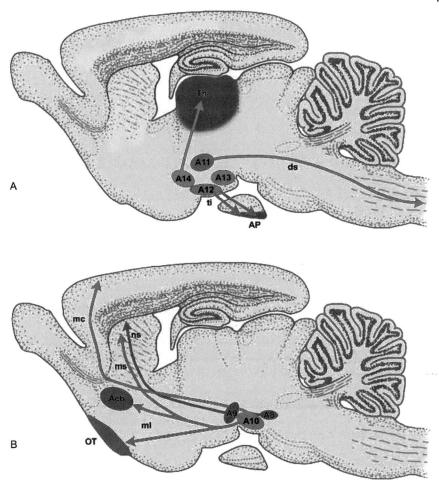

Fig. 3.6 (A) So-called short projections which mainly arise in the arcuate nucleus (A12), posterior hypothalamus (A11), the zona incerta (A13) and the paraventricular nucleus (A14). *Abbreviations for projections*: ds = diencephalo-spinal projection; ti = tuberoinfundibular hypothalamothalamic area; Th = thalamus. (B) This figure indicates the class of long projections. The cell bodies of this dopaminergic neurons have been grouped into compartments A8 to A1). *Abbreviations for projections*: mc = mesocortical projection; ml = mesolimbic projection; ms = mesostriatal projection; ns = nigro-striatal projection; Acb = *nucleus accumbens*; OT = olfactory tubercle.

The class with long projections: This class consists of neurons with very long dopaminergic projections. The projections arise either in the retrorubral field (A8), in the substantia nigra (A9) or in the ventral tegmental area (VTA; A10). From there, they project to three different brain areas (Fig. 3.6 B):
- the neostriatum;
- the limbic cortex;

- additional limbic structures (i.e. the olfactory tubercle, septal areas, the nucleus accumbens and the amygdaloid complex).

Synonyms for these projections are the nigro-striatal, mesocortical or mesolimbic system. The specific projections of the dopaminergic neurons from A8, A9 and A10 to the nucleus accumbens and the striatum are collectively called the mesostriatal dopaminergic system. The entire projection fields of the group from A8, A9 and A10 are known as the mesotelencephalic dopaminergic system.

The dopaminergic neurons of the mesotelencephalic system do not innervate exclusively the regions listed above, but innervate additional areas which are also depicted in Fig. 3.6 A, B.

3.2.3
Biosynthesis and Degradation

Dopamine, like other neurotransmitters, is not capable of crossing the blood–brain barrier. However, the precursors of dopamine, phenylalanine and tyrosine (tyrosine can be synthesized from phenylalanine by the activity of the phenylalanine hydroxylase) are able to cross the blood–brain barrier.

The biosynthesis of dopamine takes place within nerve terminals. Dopamine is synthesized from tyrosine in a two step process. The first step is catalyzed by tyrosine hydroxylase and results in the production of dihydroxyphenylalanine (DOPA; Fig. 3.7).

Tyrosine hydroxylase, which is the rate-limiting enzyme in the generation of dopamine, is present in all catecholamine-containing neurons. Tyrosine hydroxylase is an oxidase which requires tetrahydrobiopterin and oxygen as cofactors. An increase in the concentration of tyrosine does not lead to a greater production of L-DOPA, since tyrosine hydroxylase operates at its maximal catalytic activity.

Since tyrosine hydroxylase is the rate-limiting enzyme, this substrate is rendered a feasible target for physiological regulation and pharmacological manipulation.

An excess of catecholamines (dopamine, norepinephrine or epinephrine), for example, inhibits tyrosine hydroxylase activity. In contrast, some neuromodulators like VIP or growth factors like the nerve growth factor (NGF) can activate tyrosine hydroxylase.

The catalytic product of tyrosine is L-DOPA, which is then decarboxylated to dopamine by the cytosolic DOPA decarboxylase. DOPA decarboxylase is also known as aromatic amino acid decarboxylase because it catalyzes the decarboxylation of several other endogenous aromatic amino acids. DOPA decarboxylase is not rate-limiting. Thus, higher concentrations of cerebral DOPA enhance the synthesis of dopamine.

The biosynthesis of dopamine takes place in the cytosol, from which the transmitter is transported through a specific transport system into presynaptic

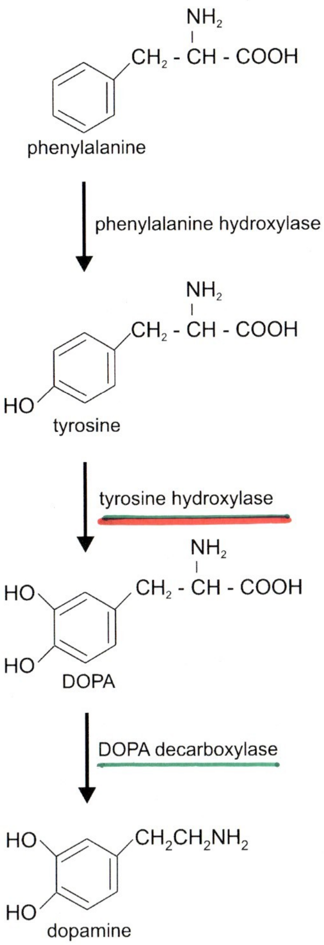

NH2
CH2 - CH - COOH
phenylalanine
phenylalanine hydroxylase
NH2
CH2 - CH - COOH
HO
tyrosine
tyrosine hydroxylase
NH2
HO
CH2 - CH - COOH
HO
DOPA
DOPA decarboxylase
HO
CH2CH2NH2
HO
dopamine

Fig. 3.7 Synthesis of dopamine from its precursors.

vesicles. This vesicular storage protects dopamine from degradation by the enzyme monoamine oxidase (MAO). The vesicular transporter is distinct from the dopamine transporter, which is found in the presynaptic membrane (see below).

Since dopamine is synthesized and stored in presynaptic vesicles, cell bodies of dopaminergic neurons contain relatively low amounts of dopamine. High concentrations of the neurotransmitter are prevalent in presynaptic terminals when visualized by the glyocylic acid technique or after immunocytochemical detection.

The dopaminergic cells give rise to a large number of dendritic processes and the axons of these neurons form highly branched collaterals. In the preterminal region of the collaterals as well as in the dendrites, a large number of varicosities have been found.

A well known example of the complexity of dopamine neurons is the nigrostriatal system, where a single dopaminergic neuron forms up to 100 000 varicosities, which make synaptic contacts with postsynaptic neurons.

3.2.4
Release, Re-uptake and Degradation

Dopamine is released by action potentials through a calcium-dependent mechanism into the synaptic cleft. The subsequent sequence of events follows the classic scheme of transmitter release. The calcium influx through voltage-dependent calcium channels triggers the fusion of the vesicle with the presynaptic membrane. A pore is formed from which dopamine is released into the synaptic cleft. Dopamine then diffuses across the synaptic cleft and binds to postsynaptic receptors. This binding induces a change in the conformation of the receptors, which in turn triggers the membrane permeability for ions and initiates a complex chain of intracellular postsynaptic events.

The outcome is an activation or inhibition of the postsynaptic neuron. The dopaminergic signal is finally terminated by removal of dopamine from the intersynaptic cleft. This removal involves specific re-uptake mechanisms into the presynaptic terminal where it can be stored and reused.

Specific dopamine transporters (DAT) support the re-uptake. This transporter system plays an important function in the inactivation and recycling of released dopamine.

The dopamine transporter is a glycoprotein of 619 amino acids (70 kDa) which shows a 12-span transmembranous motif. The function of the dopamine transporter is to reaccumulate extracellular dopamine in the presynaptic terminals and thereby to regulate the lifetime of the free intersynaptic dopamine. The uptake process, which transports the extracellular dopamine into the cells, depends on Na^+ and Cl^-. The transporter is very efficient and about 80% of the released dopamine is recaptured by this re-uptake mechanism.

The dopamine transporters are not located at the active zone of the presynaptic site but are restricted to perisynaptic areas. This implies that dopamine diffuses away from the intersynaptic cleft. Besides its physiological role, dopamine transporters provide pharmacological targets for cocaine and amphetamines.

In addition to its transporter complement, the presynaptic membrane contains some dopamine receptors. These receptors are so-called *autoreceptors*. Their functional role is to monitor the extracellular dopamine concentration and to modulate the impulse-dependent release and synthesis of dopamine. A blockade of these receptors facilitates the synthesis and presynaptic release of dopamine, while their stimulation has the opposite effect.

Dopamine can be degraded by the activity of monoamine oxidase (MAO) and aldehyde dehydrogenase to dihydroxyphenylacetic acid (DOPAC). Alternatively, it can be metabolized by the activity of catechol-O-methyltransferase (COMT) to form 3-methoxytryptamine (3-MT). DOPAC and 3-MT are then further degraded to form homovanillic acid (HVA; Fig. 3.8).

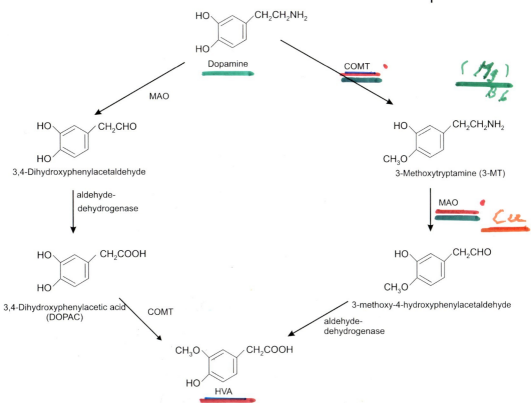

Fig. 3.8 Degradation of dopamine to homovanillic acid (HVA).

Interestingly, catechol-O-methyltransferase and monoamine oxidase do not participate exclusively in the degradation of dopamine, but are also involved in the metabolism of norepinephrine.

3.2.5
Receptors and Signal Transduction

Receptor classification
Most dopaminergic receptors are located postsynaptically, but some are located presynaptically. Stimulation of the presynaptic dopaminergic receptors results in activation or inhibition of the release and synthesis of dopamine (so-called autoreceptors, see above).

The autoreceptors play a significant role in modulating and monitoring the release and synthesis of dopamine. The autoreceptors are found at nerve terminals, as well as on the soma and on dendrites of most dopaminergic neurons. Stimulation of autoreceptors located at the nerve terminals results in an inhibition of dopamine release and synthesis, whereas stimulation of somatodendritic autoreceptors decreases the firing rate of the dopaminergic neurons. They may

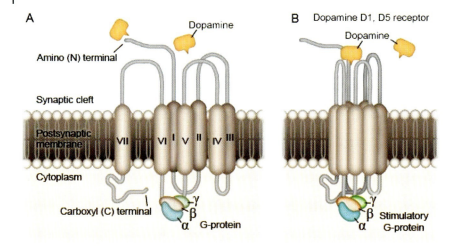
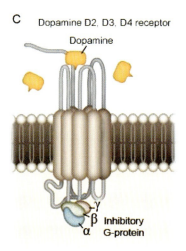

Fig. 3.9 The family of dopaminergic receptors belongs to the GPCRs. The third intracytoplasmic loop contains the G protein complex-binding site (A). D1 and D5 receptors stimulate the activity of adenylate cyclase (B), while the D2, D3, D4 group exerts an inhibitory effect on this enzyme (C).

also regulate dopamine release and synthesis. Thus, both types of presynaptic autoreceptors cooperate in the regulation of dopaminergic transmission.

The family of dopamine receptors is classified into two groups, which differ in their biochemical and pharmacological properties. The dopaminergic receptors belong to the family of receptors, which show the seven transmembrane-spanning domain motif and are all metabotropic receptors (see Fig. 3.9 A–C).

The dopaminergic receptors are divided into the D1 and D2 groups. The D1 group consists of D1 and D5 receptors, which are positively linked to adenylate

Table 3.4 Properties of dopaminergic receptors.

Type	Amino acid sequence	Location on the human chromosome	Agonists	Antagonists
D_1 family				
D_1 (1A)	446 (human, rat)	5q-34–35	A-68930 CY-208-345 Dihydrexidine Hydroxybenzazepine	Halobenzazepine (e.g. SCH-23390) Thioxanthene
D_5 (1B)	477 (human) 475 (rat)	4p-15.1–3	Hydroxybenzazepine	Halobenzazepine
D_2 family				
D_{2L} (2Aα)	443 (human) 444 (rat)	11q-22–23	Aminotetralin Ergoline (−)-Hydroxyaporphine	Benzamide Butyrophenone Phenothiazine
D_{2S} (2Aβ)	414 (human) 415 (rat)	11q	Ergolin (−)-Hydroxyaporphin	Benzamide Butyrophenone
D_3 (2B)	400 (human) 446 (rat)	3q-13.1	(+)-7-OH-DPAT (+)-PD-128-907 Dihydrexidine Pramipexol	Benzamide S-14297
D_4 (2C)	387+ (several variants in human) 368 (rat)	11p-15.5	Ergoline	(+)-Aporphine Clozapine Olanzapine

cyclase. The D2 group consists of D2, D3 and D4 receptors (Table 3.4) and each of these receptor types exists in different isoforms. The dopaminergic autoreceptors belonged to the family of D2 receptors.

Molecular topology

A very short third intracellular loop and a long C-terminal tail characterize the topology of receptors of the D1 group, while a long third cytoplasmic loop and a short C-terminal tail specifies members of the D2 family. This long third cytoplasmic loop is regarded to be essential for the interaction of the receptor with G proteins.

The human D1 family (D1 and D5 receptors) is encoded by genes located on chromosomes 5 and 4. Two D5-like pseudogenes are located on human chromosomes 1 and 2. These pseudogenes encode for a truncated and inactive form of the D5 receptor (154 amino acids, instead of 477 amino acids).

Specification of the dopamine receptor

D1 receptors

The D1 receptors appear to be most abundant in the mammalian forebrain. In addition, they are expressed (in decreasing concentrations): in the striatum, the amygdala, the thalamus, the mesencephalon, the hypothalamus and the hindbrain.

The D1 receptors have a distribution pattern which differs from that of the D2 receptors (see below). The absence of D1 receptors in adults is an indication that this receptor type lacks a role as an autoreceptor. However, it can not be excluded that they function as autoreceptors during embryonic development.

The D1 receptors, like the D5 receptors (the second member of the D1 family), stimulate the activity of the adenylate cyclase. This is the most significant respect in which they differ from members of the D2 family (Fig. 3.9 B, C).

The D1 receptor is composed of 446 amino acids (Table 3.4) and is a G protein-coupled receptor. The gene, which encodes for the D1 receptor protein, is located on human chromosome 5q35.

Several selective D1 receptor antagonists have been developed. Most of them are halogenated phenylbenzazepines, such as the compounds SCH-23390 and SCH-39166.

D2 receptors

The D2 receptors, like the D1 receptors, are widely distributed within the brain. Their relative abundance (in decreasing concentration) is as follows: striatum, mesencephalon, spinal cord, hypothalamus and hippocampus.

The D2 receptor gene was first cloned from rat tissue in 1988 by Bunzow and coworkers. A year later, the human D2 gene was cloned. The human gene, which encodes the D2 receptor, is located on chromosome 11q23. Molecular analysis of different cloned D2 receptor peptides indicates that the D2 receptors differ slightly, even within a single species. The difference is the facultative presence of a stretch of 29 amino acids in the third intracellular loop of the receptor peptide, leading to a variance of these isoforms between amino acids 415 and 444. The variants are generated by alternative splicing of the same gene (Table 3.4) during maturation of the corresponding pre-mRNA.

These isoforms are termed "long" or "short" forms (D2L and D2S, or D2A and D2B) which both inhibit adenylate cyclase activity and undergo agonist-induced desensitization.

The distribution of the D2L and D2S receptors is not uniform within in the central nervous system. D2 receptors of the short isoform are relatively more abundant in the posterior cerebral cortex, the amygdala, the hypothalamus and the brain stem as compared to the D2 receptors of the long form. However, more D2L receptors are found in extrapyramidal basal ganglia.

D2 receptors seem to exist in a monomeric as well as in a dimeric form, a feature, which enhances the diversity of the D2 receptor type. Both D1 and D2 receptors are expressed in the striatum, which raises the possibility of coexpression of D1 and D2 receptors in individual striatal neurons.

D3 receptors

The D3 receptor has structural and pharmacological similarities with the D2 receptor. D3 receptors are widely distributed in the basal forebrain, the olfactory tubercle, nucleus accumbens, striatum and substantia nigra, but they are infrequent in limbic and extrapyramidal regions, where D2 receptors have been found in high densities.

The locus of the human D3 receptor gene is on chromosome 3q13. The human receptors of the D3 type are composed of 400 amino acids and the D3 receptor of rats consists of 446 amino acids (Table 3.4).

Like the receptors of type D2, the D3 receptors occur in different isoforms (D3$_{GLY}$, and a short and long splice variant), however, it seems likely that not all of the variants form functional receptors.

The most significant difference distinguishing the D3 receptors from the other dopaminergic receptors is their lack of interaction with G proteins. However, the signal transduction pathway of this receptor is still unknown. Based on the structural similarities with the D2 receptors, it is discussed that D3 receptors are capable of inhibiting the formation of cAMP.

D4 receptors
A further member of the D2 receptor family is the D4 receptor. The gene locus of the human D4 receptor is located on chromosome 11q15. D4 receptors occur in several isoforms differing in the length of their amino acid sequences (Table 3.4).

In the brain, the mRNA for D4 receptors is expressed in the frontal cerebral cortex, the hippocampus, the thalamus, the striatum, basal ganglia and the cerebellum.

Like the other dopaminergic receptors of the D2 family, the D4 receptors inhibit the activity of adenylate cyclase.

In contrast to the D1, D2 and D3 receptors which have little affinity to clozapine (a dopamine receptor antagonist), the D4 receptors display moderate affinity to this substance.

D5 receptors
A further dopaminergic receptor has been cloned, which shares great structural homology with the D1 receptor. This dopaminergic receptor was classified following the order of its discovery, as D5 receptor. This receptor is composed of 476 amino acids (Table 3.4). Like the D1 receptors, the D5 receptors are coupled to G proteins, activate adenylate cyclase and reveal a high affinity for dopamine.

This receptor type has a very restricted distribution within the brain. The D5 receptors are expressed exlusively in the hippocampus and in the thalamus (parafacial nucleus).

3.2.6
Biological Effects

The differential distribution of the diverse dopaminergic systems indicates that dopamine influences a variety of brain functions. For instance, dopamine is involved in the modulation of arterial blood-flow, higher brain functions like cognition and learning and in anxiety-related behavior. Therefore it is not surpris-

ing that the dopaminergic systems serve as a target for antipsychotic drugs, i.e. in schizophrenia treatment. The classic domain for dopamine substitution, however, is Parkinsonism, where a degeneration of dopaminergic neurons in the substantia nigra is causal for the disease.

The nigro-striatal system is concerned with the initiation and maintenance of motor behavior. The mesolimbic and mesocortical systems appear to be involved in goal-directed and reward-mediated behavior and in motivation-dependent behavior. A dysfunction in these systems alters normal associative processes.

An enhancement of dopaminergic transmission in the mesolimbic system is linked with reinforcing effects of psychostimulant drugs. The hypothalamic–hypophyseal axis in the form of the tuberoinfundibular system plays a major role in the regulation of pituitary and hypothalamic peptides, for instance in the release of prolactin. An increase in dopamine activity in this system results in an inhibition of prolactin release. Thus, dopamine constitutes the prolaction-inhibiting factor.

It is thought that the dopaminergic tuberoinfundibular neurons are involved in the regulation of α-MSH and β-endorphin and in the release of oxytocin and vasopressin from the pituitary.

With respect to the autonomic centers of the hypothalamus, dopamine seems to be an essential neurotransmitter, which impinges on arterial blood-flow regulation. Feeding, as well as drinking activities initiated by the ventromedial and lateral hypothalamic nuclei including the zona incerta, are also modulated by dopamine.

3.2.7
Neurological Disorders and Neurodegenerative Diseases

As mentioned earlier, dopamine is an essential neurotransmitter in manifold cerebral functions. Consequently, dysfunction in the dopaminergic transmission influences a variety of neurological and psychiatric disorders. Some dysfunctions may result in hyperactivity of the dopaminergic system. Such hyperactivity leads to an accumulation of the neurotransmitter in the synaptic cleft and/or hypersensitivity of dopaminergic receptors. Dopaminergic hyperactivities have been found to be linked to some psychotic disorders, including hallucinations and manic states.

Hypoactivity of the system seems to be primarily involved with motor dysfunction, deficits in motivation-dependent behavior and imbalance of emotional perception. Thus one has to expect a combination of different clinical symptoms if the balance of the dopaminergic system is afflicted.

The basal ganglia, including the striatum, play a crucial role in the control of movement. The importance of dopamine in the striatum becomes evident in patients with Parkinson's disease (PD). PD was originally described by James Parkinson in 1817 and is characterized by a trio of cardinal symptoms:
- bradykinesia (slowed movement)
- resting tremor
- rigidity.

Normal motor function depends on the highly regulated synthesis and release of dopamine by neurons projecting from the substantia nigra to the striatum. Degeneration of the DAergic neurons located in the substantia nigra pars compacta (SNpc) and a subsequent loss of DAergic nerve terminals in the striatum are responsible for most of the movement disorders. There is a gradual decline in dopaminergic neurons of the SNpc during aging, which is accompanied by a reduction of the striatal DA content. In idiopathic PD, however, the symptoms become apparent when about 70% of the striatal dopamine and more than 50% of the nigral dopaminergic neurons are lost. Further pathological features of this disease, besides the apparent loss of neurons, are dystrophic neurites termed Lewy neurites, and characteristic round eosinophilic inclusions of about 5–25 µm in diameter, known as *Lewy bodies*, which consist of intracellular inclusions of α-synuclein. The presence of Lewy bodies in association with nerve cell loss in the substantia nigra and various other regions of the nervous system is a diagnostic hallmark of PD.

The underlying mechanisms of most cases of PD are still unknown but, recently, specific genetic defects were identified. In addition, environmental factors may also contribute to the disease. Epidemiologic studies indicated that a number of factors, including pesticides, herbicides, industrial chemicals and farming, may increase the risk of developing PD. Both lines of evidence, environmental and genetic risk factors, point to a convergence between energy metabolism and deficits in the disposal of damaged proteins in the pathogenesis of PD. Thus, the pathogenesis is thought to be multifactorial, deriving from environmental factors acting on genetically predisposed individuals with ageing. However, the relationship between genetic and environmental factors is poorly understood and most models of PD focus on single genes or toxins.

Current approaches in PD therapy include, among others:
- replenishment of dopamine by giving l-3,4-dihydroxyphenylalanine (Levadopa);
- increasing the action of remaining dopamine by reducing the destruction of dopaminergic cells;
- mimicking the action of dopamine with appropriate dopamine receptor agonists;
- electric stimulation of specific brain areas;
- manipulating the actions of neurotransmitter which interfere with the actions of dopamine in the striatum;
- transplantation of appropriate neuronal tissue into the striatum.

However, all these treatments only ameliorate the symptoms of PD, but are unable to cure the disease. Thus, a major challenge will be to identify the molecular and cellular bases which are responsible for the destruction of dopaminergic midbrain neurons.

Agents that selectively disrupt or destroy catecholaminergic systems, such as 6-hydroxydopamine (6-OHDA) and 1-methyl-4-phenyl-1,2,3,6-tetrahydropyridine (MPTP) have been used to develop animal models of Parkinson's disease, since neurotoxins reproduce specific features of this disease (Fig. 3.10).

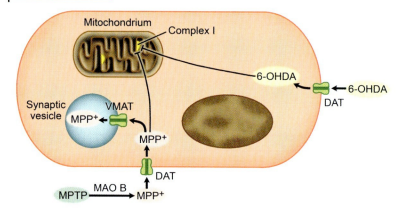

Fig. 3.10 The neurotoxins used in animal models of PD induce mitochondrial dysfunction. MPTP is converted by monoamine oxidase B (MAO B) to MPP$^+$. Like 6-OHDA, MPP$^+$ is taken up by dopamine transporters and can then be accumulated by mitochondria, leading to complex I inhibition. In contrast to 6-OHDA, MPP$^+$ can be taken up by vesicular monoamine transporters, which reduces the toxicity of MPP$^+$.

6-OHDA uses the same catecholaminergic transport system as norepinephrine and DA and produces specific degeneration of catecholaminergic neurons. 6-OHDA seems to be toxic to mitochondrial complex I and induces the generation of reactive oxygen species (ROS). 6-OHDA is usually injected unilaterally, while the intact hemisphere serves as an internal control. This unilateral 6-OHDA injection represents the so-called "hemiParkinson model", which is characterized by an asymmetric motor-circling behavior after administration of DAergic drugs, due to the physiological imbalance between the lesioned and unlesioned striatum. Therefore, animals rotate contralaterally to the hemisphere in which dopamine receptor stimulation is stronger. This means that they turn towards the side of the lesion after challenge with dopamine-releasing drugs and away from it after treatment with L-DOPA or dopamine agonists. Rotation behavior can be quantified and correlated with the degree of lesion, a major advantage of this model.

Most of the current evidence on the mechanisms of cell death in Parkinson's disease originate from studies using the MPTP model. MPTP is highly lipophilic and readily crosses the blood–brain barrier. MPTP-mediated toxicity is induced through conversion to the 1-methyl-4-phenyl-2,3-dihydropyridium ion (MPP+) in astrocytes by monoamine oxidase B. To exert its toxicity MPP+ must be transported into dopaminergic neurons by neurotransmitter transporters and, once inside the neuron, it is thought that MPP+ acts by inhibiting the electron transport system of mitochondrial complex I.

The treatment of primates or rodents with the neurotoxin MPTP represents an animal model which reflects many features of human Parkinson's disease. In primates, depending on the application paradigm used, MPTP can produce an irreversible and severe Parkinsonian syndrome that replicates nearly all features of Parkinson's disease, including rigidity, tremor, slowness of movement and even freezing.

A growing body of evidence has suggested that dopamine mechanisms are significant for aspects of affective disorders. Since dopamine concentrates in target structures of the mesolimbic system, i.e. the amygdala, medial prefrontal cortex and nucleus accumbens, it comes without surprise that this neurotransmitter is considered to be of paramount importance for a number of fear-related diseases such as phobia, panic attacks, posttraumatic stress disorders, obsessive compulsive disorders, or generalized anxiety. Common psychiatric diseases which entail fear symptoms, like depression and schizophrenia, are discerned to be related to disturbances of dopamine and its receptor metabolism. Though the ethiology of these disorders is far more complex, involving additional neurotransmitter systems like glutamate, GABA, serotonin, glycin and norepinephrine, the role of dopamine and its receptor complement in the treatment of psychotic disorders is evident.

More recent studies also indicate that mild mental disorders, including attention-deficit/hyperactivity disorder (ADHD) and restless legs syndrome, also involve the dopaminergic system.

Further Reading

Asan, E. (1997): Ultrastructural features of tyrosine-hydroxylase-immunoreactive afferents and their targets in the rat amygdala. *Cell Tissue Res.* **288**: 449–464.

Bannon, M. J. (2005): The dopamine transporter: role in neurotoxicity and human disease. *Toxicol. Appl. Pharmacol.* **204**: 355–360.

Bell, C. (1989): Peripheral dopaminergic nerves. *Pharmacol. Ther.* **44**: 157–179.

Betarbet, R., Sherer, T. B., Greenamyre, J. T. (2002): Animal models of Parkinson's disease. *Bioessays* **24**: 308–318.

Bunzow, J. R., van Tol, H. H., Grandy, D. K., Albert, P., Salon, J., Christie, M., Machida, C. A., Neve, K. A., Civelli, O. (1988): Cloning and expression of a rat D2 dopamine receptor cDNA. *Nature* **336**: 783–787.

Chase, T. N., Oh, J. D., Blanchet, P. J. (1998): Neostriatal mechanisms in Parkinson's disease. *Neurology* **51**: S30–S35.

Civelli, O., Bunzow, J. R., Grandy, D. K. (1993): Molecular diversity of dopamine receptors. *Annu. Rev. Pharmacol. Toxicol.* **32**: 281–307.

Corti, O., Hampe, C., Darios, F., Ibanez, P., Ruberg, M., Brice, A. (2005): Parkinson's disease: from causes to mechanisms. *C.R. Biol.* **328**: 131–142.

Dahlström, A., Fuxe, K. (1964): Evidence for the existence of monoamine-containing neurons in the central nervous system. I. Demonstration of monoamines in the cell bodies of brainstem neurones. *Acta Physiol. Scand.* **62**: 1–55.

Dailly, E., Cheu, F., Renard, C. E., Bourin, M. (2004): Dopmaine, depression and antidepressants. *Fund. Clin. Pharmacol.* **18**: 601–607.

Dunnett, S. B., Bjorklund, A. (1999): Prospects for new restorative and neuroprotective treatments in Parkinson's disease. *Nature* **399**: A32–A39.

Fremeau, R. T. Jr, Duncan, G. E., Fornaretto, M. G., Dearry, A., Gingrich, J. A., Breese, G. R., Caron, M. G. (1991): Localization of D1 dopamine receptor mRNA in brain supports a role in cognitive, affective, and neuroendocrine aspects of dopaminergic neurotransmission. *Proc. Natl Acad. Sci. USA* **88**: 3772–3776.

Forno, L. S. (1996): Neuropathology of Parkinson's disease. *J. Neuropathol. Exp. Neurol.* **55**: 259–272.

Fukuda, M., Ono, T., Nakamura, K., Tamura, R. (1990): Dopamine and ACh involvement in plastic learning by hypothalamic neurons in rats. *Brain Res. Bull.* **25**: 109–114.

Grace, A.A. (1993): Cortical regulation of subcortical dopamine systems and its possible relevance to schizophrenia. *J. Neural Transm.* **91**: 111–134.

Hogl, B., Poewe, W. (2005): Restless legs syndrome. *Curr. Opin. Neurol.* **18**: 405–410.

Horn, A.S., Korf, J., Westerink, B.H.C. (eds) (1979): *The Neurobiology of Dopamine*. Academic Press, London.

Hornykiewicz, O. (1993): Parkinson's disease and the adaptive capability of the nigrostriatal dopamine system. *Adv. Neurol.* **60**: 140–147.

Huang, Q., Zhou, D., Chase, K., Gusella, J.F., Aronin, N., DiFiglia, M. (1992): Immunohistochemical localization of the D1 dopamine receptor in rat brain reveals its axonal transport, pre- and postsynaptic localization, and prevalence in the basal ganglia, limbic system, and thalamic reticular nucleus. *Proc. Natl Acad. Sci. USA* **89**: 11988–11992.

Kalia, M. (2005): Neurobiological basis of depression: an update. *Metabolism* **54**[Suppl 1]: 24–27.

Kitahama, K., Geffard, M., Okamura, H., Nagatsu, I., Mons, N., Jouvet, M. (1990): Dopamine- and dopa-immunoreactive neurons in the cat forebrain with reference to tyrosine hydroxylase immunohistochemistry. *Brain Res.* **518**: 83–94.

Lisman, J.E., Grace, A.A. (2005): The hippocampal–VTA loop: controlling the entry of information into long-term memory. *Neuron* **46**: 703–713.

Madras, B.K., Miller, G.M., Fsichman, A.J. (2005): The dopamine transporter and attention-deficit/hyperactivity disorder. *Biol. Psychiatr.* **57**: 1397–1409.

Meador Woodruff, J.H., Mansour, A., Healy, D.J., Kuehn, R., Zhou, Q.Y., Bunzow, J.R., Akil, H., Civelli, O., Watson, S.J. Jr (1991): Comparison of the distributions of D1 and D2 dopamine receptor mRNAs in rat brain. *Neuropsychopharmacology* **5**: 231–242.

Pezze, M.A., Feldon, J. (2004): Mesolimbic dopaminergic pathways in fear conditioning. *Prog. Neurobiol.* **74**: 301–320.

Reith, M.E.A., Xu, C., Chen, N.-H. (1997): Pharmacology and regulation of the neuronal dopamine transporter. *Eur. J. Pharmacol.* **324**: 1–10.

Revay, R., Vaughan, R., Grant, S., Kuhar, M.J. (1996): Dopamine transporter immunohistochemistry in median eminence, amygdala, and other areas of the rat brain. *Synapse* **22**: 93–99.

Riess, O., Kruger, R. (1999): Parkinson's disease – a multifactorial neurodegenerative disorder. *J. Neural Transm. Suppl.* **56**: 113–125.

Schultz, W. (2005): Behavioral theories and the neurophysiology of reward. *Annu. Rev. Psychol.* **57**: 87–115.

Seeman, P. (1995): Dopamine receptors and psychosis. *Sci. Am.* **2**: 28–37.

Seeman, P., Van Tol, H.H.M. (1994): Dopamine receptor pharmacology. *Trends Pharmacol. Sci.* **15**: 264–270.

Sibley D.R., Monsma, F.J. (1992): Molecular biology of dopamine receptors. *Trends Pharmacol. Sci.* **17**: 61–67.

Smits, R.P., Steinbusch, H.W., Mulder, A.H. (1990): Distribution of dopamine-immunoreactive cell bodies in the guinea-pig brain. *J. Chem. Neuroanat.* **3**: 101–123.

Smythies, J. (2005): The dopamine system. Section II. *Int. Rev. Neurobiol.* **64**: 123–172.

Vanhatalo, S., Soinila, S., Kaartinen, K., Back, N. (1995): Colocalization of dopamine and serotonin in the rat pituitary gland and in the nuclei innervating it. *Brain Res.* **669**: 275–284.

Vitalis, T., Cases, O., Parnavelas, J.G. (2005): Development of the dopaminergic neurons in the rodent brainstem. *Exp. Neurol. Suppl.* **1**: 104–112.

Vizi, E.S. (2000): Role of high-affinity receptors and membrane transporters in nonsynaptic communication and drug action in the central nervous system. *Pharmacol. Rev.* **52**: 63–89.

Volz, T.J., Schenk, J.O. (2005): A comprehensive atlas of the topography of functional groups of the dopamine transporter. *Synapse* **58**: 72–94.

von Bohlen und Halbach, O., Schober, A., Krieglstein, K. (2004): Genes, proteins, and neurotoxins involved in Parkinson's disease. *Prog. Neurobiol.* **73**: 151–177.

3.3 γ-Amino Butyric Acid

3.3.1 General Aspects and History

γ-Amino butyric acid – or GABA for short – is the most ubiquitous inhibitory neurotransmitter in the brain. GABA was discovered in 1883 but it took until the late 1950s before its inhibitory function was described (Bazemore et al. 1956). Subsequently, it became evident that this small amino acid is the principal inhibitory neurotransmitter in the central nervous system. GABA is widely spread among the phyla, including invertebrates as well as vertebrates. In invertebrates, it occurs in both the central and the peripheral nervous system.

Its abundance in the central nervous system of vertebrates becomes obvious when one considers that GABA is expressed in about 30% of all synapses. It was initially thought that GABA was confined to inhibitory interneurons and was absent from projection neurons. Today, we know that there are many examples of GABAergic projection neurons as well.

The pharmacological potency of synaptic transmission by GABA becomes apparent by recollecting that some of the most widely distributed sedatives, like barbiturates and benzodiazepines, interact with GABA receptors and modulate GABAergic effects.

3.3.2 Localization Within the Central Nervous System

The distribution and localization of the GABAergic system has been the subject of extensive investigations by means of immunohistochemistry. A reliable marker for the identification of GABAergic neurons was achieved through the production of antibodies which specifically recognize the critical biosynthetic enzyme for GABA synthesis, glutamic acid decarboxylase (GAD). Alternatively, *in situ* hybridization with cDNA of GAD can be used for the identification of GABAergic neurons.

In several places, GABAergic cells occur at high densities. A prominent example is the striatum, where nearly 95% of the cell somata are GABAergic.

Additional brain areas, identified as exhibiting GABAergic cells in large amounts and densities are the globus pallidus, the substantia nigra (pars reticularis) and the cerebellum. GABAergic interneurons are most frequent in the thalamus, the hippocampus and in the cerebral cortex.

The globus pallidus, besides its abundance of GABAergic neurons, receives many GABAergic terminals, as is the case for the entopeduncular nucleus and the superior colliculus.

3.3.3
Biosynthesis and Degradation

GABA is synthesized almost exclusively from glutamate (Fig. 3.11). The critical step in GABA biosynthesis is the decarboxylation of glutamate by GAD. In addition to their localization in the central nervous system, GABA and GAD can occur also in the peripheral nervous system. The reaction, which is catalyzed by glutamate decarboxylase, requires the presence of the co-factor pyridoxal phosphate (PLP, a form of vitamin B6). Removal of the co-factor causes a loss of GAD activity, which can be recovered by the administration of the co-factor.

Glutamic acid decarboxylase exists in two different isoforms. Two genes express these isoforms (GAD65 and GAD67, according to their molecular mass). The amino acid sequences indicate that both isoforms are almost identical, including their reactive centers. However, the subcellular distribution of GAD65 and GAD67 is different, as is the case for their interaction with the co-factor pyridoxal phosphate.

GAD67 possesses a preferential somato-dendritic localization and is saturable with PLP. GAD67 is therefore more prone to elicit a tonic release of GABA and is regulated by enzyme induction at the transcriptional level.

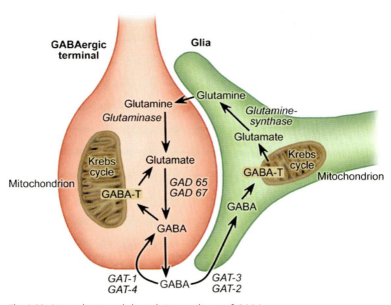

Fig. 3.11 Biosynthesis and degradation pathway of GABA. The pathway involves a GABAergic nerve terminal and an associated glial cell. *Abbreviations*: GABA-T = GABA transaminase; GAD = glutamatic acid decarboxylase; GAT = GABA transporter.

In contrast, GAD65 exhibits a preferred axonal localization and is not saturable with PLP. Consequently, elevation of the PLP level in the terminals enhances the activity of GAD65. Such a mechanism can serve as a fast regulative element for monitoring GABA levels in axon terminals.

The apparent prominent role of GAD in modulation of GABA levels becomes obvious under pathological conditions, where GAD concentrations can differ significantly from normal levels.

For instance, in models of temporal lobe epilepsy, the level of GAD65 and GAD67 is reported to be upregulated in hippocampal GABAergic interneurons.

The role of the two isoforms of GAD is not fully understood. Since both forms of GAD are expressed and regulated in different ways, it seems possible that the existence of two isoforms offers a sensitive tool for fine-tuning the GABA metabolism.

GABA is inactivated by transamination with a-ketoglutarate. This reaction is under the control of the enzyme GABA transaminase (GABA-T). GABA-T catalyzes the formation of succinic acid semialdehyde and glutamate. Succinic acid semialdehyde is further metabolized to form succinic acid, which is also an intermediate of the Krebs cycle.

In many inhibitory neurons, both GAD and GABA-T are coexpressed but, unlike GAD, GABA-T is associated with mitochondria.

3.3.4
GABA Transporters

GABA transporters (GAT) are located in the plasma membrane and mediate the removal of GABA from the extracellular space in order to terminate synaptic events evoked by released GABA. GABA transporters are found in GABAergic neurons as well as in glia.

GABA transporters are classified into two groups, which differ in their sensitivities to GABA. In the mouse, four different types of GABA transporters have been identified (GAT-1, GAT-2, GAT-3 and GAT-4). The transporters GAT-1, GAT-3, GAT-4 correspond to the transporters GAT-1, GAT-2, GAT-3 of rats. GABA transporters, like many members of the neurotransmitter transporter family, exhibit 12 transmembrane-spanning segments.

The uptake mechanism depends on Na^+ and Cl^-. The chloride ion binds close to the GABA binding site of the transporter and enhances its affinity for the substrate. Cellular uptake is driven by the electrochemical gradient of Na^+. One cycle of GABA transport carries two Na^+ ions from the extracellular space into the cell.

A reversal of the Na^+ gradient gives rise to an inverse effect, by which GABA is released from neurons and glia.

3.3.5
Receptors and Signal Transduction

GABA is released from the terminals of specific inhibitory neurons. GABA binds to its receptors and produces an increase in membrane permeability to Cl^- ions, which elicit a hyperpolarization of the postsynaptic membrane.

Three types of GABA receptors can be classified; and these are designated as $GABA_A$, $GABA_B$ and $GABA_C$ receptors. The three subtypes differ in their pharmacological properties and physiological behavior (Table 3.5).

$GABA_A$ receptors

$GABA_A$ receptors show a ubiquitous distribution throughout the central nervous system and have been identified on both neurons and glia. Outside the central nervous system, $GABA_A$ receptors are located on autonomic ganglia and on some unmyelinated nerve fibers.

The binding of GABA to the $GABA_A$ receptor increases the permeability to chloride ions, which causes a hyperpolarization of the postsynaptic membrane.

The $GABA_A$ receptor was first cloned and sequenced in 1987 by Schofield and coworkers. The $GABA_A$ receptor is a hetero-oligomeric receptor and belongs to the superfamily of ligand-gated ion channels (Fig. 3.12). The $GABA_A$ receptor consists of different subunits. To date, four different types of subunits (α, β, γ and δ) have been described, each of which encloses different members. They are depicted in Table 3.6.

Table 3.5 Properties of the different GABA receptor subtypes. The list of agonists and antagonists is not complete, but includes some of the most commonly used substances.
Abbreviations: CACA = *cis*-aminocrotonic acid;
CAMP = *cis*-2-(aminomethyl)cyclopropylcarboxylic acid;
3-APS = 3-aminopropanesulfonic acid;
TBPS = *t*-butylbicyclophosphorothionate.

Receptor	Ion channel gating	Second messenger	Agonists	Antagonists	Modulators
$GABA_A$	Direct	–	GABA, muscimol, 3-APS	Bicuculline, picrotoxin, TBPS	Benzodiazepines, barbiturates, steroids, Zn^{2+}, PKA, PKC
$GABA_B$	Indirect via G-protein	cAMP level	(–)-baclofen, GABA	CGP 36742, CGP 54626, CGP 55845	–
$GABA_C$	Direct	–	GABA, CACA, CAMP	picrotoxin	PKC

Table 3.6 Properties of recombinant GABA$_A$ receptors. Most of the subunits have been cloned in mammals. The number of amino acids of the predicted sequence is displayed, as well as the length of the mRNA.

Subunit	Amino acids	mRNA (kb)
Alpha		
α1	428	3.8, 4.2
α2	423	3.0
α3	465	4.2
α4	521	4.0
α5	433	2.8
α6	434	3.2
Beta		
β1	449	12.0
β2	450	8.0
β3	448	2.5, 6.0
Gamma		
γ1	430	3.8
γ2S	428	2.8, 4.2
γ2L	436	–
γ3	450	–
Delta		
δ	433	2.0

Alternative splicing of the RNAs of some subunits causes a further diversification of GABA$_A$ receptors. The splicing mechanism can generate long and short forms of the subunits. Coexpression of members of the α-, β-, γ-subunits seems to be necessary for the assembly of the functional, neuronal GABA$_A$ receptor.

In addition, functional GABA$_A$ receptors can also be formed by binary complexes (for example α- and β-subunits), but it seems unlikely that these forms are present *in vivo*.

The most common model of the GABA$_A$ receptor consists of a pentameric structure, which forms an ion pore and is selective for Cl$^-$ (Fig. 3.12). The pentameric structure is composed of two α-subunits, one β-subunit and one γ-subunit. The fifth unit in the pentamer is variable and can be provided either by one of the α- or γ-subunits or a δ-subunit. Variation in the oligomers leads to pharmacologically different GABA$_A$ receptor subtypes. GABA$_A$ receptors, which contain α1-, α2-, α3- or α5-subunits, are sensitive to diazepam (a benzodiazepine derivative), whereas receptors containing α6-subunits are diazepam-insensitive.

However, the precise stochiometry and composition of GABA$_A$ receptors is still under investigation.

The GABA$_A$ receptors have been characterized by their pharmacology. Classic agonists of these receptors include GABA itself, as well as muscimol.

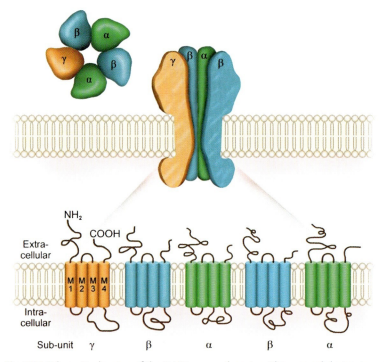

Fig. 3.12 Schematic drawing of the GABA$_A$ receptor which comprises an ionotropic receptor. The receptor is composed of five subunits which form four transmembrane domains. This general design is common for all ionotropic receptors. The heteromeric complex consists of 2α, 2β and one γ subunit.

Bicuculline is a specific antagonist of the GABA$_A$ receptor. Picrotoxin is a non-competitive antagonist as well as t-butylbicyclophosphrothionate (TBTS).

The functional behavior of GABA$_A$ receptors can also be influenced by certain ions, like Zn^{2+}, H^+ and some polycations. With the exception of La^{3+} and H^+, most of these ions lead to inhibition of the GABA$_A$ receptor.

La^{3+} enhances GABA$_A$ receptor-mediated effects, while variations in H^+ concentrations have different effects, ranging from inhibition to enhancement of GABA$_A$ receptor-mediated functions.

More recent studies indicate that naturally occurring pregnane steroids, so-called neurosteroids, can potentially and specifically enhance GABA$_A$ receptor activity in a direct (non-genomic) manner and have anxiolytic, analgesic, anti-convulsant, sedative, hypnotic and anaesthetic properties. Among these are progesterone and deoxycorticosterone metabolites, which act as potent stereoselective positive allosteric modulators on GABA$_A$ receptors.

The GABA$_A$ receptor possesses three different binding sites. The first binding site binds the neurotransmitter GABA. A second binding site on the receptor is a specific binding site for benzodiazepines and a third binding site is specific for barbiturates (Fig. 3.13).

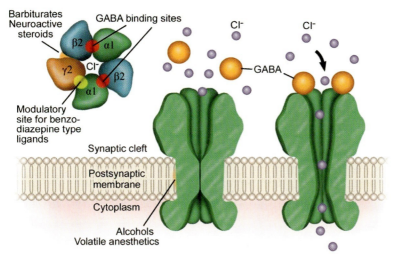

Fig. 3.13 The GABA$_A$ receptor and its binding and modulatory sites. The left drawing depicts the closed state and the right drawing the open state of the channel.

GABA$_B$ receptors

The first GABA$_B$ receptor was cloned in 1997 and, soon after, a second GABA$_B$ was identified. Cloning of both receptors provided evidence that native GABA$_B$ receptors are heterodimers of two subunits, GABA$_{B(1)}$ and GABA$_{B(2)}$. Isoforms of the GABA$_{B(1)}$ gene (Gabbr1) have also been described, GABA$_{B(1a)}$ and GABA$_{B(1b)}$, which seem to be conserved in different species, including humans. Rats show an age-related pattern of expression with high levels of GABA$_{B(1a)}$ at birth and increasing levels in adults. Both isoforms have been found to be transcribed by different promoters and further evidence showed that the promoters are regulated differentially by cAMP-response element-binding protein (CREB), activating transcrioption factor 4(CREB2) and depolarization-sensitive upstream stimulatory factor. Heterodimerization with GABA$_{B(2)}$ of both isoforms at presynaptic and postsynaptic sites are likely to occur.

Additional splice variants of Gabbr1 have been identified (GABA$_{B(1c-g)}$), but their functional significance is unclear.

Functional expression of the GABA$_B$ receptors provided evidence that the GABA$_B$ receptor belongs to the class of metabotropic receptors and thus is a member of the superfamily of G protein-coupled receptors (GPCRs). The cloning work on GABA$_B$ receptor gained a substantial breakthrough in our understanding of GPCRs. In general, it is now well accepted that GPCRs can exist in heterodimers and that concerted efforts in cloning work on the GABA$_B$ receptor fundamentally changed our view on the structure and functioning of this important class of metabotropic receptors. Figure 3.14 depicts some of the essential structural features of the GABA$_B$ receptor complex.

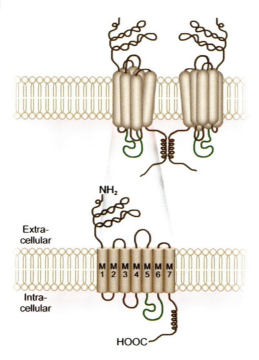

Fig. 3.14 Schematic drawing of the GABA$_B$ receptor which belongs to the GPCR family. The native GABA$_B$ receptor is a heterodimer consisting of two subunits, GABA$_{B(1)}$ and GABA$_{B(2)}$. The subunits dimerize in the coiled-coiled domains of the carboxyl termini to form an active receptor.

In contrast to GABA$_A$, the metabotropic GABA$_B$ receptor is insensitive to the GABA$_A$ receptor antagonist bicuculline. Baclofen is a specific agonist of GABA$_B$ receptors, while compounds like CGP 36742, CGP 54626 and CGP 55845 are selective antagonists.

GABA$_B$ receptors are not directly associated with ion channels and their signal transduction is mediated through G protein-coupled processes (G$_o$ or G$_i$). The G$_o$ protein can bind to a Ca^{2+} or a K$^+$ channel. Figure 3.15 depicts a scheme for the receptor–ligand binding induced cascade of G protein interaction in the case of the GABA$_B$ receptor.

Presynaptically located GABA$_B$ receptors modulate neurotransmitter release by depressing Ca^{2+} influx through voltage-activated Ca^{2+} channels of the N type (Ca$_v$2.2). Both types of presynaptic GABA$_B$ receptors are expressed: autoreceptors that control GABA release and heteroreceptors that inhibit all other neurotransmitter release. Presynaptic inhibition at GABAergic synapses is considered to be involved in the induction of LTP. Postsynaptic GABA$_B$ receptors induce a slow inhibitory postsynaptic current through activation of inwardly rectifying K$^+$ channels (GIRK or Kir3). The physiological effect of Kir3 channel activation is a K$^+$ efflux, resulting in hyperpolarization.

GABA$_B$ receptors are particularly prominent in the cerebral cortex, the thalamus, the superior colliculus, the cerebellum and the dorsal horn of the spinal chord. In the peripheral nervous system, they have been identified on autonomic ganglia, unmyelinated fibers and some smooth muscle cells.

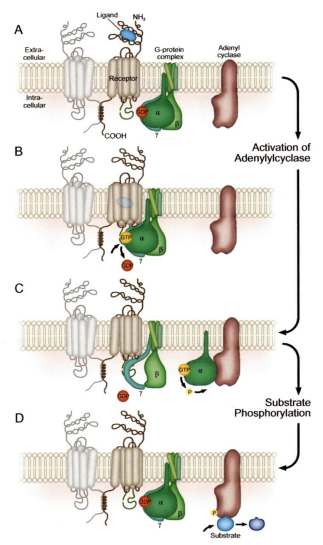

Fig. 3.15 Diagrammatic presentation of the cAMP cycle in the case of GABA$_B$ receptor activation. Binding of the transmitter alters confirmation of the receptor and exposes the binding site for G protein-coupling (A). The transmitter–receptor complex associates with the G protein complex (G$_s$), thereby activating it for GDP–GTP exchange (B). Release of GDP and binding of GTP causes the dissociation of the α-subunit from the G$_s$ complex and exposes a binding site for adenylyl cyclase (C). Activated adenylyl cyclase produces cAMP for further substrate phosphorylation (D).

GABA$_C$ receptors

An analogue of GABA, cis-aminocrotonic acid (CACA), is able to bind to a GABA receptor, which is different from either GABA$_A$ or GABA$_B$ receptors. However, this receptor is insensitive to bicuculline or baclofen. Following this specific pharmacology, the receptor was classified as the third or GABA$_C$ receptor subtype. The GABA$_C$ receptor is coupled to a Cl$^-$ selective ion channel. GABA$_C$ receptors are hetero-oligomers, which exhibit another subunit family, the so-called ρ-subunits (ρ1 and ρ2).

An allosteric modulation side for benzodiazepines or barbiturates seems to be absent from the GABA$_C$ receptor, but picrotoxin acts as an antagonist.

Attempts have failed to demonstrate GABA$_C$ receptors in the forebrain, but they have been identified in the pituitary and in horizontal and bipolar neurons of the retina.

3.3.6
Biological Effects

General aspects

GABA mediates inhibitory effects on the postsynaptic membrane. These effects are a consequence of the hyperpolarization of the neuron. In cases of high intracellular concentrations of chloride, GABA can also reduce the membrane potential and thus lead to a depolarization of the neuron. This has been described in immature neurons and in mature dorsal root ganglion cells (DRG). The mechanisms responsible for this reverse effect were recently elucidated. The excitatory effect of GABA seems to depend on the presence of cation–chloride cotransporters that tighty control the chloride gradient across neurons. Apparently the differential expression of two chloride cotransporters during development is responsible for the bipolar behavior of the GABA$_A$ receptor. Immature neurons express a higher level of the Na$^+$–K$^+$–2Cl$^-$ cotransporter (NKCC1) which results in a high concentration of intracellular Cl$^-$. GABA$_A$ receptor activity thus leads to an efflux of Cl$^-$ and subsequent depolarization. In mature neurons, expression of a second Cl$^-$ cotransporter (KCC2) prevails which lowers Cl$^-$ below that achieved by passive diffusion. This results in a low intracellular Cl$^-$ concentration. Consequently, GABA$_A$ activation causes a Cl$^-$ influx and hyperpolarization.

It is assumed that most of the physiologically relevant functions of GABA are mediated by GABA$_A$ receptors and that these receptors are involved in postsynaptic inhibition of signal transduction in mature brains.

Iontophoretic activation of GABA$_A$ receptors results in fast opening (about 1 ms) of the associated chloride channel. The influx of chloride into the neuron hyperpolarizes the cell membrane of the postsynaptic neuron and inhibits the cell. Activation of this receptor type gives rise to the classic inhibitory postsynaptic potentials.

Binding of GABA to either the GABA$_A$ or the GABA$_B$ receptors depends on the presence of calcium and magnesium.

Glutamat-Decarboxylase Gaba

(B6) - abhängig (exzit. Tsg.)
                    Löffler

↑ → Glutamat
    Excitatory Neurotr. Tsg

↓ GABA
  Inhibitor. Neurotr. (u. ü. Tsg

Binding of GABA
to either GABA A or
the   GABA B
receptors depends
of Ca and Mg

Since GABA$_B$ receptors are located mainly presynaptically, binding of GABA (or baclofen) to these receptors inhibits the release of neurotransmitters from the terminals. Such an effect has been shown to occur in several dopaminergic, glutamatergic, noradrenergic, as well as serotoninergic neurons, indicating that GABA$_B$ receptors can function as heteroceptors.

GABA is present not only in the central nervous system, but also in the peripheral nervous system, including the enteric nervous system of the gastrointestinal tract.

In the enteric nervous system, GABA shows some effects which are similar to those occurring in other parts of the peripheral nervous system, but also shows other effects which are notably different. Like the cell bodies of other autonomic and sensory neurons, the cell bodies of enteric neurons possess bicuculline- and picrotoxin-sensitive GABA receptors. However, some evidence suggests that, unlike other parts of the peripheral nervous system, the enteric ganglia may also contain a population of GABAergic neurons. It was assumed for a long time that such neurons in vertebrates are exclusively present in the brain and spinal cord.

GABA$_A$ and memory storage

Neuronal processes suggested GABA$_A$ receptors being involved in memory storage and long-term potentiation (LTP). LTP are blocked at the time of their initiation by antagonists of glutamate, NMDA, or by GABA$_A$ receptor agonists. The GABA$_A$ receptor antagonists picrotoxin and bicuculline, for instance, enhance memory functions, whereas benzodiazepines and the GABA$_A$ agonist muscimol depress memory functions. The discovery of naturally occurring benzodiazepines in the brain prompted a recent investigation of whether these compounds could act as physiological regulators of the GABA$_A$ receptors involved in memory modulation. Different forms of learning cause a rapid reduction of benzodiazepine-like immunoreactivity in the septum area, amygdala and hippocampus; and microinjection of the benzodiazepine antagonist flumazenil into these regions, at the time that consolidation is taking place, enhances memory.

Ivan Izquierdo and Jorge Medina (1991) suggested that these findings indicate that benzodiazepines released into the septum, amygdala and hippocampus do indeed downregulate memory storage processes. Moreover, benzodiazepine release could be modulated by the anxiety and/or stress associated with learning.

3.3.7
Neurological Disorders and Neurogenerative Diseases

Specific functional and pharmacological aspects

Anxiety response
Blockade of GABAergic inhibition in the basolateral amygdala (BLA) of rats elicits physiologic changes associated with a defense reaction. In the BLA, injection of the GABA$_A$ receptor antagonists bicuculline or picrotoxin produces anxio-

genic-like effects. This suggests that endogenous GABA binds to GABA$_A$ receptors in the basolateral amygdala and inhibits anxiety responses.

Alterations in GABAergic transmission can result in severe disturbances in brain activity and a deficit in GABAergic transmission can lead to epileptogenesis.

GABA and seizure induction

The administration of certain GABA$_A$ receptor antagonists can produce seizures and convulsions in animals.

Studies of experimental epilepsy in animals have shown a significant reduction in numbers of GABAergic terminals in different brain areas as well as a significant loss of inhibitory interneurons. Concurrently, the remaining hippocampal GABAergic neurons show an up-regulation of GAD65 and GAD67, which can be interpreted as a compensation effect.

Neuroactive compounds, which reduce the GABAergic transmission, can promote seizures and convulsions. These effects can be explained in different ways:
- The compounds influence the biosynthesis of GABA by inhibiting the activity of glutamic acid decarboxylase or, alternatively, they inhibit the formation of its co-factor PLP.
- They interact directly with the GABA receptors and inhibit them.
- They change the permeability of the plasma membrane for chloride ions by a perturbation of the channel.
- They inhibit the release of GABA from the nerve terminals. One example for this mode of action is the tetanus toxin.

In contrast, inhibitors of the GABA transporters as well as benzodiazepines and barbiturates, which facilitate GABAergic transmission, show anticonvulsive properties.

Several studies show that administration of GABA$_B$ receptor antagonists can elicit absences. This indicates that activation of GABA$_B$ receptors is also involved in some specific forms of epilepsy. Baclofen currently is the only GABA$_B$ medication and is used to treat spasticity and skeletal muscle rigidity in patients with spinal cord injury, multiple sclerosis, amyotrophic lateral sclerosis, and cerebral palsy.

Moreover, some metabolic alterations in GABA levels in the brain occur coincidentally with some degenerative brain diseases, which include Huntington's chorea (associated with a degeneration of GABAergic nigro-striatal neurons) and Parkinsonism.

GABA and benzodiazepines

Benzodiazepines and barbiturates bind to GABA$_A$ receptors. Their binding sites differ from that of GABA. The principal effect of both drugs is a change in Cl$^-$ conductance (see Fig. 3.13).

Barbiturates are used therapeutically as sedatives and hypnotics. Benzodiazepines are used to treat a variety of clinical symptoms, including anxiety, convulsion and muscle tension. However, like barbiturates, their most common use is as sedatives and hypnotics.

Benzodiazepines, like chlordiazepoxid (Librium) or diazepam (Valium) were introduced into clinical practice at the beginning of 1960.

More than 50 different benzodiazepine derivatives have since been engineered, revealing a wide range of physiological and pharmacological properties.

The benzodiazepines show anticonvulsive, anxiolytic, sedative and muscle-relaxant properties. They are also used for the treatment of anxiety and sleep disorders. Since they exhibit anticonvulsant properties, the possibility was considered that benzodiazepines might be useful as anti-epileptic drugs. However, since clinical trials have shown that patients develop tolerance to the anticonvulsive effects, the benefits of benzodiazepines in the treatment of epilepsy are limited.

Benzodiazepines have been classified into three groups, according to their effects on GABA transmission. Some benzodiazepines work like antagonists, others like agonists.

Typical anxiolytic compounds like diazepam show agonistic effects on $GABA_A$ receptors. Their binding to the allosteric site of the receptor enhances the affinity of the GABA-binding site for GABA, which results in an enhancement of chloride influx.

Flumazenil, in contrast, (Fig. 3.16) mediates antagonist-like effects by blocking the effects of agonists, but shows no direct effect itself.

The agonistic-like properties mediate the classic effects of the benzodiazepines; however some compounds initiate completely contrary effects. These compounds with diametrically opposite effects were termed inverse agonists. Most of these compounds belong to the β-carboline group.

One member of this group is ethyl-β-carboline-3-carboxylate (β-CCE), which has a high affinity for the benzodiazepine specific binding site. However, it shows no anticonvulsive behavior, like the classic benzodiazepines, but exhibits proconvulsive effects (Fig. 3.17).

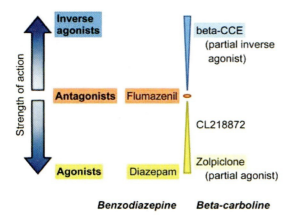

Fig. 3.16 Benzodiazepines and β-carboline act like antagonists, agonists or inverse agonists.

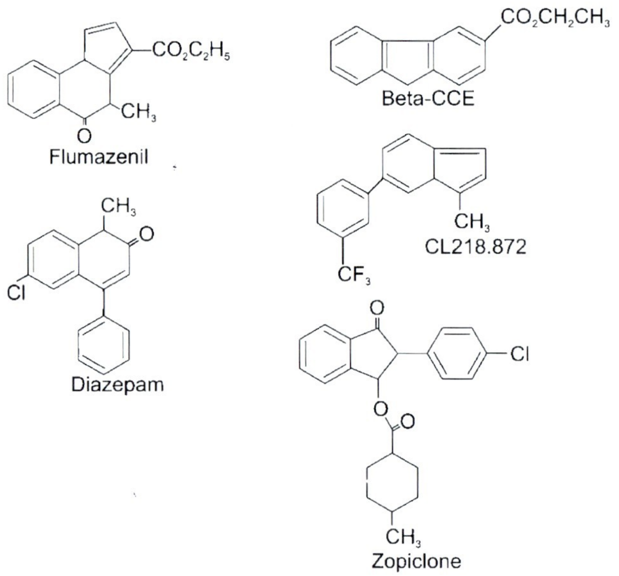

Fig. 3.17 Structure of some benzodiazepines which work as agonists, antagonists or as inverse agonists.

Neurosteroids: endogenous regulators of the $GABA_A$ receptor

Brain-derived pregnane steroids can potently and specifically enhance $GABA_A$ receptor function (see Fig. 3.13). This effect occurs fast in a non-genomic (direct) manner, different from the steroid hormones that are produced in peripheral endocrine glands, cross the blood–brain barrier and produce changes at the genomic level. The latter effects develop slowly and can persist long after the disappearance of the steroids from the brain. Selye (1942) was the first to show that pregnane steroids possess rapid sedation and anaesthetic effects. 5α-Pregnan-3α-OL-20-one (3α, 5α-THPROG), 5β-pregnan-3α-ol-20-one (3α, 5β-THPROG) and the deoxycorticosterone metabolite 5α pregnan-3α,21-diol-20-one are potent selective allosteric modulators of the $GABA_A$ receptor. Patch-clamp studies indicate that these steroids promote the open state of the GABA-gated ion channel. At concentrations of ≥ 100 nM, these steroids directly activate the $GABA_A$ receptor–channel complex. This GABA-mimetic effect is sufficient to suppress excitatory neurotransmission. Besides their anaesthetic effects, lower doses produce anxiolytic, analgesic, anticonvulsant, sedative and hypnotic effects. Neurosteroids can either be synthesized locally by neurons and glial cells which express enzymes that are required for de-novo synthesis or they can be recruited from peripheral steroid sources. Their general action seems to be the fine-tuning of the $GABA_A$ receptor activity.

Recent evidence suggests that neurosteroids are involved in brain development and plasticity. This was concluded from the relative high levels of neurosteroids just before parturition. Further evidence indicates that the steroid

3α,5α-THPRO is important in the Niemann-Pick type C (NP-C) neurodegenerative disorder. In a transgenic mouse model which carries an insertion in the NPC$_1$ gene, which is responsible for the NP-C phenotype, characteristic neuropathological features occur: Purkinje cell degeneration, demyelination and loss of motor function. Interestingly, the neurosteroid 3α,5α-THPRO cannot be detected in the brains of these mice. Neonatal application of this neurosteroid doubles the lifespan, delays the appearance of neurological symptoms and increases the Purkinje cell survival.

Patients with Alzheimer's disease and Parkinsonism also show reduced levels of 3α,5α-THPRO, compared with age-matched controls.

Further Reading

Bazemore, A.W., Elliott, K.A.C., Florey, E. (1956): Factor I and gamma-amino-butyric acid. *Nature* **178**: 1052–1053.

Belelli, D., Lambert, J.J. (2005): Neurosteroids: endogenous regulators of the GABA(A) receptor. *Nat. Rev. Neurosci.* **6**: 565–575.

Bettler, B., Kaupmann, K., Mosbacher, J., Grassmann, M. (2004): Molecular structure and physiological functions of GABA$_B$ receptors. *Physiol. Rev.* **84**: 835–867.

Ben-Ari, Y. (ed) (2005): Multiple facets of GABAergic synapses. *Trends Neurosci.* **28**: 277–340.

Betz, H. (1990): Ligand-gated ion channels in the brain: the amino acid receptor superfamily. *Neuron* **5**: 383–392.

Boue-Grabot, E., Taupignon, A., Tramu, G., Garret, M. (2000): Molecular and electrophysiological evidence for a GABAc receptor in thyrotropin-secreting cells. *Endocrinology* **141**: 1627–1632.

Bowery, N.G. (1993): GABA$_B$ receptor pharmacology. *Annu. Rev. Pharmacol. Toxicol.* **33**: 109–147.

Brandao, M.L., Borelli, K.G., Nobre, M.J., Santos, J.M., Albrecht-Souza, L., Oliveira A.R., Martinez, R.C. (2005): Gabaergic regulation of the neural organization of fear in the midbrain tectum. *Neurosci. Biobehav. Rev.* **29**: 1299–1311.

Cyran, J.F., Kaupmann, K. (2005): Don't worry "B" happy!: a role for GABA (B) receptors in anxiety and depression. *Trends Pharmacol. Sci.* **26**: 36–43.

Dalison, M.G., Pahal, I., Thode, C. (2005): Consequences of the evolution of the GABA(A) receptor gene family. *Cell Mol. Neurobiol.* **25**: 607–624.

Farrant, M., Nusser, Z. (2005): Variations on an inhibitory theme: phasic and tonic activation of GABA(A) receptors. *Nat. Rev. Neurosci.* **6**: 215–229.

Izquierdo, I., Medina, J.H. (1991): GABA$_A$ receptor modulation of memory: the role of endogenous benzodiazepines. *Trends Pharmacol. Sci.* **12**: 260–265.

Jessen, K.R. (1981): GABA and the enteric nervous system. A neurotransmitter function? *Mol. Cell Biochem.* **38**[Spec No]: 69–76.

Johnston, G.A. (2005): GABA(A) receptor pharmacology. *Curr. Pharm. Des.* **11**: 1867–1685.

Laurie, D.J., Wisden, W., Seeburg, P.H. (1992): The distribution of 13 GABAA receptor subunit messenger RNAs in rat brain. 3. Embryonic and postnatal development. *J. Neurosci.* **12**: 4151–4172.

Levitan, E.S., Schofield, P.R., Burt, D.R., Rhee, L.M., Wisden, W., Köhler, M., Fujita, N., Rodriguez, H.F., Stephenson, A., Darlison, M.G., Barnard, E.A., Seeburg, P.H. (1988): Structural and functional basis for GABA$_A$ receptor herterogenity. *Nature* **335**: 76–79.

Lewis, D.A., Hashimoto, T., Volk, D.W. (2005): Cortical inhibitory neurons and schizophrenia. *Nat. Rev. Neurosci.* **6**: 312–324.

Liu, Q.R., Lopez-Corcuera, B., Nelson, H., Nelson, N. (1993): Molecular characterization of four pharmacologically distinct aminobutric acid transporters in the mouse brain. *J. Biol. Chem.* **268**: 2106–2112.

McDonald, A. J., Augustine, J. R. (1993): Localization of GABA-like immunreactivity in the monkey amygdala. *Neuroscience* **52**: 281–294.

MacDonald, R. L., Olsen, R. W. (1994): GABA$_A$ receptor channels. *Annu. Rev. Neurosci.* **17**: 569–602.

Nayeem, N., Green, T. P., Martin, I. L., Barnard, E. A. (1994): Quarternary structure of the native GABA$_A$ receptor determined by electron microscopic image analysis. *J. Neurochem.* **62**: 815–818.

Olsen, W. O., Tobin, A. J. (1990): Molecular biology of GABA-A receptors. *FASEB J.* **4**: 1469–1480.

Polc, P., Möhler, H., Haefely, W. (1974): The effect of diazepam on spinal cord activity: possible sites and mechanisms of actions. *N.S. Arch. Pharmacol.* **284**: 319–337.

Pratt, J. A., Brett, R. R., Laurie, D. J. (1998): Benzodiazepine dependence: from neural circuits to gene expression. *Pharmacol. Biochem. Behav.* **59**: 925–934.

Roberts, E. (1984): GABA-related phenomena, models of nervous system function and seizures. *Ann. Neurol.* **16**: S77–S89.

Roy-Byrne, P. P. (2005): The GABA-benzodiazepine receptor complex: structure, function, and role in anxiety. *J. Clin. Psychiatry* **66**[Suppl 2]: 14–20.

Sanders, S. K., Shekhar, A. (1995): Regulation of anxiety by GABAA receptors in the rat amygdala. *Pharmacol. Biochem. Behav.* **52**: 701–706.

Selye, H. (1942): Anestetics of steroid hormones. *Proc. Soc. Exp. Biol. Med.* **46**: 116–121.

Squires, R. F., Braestrup, C. (1977): Benzodiazepine receptors in the brain. *Nature* **266**: 732–734.

Stein, V., Nicoll, R. A. (2003): GABA generates excitement. *Neuron* **37**: 375–378.

Yonkov, D., Georgiev, V., Kambourova, T. (1989): Further evidence for the GABAergic influence on memory. Interaction of GABAergic transmission with angiotensin II on memory processes. *Methods Find. Exp. Clin. Pharmacol.* **11**: 603–606.

3.4
Glutamate and Aspartate

3.4.1
General Aspects and History

The amino acids L-glutamate and L-aspartate are the most abundant excitatory neurotransmitter in the central nervous system. Numerous studies have confirmed the excitatory effects of both amino acids and corroborated their role as key excitatory amino acids in brain tissues. Several related amino acids, like homocysteic acid and N-acetylaspartylglutamate, may also serve a neurotransmitter function. As in the case of GABA, the excitatory amino acid neurotransmitters participate in intermediary metabolism as well as in neuronal signal transmission, so that there is often a problem of distinguishing between their neurotransmitter and metabolic functions. The following experimentally collected data clearly document the excitatory role of glutamate and aspartate:

- Stimulation of afferent fibers in the hippocampal formation is coupled to calcium-dependent release of glutamate or aspartate.
- High-affinity recapture systems are present in nerve terminals.
- Nerve terminals contain glutamate or aspartate as well as the enzymes necessary for synthesis of both excitatory amino acids.

- Postsynaptic depolarization elicited through the excitatory amino acids can be inhibited by specific antagonists.
- Specific receptors, capable of binding the excitatory amino acids are present in the central nervous system.
- Iontophoretically injected glutamate enhances postsynaptic responses.
- These responses are comparable to the effects of endogenous glutamate.

3.4.2
Localization Within the Central Nervous System

Since glutamate is the best known member of the excitatory amino acid neurotransmitter family, we will concentrate our survey on this amino acid. Glutamatergic neurons are particularly prominent in the cerebral cortex. They project to a variety of subcortical structures: these include the hippocampus, the basolateral complex of the amygdala, the substantia nigra, the nucleus accumbens, the superior colliculus, the caudate nucleus, the red nucleus (nucleus ruber) and the pons.

Various intrinsic glutamatergic pathways have been described in the hippocampus and also glutamatergic projections from the hippocampal formation to the hypothalamus, the nucleus accumbens and the lateral septum. Glutamatergic neurons are located in some additional brain areas: their location and projection are depicted in Fig. 3.18.

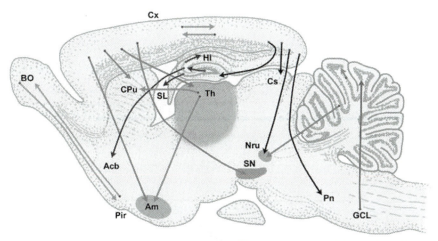

Fig. 3.18 Schematic representation of the glutamatergic pathways in the central nervous system. *Abbreviations*: Acb=*nucleus accumbens*; BO=olfactory bulb; CPu=caudate-putamen; Cs=*superior colliculus*; Cx=cortex; GCL=granular cell layer of the cerebellum; HI=hippocampus; Nru=*nucleus ruber* (red nucleus); Pn=pons nuclei; Pir=piriform cortex; SL=lateral septum; SN=*substantia nigra*; Th=thalamus.

3.4.3
Biosynthesis and Degradation

Brain glutamate and aspartate are derived solely by local synthesis, as neither amino acid can cross the blood–brain barrier. Glutamate is formed through the Krebs cycle by the activity of glutamate dehydrogenase or by transamination of α-ketoglutarate. The neurotransmitter glutamate is synthesized in nerve terminals and accumulates in synaptic vesicles. Interestingly, glutamate serves as the donor for GABA, the most potent inhibitory amino acid neurotransmitter, through decarboxylation by glutamic acid decarboxylase (see Section 3.3). Released glutamate is restored to presynaptic terminals by a specific re-uptake mechanism or cleared from the synaptic region by high-affinity membrane transporters through adjacent astrocytes. The pathway through astrocytes serves a recycling function since the imported glutamate is metabolized into glutamine – under the enzymatic control of glutamine synthetase – and released for subsequent uptake into neuronal terminals. The glutamine in the neurons is then transformed into glutamate.

3.4.4
Transporters

Sets of specific transporters regulate this uptake from the extracellular space into neurons or glia. To date, a family of four different members of the excitatory amino acid transporters (EAAT) has been identified: EAAT1-4, EAAT-1, EAAT2 and EAAT3. These transporters have a molecular mass between 57 kDa and 64 kDA. All four types of glutamate transporter are expressed in the brain but are distributed differentially: EAAT1 and EAAT2 transporters appear to occur exclusively in glia, whereas EAAT3 seems to be the neuronal transporter. These three transporters share a putative amino acid homology of about 50% among themselves, but little homology with other neurotransmitter transporters.

The cellular uptake of glutamate is driven by the electrochemical gradients of Na^+ and K^+ and is accompanied by pH and voltage changes. Some specific uptake blockers, e.g. dihydrokainate (DHK), β-threohydroxyaspartate (β-THA) and L-trans-3,4-pyrolidine-dicarboxylate (L-trans-2,4-PDC) are able to block the function of glutamate transporters.

The transporters are essential in terminating the excitatory signal and they lower the extracellular glutamate levels below thresholds that could induce excitotoxic damage (see below). Furthermore, they are a prerequisite to the recycling of the transmitter. A reduction in the transport capacity is associated with a concomitant increase in neurotoxicity. A prominent example where neuronal degeneration is accompanied by a reduction of glutamate transporters is the amyotrophic lateral sclerosis (ALS). This is characterized by cellular death of motor neurons in the spinal cord, although other factors, such as impairment of superoxyde dismutase seem to play a key role in the initiation of this disease.

3.4.5
Receptors and Signal Transduction

The differential distribution of the excitatory amino acids goes hand in hand with a diversity of receptors.

The excitatory amino acid receptors include ionotropic (ligand-gated ion channels) and metabotropic (receptors involving second messenger systems) receptor subtypes (Table 3.7). Since some prominent pharmacologically active substances have been used to classify the subtypes of excitatory amino acid receptors it seems important to look at the essential molecular features which make a drug capable of interacting with the receptors.

Such a drug should exhibit:
- one acidic group (usually a carboxyl group)
- an amino group on the same carbon as the acidic group
- a second group (polar or acidic) located two or four carbon units apart from the other acidic group. Apparently, this conformation is necessary to allow interaction with the active center of the receptors.

The naturally occurring agonists of the excitatory amino acid receptors are glutamate and aspartate. Pharmacologically selective agonists are substances like kainate, quisqualate, N-methyl-D-aspartate (NMDA) and α-amino-3-hydroxy-5-methyl-4-isoxazolepropionate (AMPA). As can be seen in Fig. 3.19, the endogenous excitatory amino acid and the drugs share the molecular features for receptor affinity described above.

The glutamate receptors, which could occur as homomeric or heteromeric structures (Table 3.8) are classified according to the binding of the most common agonists (1), or alternatively according to their functional properties (2), reflecting the pharmacology of the receptors. Four groups are differentiated: the NMDA receptors, the AMPA receptors, the kainate receptors and receptors which are activated by quisqualat.

Based on their general functional properties, two groups can be distinguished: the group of ionotropic receptors and the group of metabotropic receptors. The receptors which are activated by quisqualate belong to the group of metabotropic receptors. They involve second-messenger systems in signal transduction and they are coupled to G proteins. The remaining receptors (NMDA receptors, AMPA receptors and kainate receptors) constitute ligand-gated cation-selective ion channels and thus belong to the ionotropic receptor family (Fig. 3.20).

The ionotropic receptors include a glutamate-sensitive ion channel that opens upon activation and allows the influx of Na^+ and K^+ and/or Ca^{2+}, which subsequently elicit fast excitatory responses, measurable in the form of excitatory postsynaptic potentials (EPSP). However, the effects of the ionotropic receptors are not confined to these fast postsynaptic responses, but include post- and presynaptic processes that are involved in long-lasting changes in synaptic activity (see below).

Table 3.7 Glutamatergic receptors and their general properties. The table indicates the different receptor types and their properties. Some common agonists and antagonists are shown. *Abbreviations*: L-AP4 = L-2-amino-4-phosphonobutanoic acid; 2-AP5 = D-2-amino-5-phosphonopentanoic acid; 2-AP7 = D-2-amino-7-phosphonoheptanoic acid; CCG = 2-carboxycyclopropylglycine; CGP39653 = 2-amino-4-propyl-5-phosphonopentenoic acid; 4C3HPG = (S)-4-carboxy-3-hydroxyphenylglycine; CNQX = 6-cyano-7-nitroquinoxaline-2,3-dione; 4CPG = (S)-4-carboxyphenylglycine; CPP = {3-[(+)-carboxypiperazin-4-yl]prop-1-yl}phosphonic acid; 5,7-di-Cl-Kyn = 5,7-dicholorokynurenic acid; DNQX = 6,7-dinitroquinoxaline-2,3-dione; GAMS = glutamylaminomethylsulfonate; 3HPG = (S)-3-hydroxyphenylglycine; JST = Joro spider toxin; MK-801 = (+)-5-methyl-10,11-dihydro-5H-dibenzo[a,d]cyclohepten-5,10-imine; NBQX = 2,3-dihydroxy-6-nitro-7-sulfamoyl-benzo[f]quinoxaline.

	Ionotropic receptor			Metabotropic receptor
Receptor subtype	NMDA	AMPA	Kainate	mGluR
Selective agonists	NMDA	AMPA	Kainate	Quisqualate, L-AP4
Functional characteristics	Activation of Na^+, K^+ and Ca^{2+} channels	Activation of Na^+ and K^+ channels	Activation of Na^+ and K^+ channels	Activation of phospholipase C; inhibition of adenylate cyclase
Other agonists	Ibotenate, quinolinate	Quisqualate, kainate, domoate	Domoat, acromelic acids A and B	Quisqualat, L-serine O-phosphate, ibotenat, CCG
Competitive antagonists	2-AP5, 2-AP7, CPP, CGP39653	CNQX, NBQX, DNQX	GAMS, glutamyl-glycine	Phenylglycine analogs (3HPG, 4CPG, 4C3HPG)
Allosteric modulators	Glycin, D-serin, spermin	Benzothiazide	Concanavalin A (Con A)	
Antagonists on allosteric binding sites	5,7-diCl-Kyn, HA-966, CNQX			
Inhibitors of the ion channel	PCP, MK-801, Ketamine	Jorospider toxine (JST)		

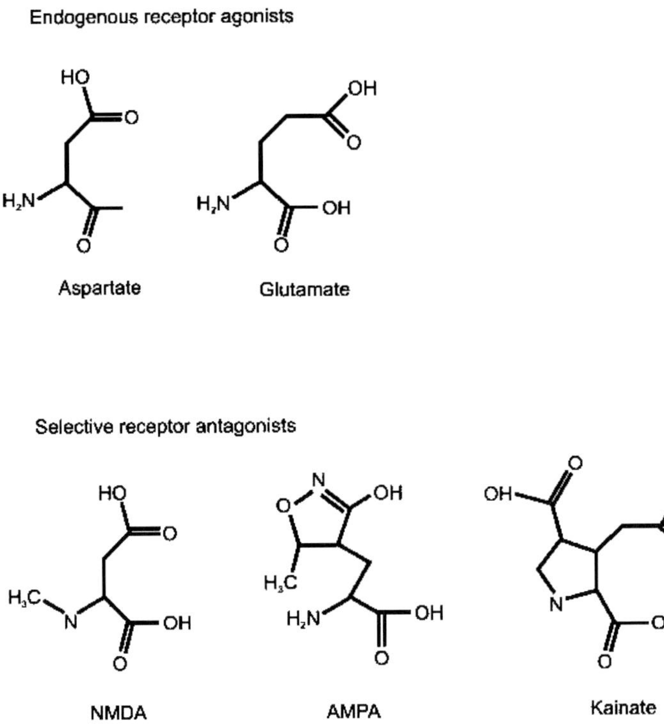

3.4 Glutamate and Aspartate

Endogenous receptor agonists

Aspartate   Glutamate

Selective receptor antagonists

NMDA   AMPA   Kainate

Fig. 3.19 Structures of the endogenous agonists aspartate and glutamate and of the receptor subtype-selective agonists NMDA, AMPA and kainate.

Table 3.8 Some characteristics of the exitatory amino acid receptors.

Receptor gene/ subunit	Structur	Characteristics
NMDA		
NR1	Homo- or heteromer	Eight splice-variants of a single gene
AMPA		
GluR1–4	Homo- or heteromer	Splice-variants, named "flip" and "flop"
Kainate		
GluR5	Homo- or heteromer	Low-affinity kainate-binding site
GluR6		
GluR7	Heteromer only	Low-affinity kainate-binding site
KA-1	Heteromer only	High-affinity kainate-binding site
KA-2		
Metabotropic		
mGluR1–8	Homomer only	Does not form an ion channel

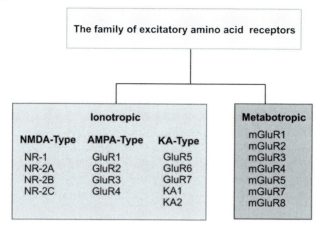

Fig. 3.20 The family of excitatory amino acid receptors and their subfamilies. The class of the ionotropic receptors includes NMDA, AMPA and kainate receptors. The class of the metabotropic receptors includes receptors linked to the activation of phospholipase C (PLC) or the inhibition of adenylate cyclase. Each receptor type is known to be composed of different subtypes.

The metabotropic receptors are coupled to G proteins. The activation of the metabotropic receptors impinges on a second messenger cascade which produces a delayed synaptic response.

NMDA receptors

The NMDA receptors seem to be most prominent in mammals. NMDA receptors are selectively activated by the drug NMDA and they are less selectively activated by glutamate, aspartate or homocysteate (HC).

The NMDA receptors respond to glutamate in a relatively slow fashion (in comparison to the other ionotropic excitatory amino acid receptors). This slow responsiveness is thought to be due to the fact that Mg^{2+} tonically inhibits NMDA receptors.

The current topological model of the NMDA receptor consists of a complex of five transmembrane proteins with different specific binding sites associated with an ion channel which is permeable to Ca^{2+}, Na^+ and K^+ (Fig. 3.21).

The different binding sites of the NMDA receptors can be divided into four categories:

- A main binding site for agonists like NMDA, glutamate and competitive antagonists, including D-2-amino-5-phosphonopentanoic acid (AP5), cis-4-(phosphonomethyl)piperidine-2-carboxylic acid (CGS 19755), and 3-[(+)-2-carboxypiperazin-4-yl] propyl-1-phosphonic acid (CPP).
- A ligand-gated ion channel. This channel can be blocked at two specific sites: a voltage-dependent cation site, which is blocked by Mg^{2+} and a binding site, which is a target for some substances acting as non-competitive NMDA an-

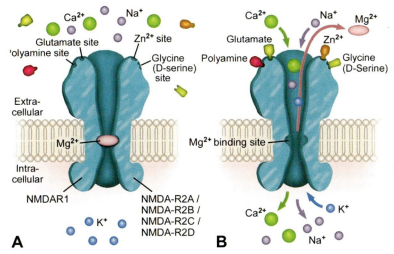

Fig. 3.21 NMDA receptor with its binding and modulatory sites. Part (A) depicts the closed receptor with Mg^{2+} block. Upon glutamate binding and Mg^{2+} release, the channel is open for Ca^{2+} and Na^+ entrance and K^+ efflux (B). The different modulatory binding sites are indicated at the extracellular extremities of the receptor and the subunits at the cytoplasmic aspect.

tagonists: 1-(1-(2-thienyl)cyclohexyl)piperidine (TCP) and (5-methyl-10,11-dihydro-5H-dienzocyclohepten-5,10-imine maleate (MK-801).
- Different regulatory binding sites such as an allosteric binding site for glycine and a polyamine site as well as Mg^{2+} and Zn^{2+} selective sites. These sites exert either negative or positive control over the function of the NMDA receptors.

The NMDA receptors play an important role in the excitatory amino acid induced transmission and in synaptogenesis. Under resting potential conditions, the NMDA receptors are not activated and the ion channel is blocked by Mg^{2+}. On depolarization, the magnesium block is released and the channel opens, thereby allowing the exchange of ions through the channel pore. The opening of the NMDA receptor ion channel increases the permeability for Na^+, K^+ and Ca^{2+}. One essential consequence of the entry of extracellular calcium through the channel is the activation of a variety of processes which alter the properties of the neuron. Excess of intracellular Ca^{2+} is also toxic to neurons; and hyperactivation of the NMDA receptor is thought to play an important role in several neurodegenerative disorders (see below).

Experiments using cDNA cloning have revealed the molecular diversity of NMDA receptor channels. The identification of multiple subunits with distinct distributions, properties and regulation implies that NMDA receptor channels are heterogeneous in their pharmacological properties, depending on the brain region and the developmental stage.

The deduced primary structures of the NMDA receptors share a homology of about 22–26% with the AMPA/kainate receptors. According to hydropathicity analyses, the ionotropic NMDA receptor forms four transmembrane segments. The second transmembrane section seems to be responsible for the formation of the ion channel and is flanked by negatively charged groups. In the original model, the transmembrane domains were proposed to be organized in a manner similar to the nicotinic ACh receptor, with the N-terminus and the C-terminus residing extracellularly and the loop between the third and fourth transmembrane domains extending into the cytoplasm. The currently proposed model differs insofar as the second membrane-spanning segment is considered to form a kink within the membrane and runs back into the cytoplasm akin to the model proposed for the pore-forming domain of voltage-activated K^+ channels. Both alternative models affect the consideration of which parts of the receptor are accessible for intracellular processing like phosphorylation or glycosylation. The intracellular domains provide consensus sequences, which are targets for protein kinase C and calmodulin-dependent kinases.

The NMDA receptors are heteromeric complexes, which consist of two different subunits:

- *NR1 subunits:* The NR1 family is encoded by a single gene, but alternative splicing generates at least eight different splice variants: NR1-1a, NR1-1b, NR1-2a, NR1-2b, NR1-3a, NR1-3b, NR1-4a, NR1-4b. A further isoform, NR1-5, seems to be biologically inactive. The NR1 subunits, when expressed in *Xenopus* oocytes, are able to form functionally active homomeric NMDA receptors.
- *NR2 subunits:* The predicted amino acid sequence of the NR2 subunits shares a homology of about 15% with the NR1 subunits. The NR2 subfamily is encoded by four different genes (NR2A-NR2D) and gives rise to five isoforms, which are named NR2-a, NR2-b, NR2-c, NR2-d1 and NR2-d2. The NR2 subunits, when expressed in oocytes, are unable to form a functional homomeric receptor.

The NR1 subunit seems to be an essential constituent of functional NMDA receptors and the subunit NR2 (NR2a-NR2d) is associated with one or more subunits of the NR1 type. The NR1 subunits are considered to form the molecular backbone of the receptor, while the subunit NR2 is assumed to be responsible for its physiological and pharmacological properties.

NMDA receptors are found in high densities in the cerebral cortex, the hippocampus, the basal ganglia, the hypothalamus and the olfactory bulb. The distribution pattern of NMDA receptors seems to be consistent with the expression of heteromeric receptors made up of one NR1 subunit and different NR2 subunits. The mRNA of the NR1 subunit is expressed throughout the brain, while the mRNA of the NR2 subunits exhibits a more restricted spatial and temporal expression.

Historically, the ionotropic glutamate receptors have been described as either NMDA or non-NMDA subtypes, depending on their abiliy to bind NMDA. AMPA and kainate receptors have been grouped to the non-NMDA receptor

3.4 Glutamate and Aspartate

family. Unlike the NMDA receptors, the non-NMDA receptors mediate fast excitatory synaptic transmission and they are primarily associated with voltage-independent channels.

AMPA receptors

AMPA receptors are ionotropic receptors and they belong to the group of non-NMDA-receptors. From the pharmacological point of view, they are not as well characterized as the NMDA receptors.

The AMPA receptors can be activated by the agonists a-amino-3-hydroxy-5-methyl-4-isoxazolleproprionat (AMPA), quisqualate and glutamate. Despite the absence of a specific antagonist for the AMPA receptors, several substances show relatively selective effects. Among these substances are glutamyl-aminoethyl-sulfonate (GAMS), glutamylglycine (DGG) and 6-cyano-7-nitroquinoxaline-2,3-dion (CNQX).

The AMPA receptors are associated with a cation-selective ion channel which is permeable for monovalent cations, like Na^+ and K^+. Under certain combinatorial conditions of the receptor subunits, it also becomes permeable to Ca^{2+}.

Molecular biological techniques have allowed the cloning of several cDNAs coding for at least four different subtypes of AMPA receptors. These subtypes were named $GluR_A$, $GluR_B$, $GluR_C$, $GluR_D$, or alternatively GluR1, GluR2, GluR3, GluR4. The different subfamilies of glutamate receptors share a homology in their predicted amino acid sequence of about 70% and each subunit is about 900 amino acids in length. The ensemble of the subunits which form the receptor and its associated ion channel provides, to a considerable degree, a combinatorial freedom that strongly influences the functional properties of the channel.

Like $GABA_A$ or nicotinic receptors, a combination of different subunits can form a functional receptor. Functional AMPA receptor channels can be assembled by expression of one type of subunit or by co-expression of two or four subunits. However, the homomeric receptors differ in their permeability to Ca^{2+}. For instance, when homomeric channels consisting of GluR2 subunits are functionally expressed in *Xenopus* oocytes, they exhibit little current in the presence of an agonist, unlike the large inwardly rectifying currents obtained when GluR1 or GluR3 are expressed.

Apparently, homomeric GluR2 channels form poorly conducting receptors. However, when GluR2 is coexpressed with either GluR1 or GluR3, the properties of the channel are distinctly different. In this case, the channel shows little rectification and a nearly linear I/V plot. Closer examination reveals that GluR1 and GluR3, either independently or under coexpression conditions, show channels permeable to Ca^{2+}, unlike channels which contain GluR2 subunits. In this case, any subunit combination makes the receptor impermeable to Ca^{2+}. Obviously, the physiological properties of the glutamate receptors can be quite different and elicit different responses depending on the subunit composition expressed in a particular type of neuron.

In addition, all subtypes can occur in two forms. These two forms are products of alternative splicing and they can be distinguished by the presence or the absence of a segment in the last transmembrane domain. These two splice variants are named "flip" and "flop" and they have different effects upon the kinetics and the amplitude of agonist-induced responses of the channels. Channels formed by the GluR-D "flip" variant, for example, show a slower desensitization kinetic than receptors formed by GluR-D "flop" variant.

Furthermore, both variants show different expression patterns in the adult and in the developing brain. An additional variation of the functional properties of recombinant AMPA receptors can be achieved by posttranscriptional RNA editing of the subtypes. RNA editing specifically permutates the encoded amino acid residues by deletion or insertion of single nucleotides, which usually leads to modified functional properties of the protein. In glutamate receptors, conversion of specific adenosine (A) residues to inosine (I) causes either a CAG → CIG conversion, with the consequence that arginine is changed to a glycine residue, or an AGA → IGA conversion having the same consequence. The most dramatic functional effects of RNA editing have been shown with edited sites located within the ion permeation pathway. When a glutamine (Q) residue at position 586 (for example present in all AMPA receptor protein except GluR2) is altered to an arginine (R) via RNA editing – a change that has been coined the Q/R site – the permation pore shuts off the calcium permeability and causes a linear current–voltage relationship instead of the normally inwardly rectifying relationship. AMPA receptors show an alteration of the GluR2/R3 expression during development [increasing to a peak between postnatal day (PND) 10 and PND 15 and then decreasing to a plateau at PND 30]; and this expression is subject to changes in the adult brain under pathological conditions. It is believed that an altered expression of GluR2 mRNA precedes neuronal degeneration.

The AMPA receptors are widely distributed in the central nervous system and their pattern is different from that of the NMDA receptors. High densities of AMPA receptors have been identified in the neocortex, the hippocampus, the lateral septum, the basolateral nucleus and the lateral nucleus of the amygdala, the caudate-putamen, the nucleus accumbens, the bulbus olfactorius and in the molecular layers of the cerebellum.

In the hippocampal formation, AMPA receptors are located preferentially in the pyramidal layer of the cerebral cortex, where they are primarily found, like NMDA receptors, in layer II and layer III.

GluR1–4 are widely distributed in the brain; however, the expression pattern and relative abundance of the different GluR subunits differ among various brain areas. In the hippocampus, for example, GluR1–3 mRNAs are highly expressed in the fields CA1, CA2 and CA3 and in the granule cells of the dentate gyrus. A low expression of GluR2–4 has been shown in the striatum, accompanied by a high expression of GluR1. The cerebral cortex exhibits high levels of GluR2 and GluR3, but shows low expression of GluR4.

GluR2 and GluR4 are abundant in the cerebellum and in the "Bergmann" glia.

Kainate receptors

Kainate receptors can be activated by kainate and glutamate. The substances glutamylaminomethylsulfonate (GAMS), 6-cyano-7-nitroquinoxaline-2,3-dione (CNQX) and 5-nitro-6,7,8,9-tetrahydrobenzo[G]indole-2,3-dione-3-oxime (NS102) show weak antagonizing effects on the kainate receptor, but no specific antagonists have been found yet.

Selective antagonists are currently not available. For this reason, it was thought for a long time that AMPA and kainate receptors were identical (and were thus termed the AMPA/kainate receptor).

Like the AMPA receptors, the kainate receptors are associated with an ion channel which is permeable for the monovalent cations Na^+ and K^+ and for Ca^{2+}.

High-affinity kainate receptors are formed by different subunits, which belong to two structural classes. These two classes consist of the subtypes GluR5, GluR6, GluR7 and KA-1 and KA-2. Sequence homology of both classes is less than 50%.

One class comprises GluR5, GluR6 and GluR7, which share a homology of about 80% with one another, but they share a homology of only 40% with the subunits of the AMPA receptors. The homomeric expression of GluR5 or GluR6, but not GluR7, yields functional channels with a low affinity for kainate.

In addition, GluR5 and GluR6 occur in two isoforms. These differ in their Q/R site (see above) due to RNA editing.

The second class comprises KA-1 and KA-2. Homomeric expression of both subtypes yields channels exhibiting a high affinity for kainate, but the homomeric expression does not qualify for agonist-sensitive ion channels.

Functional channels with affinity for kainate are also formed by coexpression of KA-1 or KA-2 in combination with GluR5 or GluR6.

Kainate receptors have been found in the neocortex, the piriform cortex and the hippocampal formation as well as in the caudate-putamen, the reticular nucleus of the thalamus and in other brain areas. The distribution patterns of kainate receptors depend on the configuration of the subtype. By *in situ* hybridization it has been demonstrated that the mRNAs of different kainate receptor subunits are differentially expressed.

The GluR5 mRNA is localized in the piriform and cingulate cortex, the medial nucleus of the amygdala, the medial habenula and in the Purkinje layer of the cerebellum.

The mRNA coding for GluR6 is found in the hippocampus, the caudate-putamen and in the granular layer of the cerebellum, while the distribution of KA-1 seems to be limited to the hippocampal formation (region CA3 and dentate gyrus).

The KA-2 subunit is extensively distributed in the brain with the exception of some thalamic brain nuclei. High levels of KA-2 mRNA are present in the cerebral cortex, the hippocampal formation and the cerebellum.

The precise function of the kainate receptors has not been clarified in detail. It has been shown that the high-affinity binding sites for kainate are located preferentially on presynaptic membranes. It is possible, therefore, that these receptors are mainly involved in modulating the release of excitatory amino acids and additional neurotransmitters or neuromodulators.

The metabotropic glutamate receptors

The metabotropic glutamate receptors (Qp) are coupled to G proteins and the signal transduction involves different second-messenger systems. The metabotropic receptors generate slow postsynaptic responses after an adequate stimulus.

The metabotropic glutamate receptors reveal a considerable diversity in signal transduction pathways. They either interact with the adenylate cyclase system or with the protein kinase C system. Additionally, they are capable of activating the phospolipase C-DAG-IP3 pathway.

Moreover, a direct coupling of G proteins to cation channels may also occur, since activation of the metabotropic receptors can influence voltage-gated K^+ and Ca^{2+} channels.

The metabotropic receptors are activated by glutamate, quisqualate, 1-aminocyclopentane-(1S,3R)-dicarboxylic acid (ACPD) or L-serine-O-phosphate and ibotenic acid. However, they are resistant to activation by NMDA, AMPA or kainate.

Eight different metabotropic glutamate receptors (mGluR1-mGluR8) have been described. They share about 40% homology with each other and they all belong to the family of receptors sharing the seven membrane-spanning motif (Fig. 3.22). The eight receptors have been classified into three groups based on their linkage to second-messenger systems and their pharmacology: group I (mGluR1 and mGluR5) act via the phospholipase C system, whereas group II (mGluR2 and mGluR3) and group III (mGluR4, mGluR6, mGluR7 and mGluR8) inhibit adenylate cyclase. Groups II and III differ in their affinity to several agonists. Metabotropic receptors of group II can be activated by CCG, whereas receptors of group III bind the agonists L-serine-O-phosphate and 2-AP4.

The metabotropic glutamate receptors are widely expressed throughout the central nervous system, but the different subtypes are differentially distributed.

The mGluR1 receptors were found in high densities in the following areas: the hippocampus, the thalamus, the lateral septum, the olfactory bulb and in the Purkinje cells of the cerebellum. They appear to be localized postsynaptically.

The mGluR2 receptors have been shown in the entorhinal cortex, the parasubicular cortex, the dentate gyrus, the accessory olfactory bulbs and in Golgi cells of the cerebellum, where they seem to be primarily located at presynaptic sites.

The mGluR3 receptors are more widely distributed than mGluR2 in the brain. They have been demonstrated in the cerebral cortex, the dentate gyrus, the nucleus reticularis of the thalamus, the caudate-putamen and in the supraoptic nucleus. In addition, mGluR3 receptors have been found in glia.

Expression of mGluR4 receptors is prominent in the dentate gyrus, the hippocampal area CA3, the entorhinal cortex, the lateral septum, the thalamus, the olfactory bulb, the pontine nucleus and in the granule cells of the cerebellum.

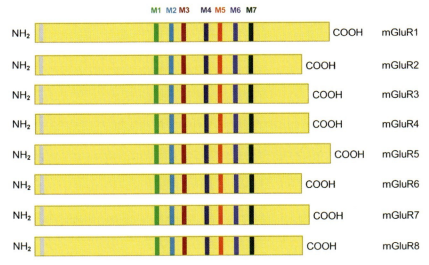

Fig. 3.22 The metabotropic glutamate receptors and their putative amino acid sequences.

The mGluR5 receptors occur in the cerebral cortex, the hippocampus (areas CA1–CA4, subiculum, dentate gyrus), the lateral septum, the caudate-putamen, the nucleus accumbens and in the olfactory bulb.

The mGluR6 receptors seem to be absent from brain tissues, but are expressed in bipolar cells of the retina.

Expression of mGluR7 has been found in the cerebral cortex, the hippocampus, the thalamus, the caudate-putamen and in the olfactory bulb.

The mGluR8 receptors are more locally restricted and occur in the olfactory bulb, the olfactory tubercle and in the mammillary bodies.

The functional significance of the metabotropic receptors has been subject of extensive investigations. They are considered to contribute to delayed neuronal responses and to synaptic plasticity. Since application of agonists of metabotropic glutamate receptors can potentiate long-term potentiation (LTP), it is believed that these receptors are involved in processes coupled to learning and memory storage.

3.4.6
Biological Effects

Glutamate is involved in fast synaptic transmission, eliciting a postsynaptic depolarization, as indicated above.

Besides the fast excitatory effect, which occurs in the millisecond range, glutamate can produce long-lasting activity-dependent changes of neuronal excitability, as is the case in long-term potentiation (LTP) of synaptic transmission.

Additional functional features comprise:

- neuroendocrine regulatory functions by influencing the secretion of pituitary hormones, which play a role in the reproductive cycle;
- participation in neuronal migration in some brain areas during development;
- involvement in the reception and processing of environmental stimuli and motor behavior.

By enumerating the diverse function of glutamate, it becomes obvious that the function of glutamatergic synapses is not confined to serving as electrochemical relays for fast interneuronal signal transmission, but that they also participate in processes related to storage of information. The remarkable capacity of the brain to translate transient experience into almost infinite numbers of memories has been attributed to long-lasting, activity-dependent changes in synaptic activity. This storage is believed to be electrophysiologically translated as a modification of synaptic strength, either by an augmentation of synaptic efficacy (which is obvious in long-term potentiation; LTP) or by its reduction (obvious in long-term depression; LTD).

In the central nervous system, long-lasting synaptic modifications of the LTP type are best studied in the hippocampus, specifically on the synapses of the Schaffer collateral and commissural axons and the apical dendrites of CA1 pyramidal cells.

Long-term potentiation (LTP) is defined as an increase in the efficacy of the synaptic responses, which can be experimentally induced by repetitive electrical stimulation or by the application of some drugs, particularly NMDA. After LTP is triggered rapidly (within seconds), it can last for hours within *in vitro* preparations and for days *in vivo*.

Glutamate appears to play a crucial role in these processes since glutamate release can be observed during LTP production throughout the mammalian brain, including the cerebral cortex. It is well accepted that postsynaptic NMDA receptors play an important role in the generation of LTP. Application of NMDA receptor antagonists, like AP7 and AP5, prior to the tetanic stimulation, can completely inhibit the induction of LTP. Preventing the rise in postsynaptic Ca^{2+} with Ca^{2+} chelators also blocks LTP, whereas directly raising the postsynaptic Ca^{2+} concentration by photolysis of caged Ca^{2+} can mimic LTP. In agreement with these functional observations is the fact that NMDA receptors are found in brain areas which are believed to be involved in memory formation and learning.

LTP may also involve a long-lasting increase in the release of transmitters (presynaptic effect) as well as the sensitization of postsynaptic membranes to the transmitter.

The detailed electrophysiological and molecular mechanisms underlying LTP are an actively debated topic. An important role in the activation mechanism of the NMDA receptor is conveyed by the calcium/calmodulin-dependent kinase II (CaMKII). Evidence from clinical studies and knockout animals show that LTP induced by glutamate contributes substantially to specific forms of memory storage, in particular spatial memory formation.

Furthermore, *in vitro* studies have shown that glutamate exerts positive effects on the survival, growth and development of neurons in the cerebellum and that glutamate can stimulate the outgrowth of dendrites of pyramidal cells in the hippocampus. These data are in agreement with findings on the density of NMDA receptors within the caudate-putamen and the hippocampus, which show a significantly higher expression in the young compared to adult animals.

The allosteric binding sites (like the glycine-binding site: see previous section, under *NMDA receptors*) have a higher density in adult animals. This indicates that there is a specific morphogenetic regulation of different components of the NMDA receptor during development.

The NMDA receptor as a target for D-amino acids

The existence of two glycine-binding sites, strychnine-sensitive and strychnine-insensitive (see Section 3.5), in mammalian brain centers have been demonstrated by radioligand-binding techniques. Glycine has been found to increase the NMDA response in a strychnine-independent manner. It also increases the frequency of channel opening without altering their conductance or mean open time in the presence of NMDA under patch clamp conditions. Occupation of the NMDA glycine site seems to be an absolute requirement for receptor activation. Support for the absolute requirement of glycine site agonist for NMDA receptor activation comes from blocking experiments exploiting the selective glycine site antagonist 7-chlorokynurenate (7-Cl-Kyn), which results in a complete block of NMDA responses after exposure to glycine. D-serine, the D-isomer of the non-essential amino acid L-serine, was found to act in a similar manner. The ED_{50} values of D-serine were found to be even lower in functional expression systems using various heteromeric combinations of NMDA receptor subunits. In support of the concept that D-serine constitutes an endogeous co-activator at the glutamatergic NMDA receptor is the finding that: (1) D-serine is confined predominantly to forebrain structure, the hippocampus and the striatum in the adult brain and (2) the developmental expression pattern corresponds well with those of the NMDA receptor. Interestingly the D-isomer of L-aspartate, D-aspartate, has also been found to activate the NMDA receptor at its glutamate-binding site. Unlike D-serine, the concentration of D-aspartate yields highest levels during early embryogenesis and declines dramatically during late gestation, indicating that this D-amino acid may play a role in developmental processes rather than being an important physiological effector in the mature brain. Since the NMDA subtype of glutamate receptors is essential for many developmental, learning and memory storage processes in mammals, the existence of endogenous agonists acting at its glycine-binding site are of considerable importance in respect to pathological disorders where the glutamatergic system is involved (see below).

3.4.7
Neurological Disorders and Neurodegenerative Diseases

The biological effects of excitatory amino acids as described above are consistent with the idea of glutamate being a Janus-faced neurotransmitter: on one hand, it provides beneficial effects in the regulation of neuronal growth and differentiation; on the other, excitatory amino acids can be harmful to brain tissue. This happens when large amounts of glutamate or aspartate are released into the extracellular space with the consequence of hyperactivation of glutamate receptors, an effect which has been termed "excitotoxicity".

This can be observed in ischemia, hypoglycemia, epileptic seizures and in neurodegenerative diseases such as Alzheimer's disease, Parkinsonism and amyotropic lateral sclerosis.

The excitotoxic effect is related to the massive entry of Ca^{2+} into the cells as a consequence of the sustained activation of glutamate receptors. An excessive raise in intracellular Ca^{2+} accounts for multiple cytotoxic damage to the neurons such as perturbation of cytoskeletal proteins and activation of proteases and phospholipases. In addition to their proteolytic and lipolytic acitivity these enzymes result in the formation of free radicals which damage the cells. It is generally believed that the most important mechanism mediating the toxic influx of Ca^{2+} into neurons, is the ionotropic channel of the NMDA receptors. The following arguments support this view:

- Cerebral ischemia is coupled to a massive increase in the extracellular concentrations of aspartate and glutamate.
- Prior to the accumulation of excitatory amino acids in the extracellular space, stimulation of AMPA receptors takes place; this is followed by a depolarization of neurons, which in turn allows the activation of NMDA receptors.
- Treatment with competitive antagonists, like APV, CPP, CGS 19755 or noncompetitive antagonists (phencyclidine, MK-801) of the NMDA receptor, can protect neurons from the neurotoxic effects.

In fact, some experimental data provide evidence for the neuroprotective effects of NMDA antagonists, thus corroborating this concept. For instance, cell death, induced by a transient ischemia of the middle cerebral artery occurs after a delay of 24–48 h. Administration of NMDA antagonists shortly after induction of the ischemic stress can reduce the number of damaged cells. One such drug is MK801. For this reason, the use of NMDA receptor antagonists could be advantageous in the therapy of ischemic insults, traumatic lesions and epilepsy. Clinical trials have yet to substantiate these therapeutic possibilities.

NMDA receptors have also been implicated in the pathophysiology of schizophrenia. In particular, the glycine modulatory site of the NMDA receptors is considered to participate in clinical manifestation of schizophrenia and is currently a favored theraupeutic target. The glycine site at the NMDAR affects channel open time and desensitization rate in the presence of agonist (glutamate), but does not, of itself, induce channel opening. The glycine binding site is also sensitive for D-serine and D-cycloserine; and it represents the strychnine-insensitive target of the glycine receptors. The strongest clinical data validating glutamatergic mechanisms behind schizophrenic symptoms come from studies with NMDAR glycine-site agonists, which function as positive allosteric modulators of the NMDAR complex. Clinical trials with NMDA receptor agonists (glycine, D-serine and D-cycloserine) together with typical antipsychotic drugs revealed large effect size improvements in negative and cognitive symptoms of schizophrenia.

Other neuropsychiatric disorders may also represent appropriate targets for glutamatergic agents. Stimulation may be beneficial in disorders associated with primary memory deficits, whereas inhibition may exert positive effects in disorders with neurodegeneration. Effects of glutamatergic agents have been studied more extensively with regard to Alzheimer's disease (AD), anxiety and posttraumatic stress disorder (PTSD). NMDAR antagonists have been used to attempt to slow down excitotoxic neurodegeneration in AD. Memantine, a weak NMDAR channel blocker, has shown safety and efficacy in slowing the decline in moderate to advanced AD. The effect of memantine has been contributed to mimicking the voltage-dependent Mg^{2+} blockade of the NMDAR, which is found to be reduced during the progression of AD. Thus memantine seems to take over the physiological role of Mg^{2+}.

Benzodiazepines and barbiturates are by far the most common group of drugs for the treatment of anxiety. The rational behind this therapeutic paradigm is to increase the inhibitory impact of GABAergic transmission. On the simplest level, a similar neurochemical effect could be achieved by reducing excitatory glutamatergic neurotransmission. AMPA receptor agonists, or group II/III metabotropic agonists could in theory achieve such an effect. Preclinical studies gained some promising outcome of some of the available drugs, but further follow-up studies need to be done in order to substantiate the therapeutic efficacy of this concept.

A further focus of glutamatergic therapies is related to neuropathic pain. Here the NMDA receptor and, in particular, glycine-site antagonists have shown much promise in this arena.

Further Reading

Arai, Y., Mizuguchi, M., Takashima, S. (1997): Developmental changes of glutamate receptors in the rat cerebral cortex and hippocampus. *Anat. Embryol.* 195: 65–70.

Buller, A.L., Larson, H.C., Schneider, B.E., Beaton, J.A., Morrisett, R.A., Monaghan, D.T. (1994): The molecular basis of NMDA-receptor subtypes: Native receptor diversity is predicted by subunit composition. *J. Neurosci.* 14: 5471–5484.

Coyle, J.T., Tsai, G. (2004): NMDA receptor function, neuroplasticity, and the pathophysiology of schizophrenia. *Int. Rev. Neurobiol.* 59: 491–515.

Hashimoto, A., Tetsuo, O. (1997): Free D-aspartate and D-serine in the mammalian brain and peripherie. *Prog. Brain Res.* 52: 325–353.

Hollmann, M., Heinemann, S. (1994): Cloned glutamate receptors. *Annu. Rev. Neurosci.* 17: 31–108.

Izquierdo, I. (1994): Pharmacological evidence for a role of long-term potentiation in memory. *FASEB J.* 8: 1139–1145.

Jansen, M., Dannhardt G. (2003): Antagonists and agonists at the glycince site of the NMDA receptor for therapeutic interventions. *Eur. J. Med. Chem.* **38**: 661–670.

Javitt, D.C. (2004): Glutamate as a therapeutic target in psychiatric disorders. *Mol. Psychiatry* **9**:979–984.

Jonas, P., Monyer, H. (eds) (1999): *Ionotropic Glutamate Receptors in the CNS*. Springer, Berlin, Heidelberg.

LoGrasso, P., McKelvy, J. (2003): Advances in pain therapeutics. *Curr. Opin. Chem. Biol.* **7**: 452–456.

Malenka, R.C., Nicoll, R.A. (1999): Long-term potentiation – a decade of progress? *Science* **285**: 1870–1874.

McDonald, A.J. (1996): Glutamate and aspartate immunoreactive neurons of the rat basolateral amygdala: colocalization of excitatory amino acids and projections to the limbic circuit. *J. Comp. Neurol.* **365**: 367–379.

Müller, D., Buchs, P.A., Stoppini, L., Boddeke, H. (1991): Long-term potentiation, protein kinase C, and glutamate receptors. *Mol. Neurobiol.* **5**: 277–288.

Nakanishi, S. (1992): Molecular diversity of glutamate receptors and implications for brain functions. *Science* **256**: 597–603.

Nicoll, R.A. (2003): Expression mechanisms underlying long-term potentiation: a postsynaptic view. *Phil. Trans. R. Soc. Lond. B* **358**: 721–726.

Nishikawa, T. (2005): Metabolism and functional roles of endogenous D-serine in mammalian brains. *Biol. Pharm. Bull.* **28**: 1561–1565.

Parsons, C.G., Danysz, W., Quack, G. (1998): Glutamate in CNS disorders as a target for drug development. An update. *Drug News Perspect.* **11**: 523–569.

Pin, J.-P., Duvoisin, R. (1995): The metabotropic glutamate receptors: structure and functions. *Neuropharmacology* **34**:1–26.

Schell, M.J. (2003): The N-methyl D-aspartate receptor glycince site and D-serine metabolism: an evolutionary perspective. *Phil. Trans. R. Soc. Lond. B* **359**: 943–964.

Sugita, S., Uchimura, N., Jiang, Z.G., North, R.A. (1991): Distinct muscarinic receptors inhibit release of gamma-aminobutyric acid and excitatory amino acids in mammalian brain. *Proc. Natl Acad. Sci. USA* **88**: 2608–2611.

Vizi, E.S. (2000): Role of high-affinity receptors and membrane transporters in nonsynaptic communication and drug action in the central nervous system. *Pharmacol. Rev.* **52**: 63–89.

3.5
Glycine

3.5.1
General Aspects and History

In mammals, glycine belongs to the non-essential amino acids. Glycine is capable of crossing the blood–brain barrier and thus can be transported from the blood into the spinal cord and brain.

Like glutamate and GABA, glycine is part of the common protein metabolism and shares with these amino acids the same difficulties in separating its metabolic and neurotransmitter function.

Until the early 1960s, glycine was considered to be of minor importance in synaptic transmission because of its simple structure and its ubiquitous distribution as a member of protein and nucleotide metabolism. Between 1960 and 1970, however, it became apparent that glycine is a potent neurotransmitter in the spinal chord and brain stem. Today, GABA and glycine are known to be the main inhibitory transmitters in the central nervous system acting through ionotropic receptors. More recently, glycine has been found to play a role in the functional modulation of N-methyl-D-aspartate (NMDA) receptors (see Section 3.4).

3.5.2
Localization Within the Central Nervous System

Glycine is present at very high concentrations in the spinal cord, the pons and the medulla oblongata, as well as in suprabulbar regions of the central nervous system. At somewhat lower concentrations, glycine has been found in the cerebellum, the forebrain and the retina.

Colocalization of glycine with other neurotransmitters is also known. For example, glycine coexists with GABA in some interneurons of the dorsal horn of the spinal chord.

3.5.3
Biosynthesis and Degradation

The precise metabolic pathway for glycine in the transmitter pool is not fully understood. However, ample evidence suggests that most of the glycine in the central nervous system is synthesized from glucose via serine in mitochondria. It is presumed that the enzyme serine hydroxymethyltransferase is responsible for converting serine to glycine. An alternative, possible pathway for the synthesis of glycine is from isocitrate by the activity of the enzyme isocitrate lyase and subsequent transamination by glycine a-ketoglutarate aminotransferase.

The degradation of the neurotransmitter glycine still remains unclear, but it has been shown that glycine can be metabolized to form guanidinoacetic acid, glyoxylate, gluthathione or serine. The glycine cleavage system has been localized in the form of a mitochondrial and cytosolic enzyme complex.

Glycine can be taken up into cells by specific transporter systems (GLYT1 and GLYT2) which comprise sodium/chloride-dependent transporters. These high-affinity glycine re-uptake mechanisms reveal an affinity constant (K_m) of about 10^{-5} M. This re-uptake mechanism is thought to be responsible for the removal of glycine from the synaptic cleft after release of the amino acid. The GLYT1 transporter is found in the spinal cord, the pons, the medulla, the diencephalon and the retina and, to a lower concentration, in the olfactory bulb and the brain hemispheres. GLYT2 is more restricted to spinal cord, brain stem and the cerebellum. GLYT2 is preferentially found in neurons, whereas GLYT1 is localized in astrocytes. Recent studies have shown that both types of GLYT are expressed by oligodendrocytic precursor cells.

The accumulation of glycine in synaptic vesicles implies that a vesicular transporter must exist. While no specific vesicular glycine transporter has been described so far and vesicular GABA transporters colocalize in glycinergic neurons, it seems suggestive that a common transporter system for both inhibitory neurotransmitters exists. Observations indicating that inhibitory presynaptic boutons may contain both GABA and glycine favor this idea.

3.5.4
Receptors and Signal Transduction

The glycine receptor (GlyR) is a strychnine-sensitive glycoprotein which is composed of five subunits. Among these subunits are the α-subunit, which is the ligand-binding structure, the β-subunit and a supplementary polypeptide (Fig. 3.23). The total glycine receptor is a 250-kDa glycoprotein and, in molecular mass, the α-subunit is about 48 kDa, the β-subunit 58 kDa and the polypeptide 93 kDa.

The receptor has a pentameric structure with three α- and two β-subunits forming the ion channel. Sequence alignment of the subunits has shown that

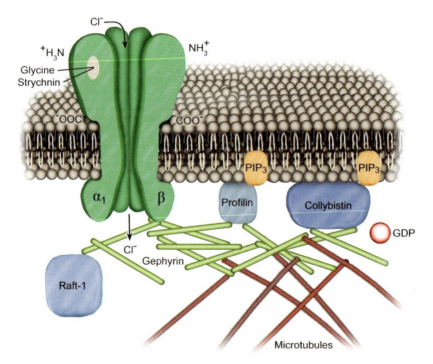

Fig. 3.23 Structure of the strychnin-sensitive glycine receptor. The receptor protein forms an ion channel which is permeable for chloride ions. The receptor is a pentameric structure which consists of three α and two β subunits. The postsynaptic scaffold is composed of gephyrin, which binds the receptor to microtubules. Gephyrin interacts with associated proteins (profilin and collibistin) involved in actin dynamics and downstream signaling. It also interacts with Raft-1, a candidate regulator of dendritic protein synthesis.

the glycine receptor belongs to the superfamily of ligand-gated ion channels, like the $GABA_A$ receptor or the nicotinergic acetylcholine receptor. Molecular cloning has revealed the existence of several subtypes of the ligand-binding subunit. This heterogeneity is responsible for the distinct pharmacological and functional properties displayed by the various receptor configurations that are differentially expressed and assembled during development.

The existence of four different isoforms of the α-subunit ($\alpha1$, $\alpha2$, $\alpha3$ and $\alpha4$) has been described. These isoforms are expressed in different areas or in different stages during development. $\alpha1\beta$-heteromers are mainly found in the spinal chord of adults, whereas $\alpha2$ homomers predominate in embryonic and neonatal stages. The switch towards the adult isoform is completed by around postnatal day 20. However, the embryonic $\alpha2$-subunit pattern prevails in the retina and in the auditory brain stem.

To date, no isoforms of the β-subunit have been described. The α-subunits share a homology of 80–90% while the β-subunit has a sequence identity of about 47% with the $\alpha1$ isoform. Splice variants have been found in rat $\alpha1$-subunit, termed $\alpha1^{ins}$, which contains an eight-amino-acid insert in the large intracellular loop that contains a possible phosphorylation site. The rat $\alpha2$-subunit exhibits two splice variants, $\alpha2A$ and $\alpha2B$. The $\alpha2B$ variant differs from $\alpha2A$ by V58I and T59A amino acid substitutions. Two transcripts of the human $\alpha3$-gene, termed $\alpha3L$ and $\alpha3K$, have also been characterized. The amino acid sequence of the $\alpha3L$ is mainly identical to the $\alpha3$ isoform. The $\alpha3K$ subunit lacks the coding sequence for 15 amino acid residues located within the putative cytoplasmic loop bewteen the transmembrane domains M3 and M4. Homologous screening of brain cDNA libraries led to the discovery of a further $\alpha2$-subunit, $\alpha2^*$, which is not generated by alternative splicing. The protein sequence differs from other $\alpha2$ isoforms by an E167G substitution which results functionally in a 40-fold lower affinity for glycine and a 500-fold lower affinity for the antagonist strychnine.

The glycine receptor is presently considered to form a complex consisting of a glycine recognition site and an associated chloride channel.

It is believed that at least two glycine molecules must bind to the recognition site to open the chloride channel and that the α-subunits are responsible for ligand binding.

The properties of GlyRs seem to depend on their α-subunit composition and their stochiometry. Different GlyR subtypes possess distinct ionic conductance and kinetic properties, affinities for various agonists and/or antagonists as well as different allosteric modulatory sites. They are also regulated in distinct ways by phosphorylation processes.

Postsynaptic glycine receptors are aggregated in clusters whose formation depends on the presence of submembrane scaffolding proteins. The size of the GlyR clusters varies considerably on postsynaptic sites. The major constituent of the postsynaptic scaffold in inhibitory synapses is a 93-kDa protein named gephyrin, which has been identified as a pheriperal membrane protein. Gephyrin binds to the long cytoplasmic loop of the β-subunit and anchors the glycine receptors and the GABA receptors to the cytoskeleton (see Fig. 3.23). Evidence in favor of this role of gephyrin derived from antisense strategies and from knockout mice in which synaptic clustering of GlyRs was abolished. However, the clustering mechanism might be more complex, since homomeric GlyRs deprived of the β-subunit can be synaptically activated. Additionally, current data indicate that homomeric GlyRs can cluster at the cell surface independently of gephyrin.

Gephyrin is also known to colocalize with $GABA_A$ receptors, where it binds specifically to $\gamma 2$- and $\gamma 3$-subunits.

Further scaffolding proteins associated with the postsynaptic web of glycinergic synapses are collybistin, profilin and RAFT1. They are discussed to act in concert with gephyrin in the GlyR aggregate complex.

Glycince, taurine and β-alanine are the most common agonist at GlyRs, whereas strychnine represents a high-affinity antagonist (see below).

Modulatory effects on GlyRs acitivity can be achieved by zinc, some alcohols and anaesthetics.

Zinc seems to be an endogenous modulator. This metal ion is released together with neurotransmitters at some synaptic terminals. Zinc ions are known to modulate several voltage-dependent and ligand-gated channels. The effects of zinc on GlyRs are concentration-dependent: at low concentrations, zinc enhances glycinergic currents, while concentrations higher than 10 µM elicit a depression effect. Further allosteric modulation is conveyed by ethanol and anaesthetics in a similar way as at $GABA_A$ receptor sites.

Picrotoxin, commonly used as a $GABA_A$ receptor antagonist, also suppresses GlyR responses to glycine.

Binding sites for glycine are present within the brain, which differ from the typical inhibitory glycine receptor. These binding sites are located on receptors, which are insensitive to strychnine and are found at high densities in the cortex, the hippocampus, the amygdala and the cerebellum.

This glycine-binding site is associated with N-methyl-D-aspartate (NMDA) receptors, because glycine can influence the activity of NMDA receptors. NMDA and glutamate are selective agonists for one of the principal receptor types (NMDA receptors) involved in excitatory synaptic transmission. Glycine may serve as a co-factor for NMDA-receptors because glycine is needed for the opening of the ion channel of the NMDA receptor. Both NMDA- and glycine-binding sites are present on this receptor; and glycine mediates its effects through the strychnine-insensitive binding site. The exact mechanism of glycine action on the NMDA receptor is largely unknown, but some evidence indicates that glycine increases the opening frequency of the ion channel associated with the NMDA receptor (see Section 3.4).

Thus, glycine has two distinct functions in the central nervous system: an inhibitory function in the spinal cord and the brain stem through activation of a specific glycine receptor; and an excitatory function in the forebrain as a co-activator of NMDA receptors.

3.5.5
Biological Effects

Glycine is known as the major inhibitory neurotransmitter in the spinal cord and the brain stem. When GlyR is activated, the resulting Cl$^-$ flux moves the membrane potential rapidly toward the Cl$^-$ equilibrium potential. Depending on the value of the equilibrium potential relative to the cell resting potential, the Cl$^-$ flux may cause either a depolarization or a hyperpolarization. The inhibitory preference of the GlyR happens because the Cl$^-$ equilibrium potential is usually close to, or more negative than, the cell resting potential. However, in embryonic neurons, the intracellular Cl$^-$ concentration is raised substantially, with the effect that GlyR activation elicits a strong, suprathreshold depolarization. The switch to the mature neuron phenotype is caused by the developmentally regulated expression of a K$^+$-Cl$^-$ cotransporter, KCC2, which lowers the internal Cl$^-$ concentration, thereby shifting the Cl$^-$ equilibrium potential to more negative values and converting the action of GlyR from excitatory to inhibitory.

Neurons within the spinal cord seem more sensitive than other neurons to glycine. Strychnine, an alkaloid from the plant *Strychnos nux vomica*, is a potent convulsant poison which antagonizes the inhibitory hyperpolarizing action of glycine by a specific, competitive interaction with glycine receptors. Based on the strychnine sensitivity of glycine-binding sites, a strychnine-sensitive site, which comprises the glycine receptor, and a strychnine non-sensitive site, represented by the glycine-binding site of the glutamatergic NMDA receptor, can be differentiated (see above).

Another toxin that occupies the glycine repector is tetanus toxin released by the bacterium *Clostridium tetani*. One of the effects of the tetanus toxin is the blockade of glycinergic inhibition, which leads to the typical clinical symptom of muscle rigidity with agonizing spasms superimposed (overactivity of motor neurons). It is hypothesized that tetanus toxin presynaptically blocks the release of glycine.

A further function for glycine is postulated in epileptogenesis, since intracerebral levels of glutamate, aspartate and glycine are significantly increased in the epileptogenic cerebral cortex. In this context, glycine is considered to act via NMDA receptors.

3.5.6
Neurological Disorders and Neurodegenerative Diseases

Hyperekplexia, or startle disease, is a rare neurological disorder characterized by an exaggerated response to unexpected stimuli. The response is typically accompanied by a transient but complete muscular rigidity. Susptectibility to startle responses is increased by emotional tension, nervousness, fatigue and the expectation of being frightened. Symptoms of this disease are present from birth, a reason why it is also named stiff baby syndrome, with infants displaying severe muscular rigidity. This disorder has a history of being misdiagnosed as epilepsy, although hyperekplexia can be readily distinguished by an absence of fits and retention of consciousness during the startle episodes. The symptoms are successfully treated by benzodiazepines, with clonazepam being the current drug of choice.

Dominant and recessive forms of this rare disease result from missense mutations of the $\alpha 1$ GlyR subunit gene on chromosome 5q. The most common mutation is the R271L substitution which leads to an autosomally dominant form. This mutation and other substitutions at the R271 site cause reduced glycine sensitivity and single-channel conductance. The reduced efficiency of glycinergic neurotransmission explains the human phenotype which exhibits an increasing level of excitability of motor neurons at the spinal cord and brain stem level.

Further Reading

Aprison, M. H., Daly, E. C. (1978): Biochemical aspects of transmission at inhibitory synapses: the role of glycine. *Adv. Neurochem.* **3**: 203–294.

Betz, H. (1987): Biology and structure of the mammalian glycine receptor. *Trends Neurosci.* **10**: 113–117.

Bormann, J., Hamill, O. P., Sakmann, B. (1987): Mechanism of anion permeation through channels gated by glycine and γ-aminobutric acid in the mouse cultured spinal neurones. *J. Physiol.* **385**: 243–286.

Grillner, S. (1981): Control of locomotion in bipeds, tetrapods and fish. In: *Handbook of Physiology (Section 1, The Nervous System, vol 2: Motor Control).* ed. Brooks, V. B., American Physiological Society, Bethesda, Md., pp 1179–1236.

Kemp, J. A., Leeson, P. D. (1993): The glycine site of the NMDA receptor – five years on. *Trends Pharmacol. Sci.* **14**: 20–25.

Kneussel, M. (2002): Dynamic regulation of GABA(A) receptors at synaptic sites. *Brain Res. Brain Res. Rev.* **39**: 74–83.

Kneussel, M., Betz, H. (2000): Clustering of inhibitory neurotransmitter receptors at developing postsynaptic sites: the membrane activating model. *Trends Neurosci.* **23**: 429–435.

Langosch, D., Becker, C.-M., Betz, H. (1990): The inhibitory glycine receptor: a ligand-gated chlorid channel of the central nervous system. *FEBS Lett.* **194**: 1–8.

Legendre, P. (2001): The glycinergic inhibitory synapse. *Cell. Mol. Life Sci.* **28**: 760–793.

Lynch, J. W. (2004): Molecular structure and function of the glycine receptor chloride channel. *Physiol. Rev.* **84**: 1051–1095.

Rajendra, S., Lynch, J. W., Schofield, P. R. (1997): The glycine receptor. *Pharmacol. Ther.* **73**: 121–146.

Vannier, C., Triller, A. (1997): Biology of the postsynaptic glycine receptor. *Int. Rev. Cytol.* **176**: 201–244.

3.6
Histamine

3.6.1
General Aspects and History

Since the discovery of histamine (β-imidazolylethylamine) by Sir Henry Dale at the beginning of the 20th century, this substance has been the subject of detailed studies.

Histamine was first known to be a substance released by mast cells in response to allergen stimulation. However, data have been obtained showing that the function of histamine is not restricted to hyperergic processes, but is also a physiological mediator within different tissues, including the central nervous system.

In 1920, Popielski described the secretagogic effect of histamine on parietal cells in the stomach.

About 20 years later, histamine was identified in the central nervous system. In the last quarter of the 20th century, the idea was born that histamine could act as a neuroactive substance within the central nervous system. Since histamine is unable to cross the blood–brain barrier, it was suggested that histamine is synthesized in brain tissues and represents a neurotransmitter *per se*.

In the meantime, both postsynaptic and presynaptic histaminergic receptors have been found in the brain as well as the precursor of histamine metabolism. Histamine has many features in common with the monoamines, such as catecholamines (dopamine, epinephrine and norepinephrine) or indolamines (like serotonin), with respect to release, metabolic pathway and mode of action at the cellular level.

3.6.2
Localization Within the Central Nervous System

The principal storage sites of histamine are basophilic leucocytes and mast cells, which release histamine as part of their reaction to allergens. Other sources include the epidermis, the intestinal mucosa and the central nervous system.

Immunohistochemical methods have played a central role in the localization of histamine in the central nervous system. Somewhat striking is the fact that histamine is located not only within neurons, but also in interstitial cerebral mast cells. It is thought that these cells are involved in the regulation of vascular permeability. An interesting feature of cerebral mast cells is that the half-life of their histamine content is much longer than in neurons.

Histaminergic neurons are predominantly located in the tuberal area of the posterior hypothalamus, which is known as the tuberomammillary nucleus (Fig. 3.24). From this nucleus, the histaminergic neurons project diffusely into several brain areas; and their axonal collaterals give rise to projections, for example to the forebrain, the cerebellum and the mesencephalon. Two ascending pathways can be distinguished in the rat brain. One pathway runs at the ventral surface of the brain to the septal area; and the other runs more caudally at the lateral side of the third ventricle.

Intrahypothalamic fibers are most common within the medial forebrain bundle from which they ascend ipsilaterally into the forebrain. Another fraction crosses the midline and ascends within the contralateral side.

A dense histaminergic innervation is also obvious in thalamic areas as well as in the nucleus accumbens and in the bed nucleus of the stria terminalis. In addition, histaminergic projections have been found in the cerebral cortex and in the hippocampus.

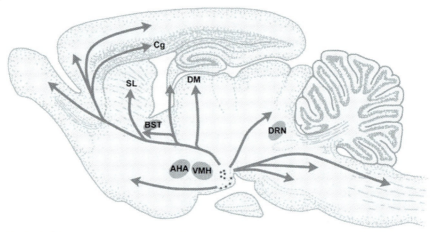

Fig. 3.24 Schematic representation of the histaminergic projection pattern within the brain of rats. *Abbreviations*: AHA=anterior hypothalamic area; BST=bed nucleus of the *stria terminalis*; Cg=cingulum; DM=dorsomedial nucleus of the thalamus; DRN=dorsal raphe nucleus; SL=lateral septum; VMH=ventromedial nucleus of the hypothalamus.

Besides the ascending projections, a long descending pathway, which originates also from the tuberomammillary nucleus, projects to various regions of the brain stem: these include the periaqueductal grey, the nucleus of the solitary tract and, more caudally, the dorsal horn of the spinal chord.

3.6.3
Biosynthesis and Degradation

Pharmacological and biochemical studies have revealed that the enzyme L-histidine-decarboxylase (HDC or HD) is responsible for the biosynthesis of histamine in the central nervous system. Histamine is formed by a single decarboxylation of L-histidine under the control of L-histidine-decarboxylase.

L-Histidine-decarboxylase shares a homology of about 50% with dopamine-decarboxylase, the rate-limiting enzyme responsible for the generation of dopamine.

The highest activity of L-histidine-decarboxylase has been detected in the posterior hypothalamus. In this area, the mRNA coding for HDC is expressed at high levels. However, the regional distribution of L-histidine-decarboxylase is extremely heterogeneous in the brain, a fact which reflects the diffuse histaminergic innervation in the different brain areas. The reason why the enzyme is found in brain areas other than the posterior hypothalamus is that L-histidine-decarboxylase, synthesized in the hypothalamus, is transported anterogradely from the soma to the nerve endings.

An important step in unraveling the function of histamine as a neurotransmitter began with the synthesis of α-fluoromethylhistidine (FMH), an inhibitor of L-histidine-decarboxylase.

With FMH, it became possible to inactivate the enzyme irreversibly and thus prevent the formation of histamine.

The decline in histamine level after administration of FMH is a reliable means of assessing the speed of renewal of neuronal histamine. Using this method, it was possible to demonstrate that histamine has a half-life of about 30 min.

The renewal of histamine can be slowed down by sedatives and enhanced by stimulation of NMDA receptors or μ-opioid receptors. This rapid adaptation after manipulation of neuronal receptor components suggests that the metabolism of the neurotransmitter histamine is modulated by neurons themselves.

Histamine synthesized in neurons is stored in synaptic vesicles by a reserpine-sensitive mechanism and is released from these stores in a calcium-dependent manner.

Histamine is inactivated by the ubiquitously distributed enzyme histamine N-methyltransferase and subsequently deaminated by monoamine oxidase B.

No evidence exists for a high-affinity re-uptake mechanism of histamine.

3.6.4
Receptors and Signal Transduction

Three different receptor subtypes, termed H1, H2 and H3, bind histamine specifically. The three receptor subtypes can be distinguished by their different binding patterns and different biological effects (Table 3.9). All three receptor subtypes seem to belong to the superfamily of G protein-coupled receptors.

Table 3.9 The three different subtypes of histamine receptors and some of their characteristic functional features.

	Subtype of histamine receptor		
	H_1	H_2	H_3
Localization on human chromosome	3p2	5	?
Specific agonist	(Trifluoromethyl phenyl)-histamine	Impromidine	(R)α-Methyl-histamine
Specific antagonist	Mepyramine	Cimetidine	Thioperamide
autoreceptor	No	No	Yes
Receptor localization	Postsynaptic	Postsynaptic	Presynaptic
G protein coupling	$G_{i/o}$	G_s	$G_{i/o}$
Highest densities	Cerebellum, thalamus, hippocampus	Caudate-putamen, cerebral cortex, amygdala	Frontal cortex, caudate-putamen, substantia nigra

H1 receptors

The H1 receptor is a glycoprotein of about 490 amino acids. Antagonists of the H1 receptor are collectively known as "antihistamines" (which are capable of crossing the blood–brain barrier). The activation of the H1 receptor gives rise to a variety of intracellular signals, which depend on the action of a G protein of the type G_q. This G protein enhances phospholipase C activation, which ultimately leads to activation of the DAG/IP3 pathway and consequent release of Ca^{2+} from intracellular calcium stores.

The activation of H1 receptors is positively coupled to an enhancement in phospholipid turnover and stimulates glycogen hydrolysis. Prolonged exposure of the receptor to its neurotransmitter can produce receptor subsensitivity (desensitization). The molecular nature of the desensitization effect is not known.

Agonists of the H1 receptor are substances like 2,3 (trifluoromethyl)-phenyl-histamine, 2-methylhistamine and 2-(pyridyl)-ethylamine. In addition, different specific antagonists have been found which interact with the H1 receptors. Some of these antagonists have been used in the treatment of allergies, but since they cross the blood–brain barrier they exhibit side-effects within the central nervous system, predominantly tranquilizing effects, by inhibition of cerebral H1 receptors. Therefore, considerable efforts have been made to develop H1 receptor antagonists which do not cross the blood–brain barrier and so are without a sedative effect.

H1 receptors are mainly found in cerebral areas, in which histaminergic terminals are present at high densities. H1 receptor-expressing cells have been identified in the pyramidal layer of the hippocampus and in cerebellar Purkinje cells.

H2 receptors

The H2 receptor is a glycoprotein of about 358 amino acids; and it shares a homology of about 40% with the H1 receptor. Like the H1 receptor, the H2 receptor belongs to the superfamily of G protein-coupled receptors. Unlike the H1 receptor, the H2 receptor displays dual signaling by means of coupling to the G_s and G_q members of the G protein family.

H2 receptors are located postsynaptically and are found in areas containing large numbers of histaminergic terminals. These include the caudate-putamen, the hippocampus and the cerebral cortex. H2 receptors have also been found in non-neuronal tissues, like small cerebral microvessels and glia.

In contrast to the H1 antagonists, most of the H2 antagonists are unable to cross the blood–brain barrier (with the exception of zolantidine, which passes this barrier).

The substance impromidine was first thought to be an agonist of H2 receptors; however, more recent studies have shown that this substance is an agonist of H3 receptors.

H3 receptors

The human histamine H3 receptor was identified by orphan GPCR cloning in 1999 (Lovenberg et al. 1999). Translation of the open reading frame revealed a 445-amino-acid coding region with low homology (20–27%) to the biogenic amine subfamily of GPCRs. The receptor shows abundant expression in the brain, particularly in the hypothalamus, the caudate nucleus, the thalamus and the cortex.

The H3 receptors function as inhibitory, presynaptic autoreceptors which regulate the synthesis and release of histamine. Based on the inhibitory function of the H3 autoreceptor, H3 antagonists elicit an increase of histamine at the extracellular site. Therefore, H3 antagonists reveal a kind of inverse agonistic effects.

However, H3 receptors may also function as heteroreceptors, inhibiting the release of other neurotransmitters, like acetylcholine, dopamine, norepinephrine and serotonin.

A controversial issue is whether the H3 receptor can bind N-methyl-D-aspartat (NMDA) on its polyamine-binding site.

The H3 receptor displays "constitutive acitivity", that is, spontaneous activativity in the absence of an agonist. This means that a proportion of the receptor population spontaneously undergoes an allosteric transition, leading to a conformation that can bind G protein. Constitutive acitivity of the H3 receptor is suggested to control histaminergic neuron activity *in vivo*.

A further human histamine receptor, H4, was cloned from human leukocytes. The deduced amino acid sequence showed about 40% identity to that of the human H3 receptor. In contrast to the H3 receptor, expression was not detected in the brain. Recent cloning work indicated that H3 receptors in different species have minor, yet significant, differences in their protein sequences that result in pharmacological heterogeneity. Three functional rat H3 receptor isoforms (H_{3A}, H_{3B}, H_{3C}) and one non-functional H_{3T} have been cloned, displaying differential expression in key areas involved in endocrine, sensory and cognitive functions. All three isoforms are generated as a result of alternative splicing. The three isoforms couple to the $G_{i/o}$ protein-dependent inhibition of adenylate cyclase. Interestingly, the activation of the H3 receptor isoforms also leads to activation of the MAP kinase pathway via PTX-sensitive G proteins. The H_{3A} isoform seems to be more effectively coupled to p44/p42 MAPK activation. Since the H3 receptor is known to be involved in learning and memory storage and histamine has been implicated in long-term potentiation, the strong expression of the H_{3A} isoform in the hippocampus and its preferential linkage to the MAPK pathway could provide a clue for these effects.

3.6.5
Biological Effects

Histamine in mast cells is a potent inflammatory substance depleted when an allergen binds to the cell surface, cross-linking specific IgE molecules. This mechanism is also present in the nervous tissue; and it seems most likely that released intracerebral histamine regulates cerebral circulation and the permeability of cerebral vessels. Anti-histaminergic drugs have been used in cases of fever but, since they produce drowsiness, which may be due to an action on histaminergic receptors in the brain, they are nowadays less frequently used.

In contrast, neuronal histamine plays a role as a neurotransmitter. Many of the physiological effects attributed to the tuberomammillary nucleus, seem to involve histamine. Histamine is a major player in the control of sleep and wakefulness. The following observations speak in favor of the crucial role of the histaminergic system in wakening:

- Lesion of the posterior hypothalamus at the site of the tuberomamillary nucleus (TM), where a high density of histaminergic neurons can be found, produces a comatose-like continous sleeping.
- TM neurons send inputs to various brain regions, including those that control the sleep-wake cycle, such as the cortex, thalamus, preoptic and anterior hypothalamus.
- Presumed histamine-containing neurons discharge tonically and specifically during wakefulness; this pattern of activity seems to be the most wake-selective neuronal pattern so far described in the brain.
- Histamine discharge in the prefrontal cortex is strictly correlated to wakening.
- Treatments that impair histamine-mediated neurotransmission increase cortical slow waves and enhance sleep.

Histamine is also involved in the regulation of the energetic balance of the body, by regulating glycogenolytic functions. The modulation of neuronal histamine, for instance by blocking H3 receptors, reduces food intake.

Furthermore, histamine exerts some endocrine functions because it influences the secretion of hormones from the pituitary. Together with some prostaglandines, histamine cooperates in the stimulation of ACTH secretion. Similar interaction may also play a role in the mediation of the ACTH response to immunochallenges. Additional evidence for an endocrine function of histamine has been obtained by experimental adrenalectomy, which enhances the plasma level of vasopression. Inhibition of histidine-decarboxylase by FMH blocks the increased vasopression response. Similar effects of FMH have been found on the secretion of ACTH and the stress-induced increase of prolactin. Taken together, these findings suggest that neuronal histamine functions to modulate the release of hormones and neuromodulators.

Blood pressure and body temperature are two further examples for a possible involvement of histamine in centrally regulated physiological processes. Intraventricular administration of histamine results in an increase in blood pressure, though a role for histamine in the development or maintenance of hypertension remains to be clarified.

The release of histamine acting on H1 and H2 receptors may be involved in radiation-induced hypothermia, since the H1 receptor antagonist, mepyramine, and the H2 receptor antagonist, cimetidine, both antagonized the hypothermia.

However, there is evidence that intracerebroventricularly applied histamine could induce biphasic changes in body temperature. Clark and Cumby (1976) observed hypothermic responses to histamine in cats at an environmental temperature of 22 °C. These responses were antagonized by the administration of H1 receptor antagonists. In some animals at higher environmental temperatures, hyperthermic responses to histamine were observed, which could be antagonized by central, but not peripheral, injection of metiamide, an H2-receptor antagonist.

The histaminergic innervation of the limbic system suggests that histamine is influential on different behavioral and cognitive functions and in the generation and maintenance of emotional states.

A competitive antagonist of the H2 receptors is the drug lysergic acid diethylamide (LSD; although it also interacts efficiently with the serotoninergic system); and some changes in the behavior induced by LSD might be coupled to the inactivation of H2 receptors. An interesting point in this context is that the stereo-isomer L-LSD is not capable of antagonizing H2 receptors.

3.6.6
Neurological Disorders and Neurodegenerative Diseases

As indicated above, histamine is an essential neurotransmitter involved in arousal and wakefulness. Drugs that improve arousal and thus modulate attention or vigilance are thought to be valuable tools in correcting cognitive deficiencies. Thus, it is not surprising that histamine is considered to be a versatile candidate for the improvement of impaired cognitive functions. In particular, the blockage of H3 receptors is the focus of current research concerning the therapeutical manipulation of histamine levels for the improvement of cognitive performances in disease states like Alzheimer's disease, schizophrenia, Parkinson's disease and trauma. Selective H3 receptor antagonists have been developed, among which thioperamide is the most widely used experimental tool. Toxicities that may be a function of the thiourea portion of the molecule have taken this drug off the list of potential clinical candidates. A further antagonist/inverse agonist, ciproxifan, is also highly potent *in vitro* and *in vivo*.

Since the tuberomammillary nucleus is afflicted in Alzheimer's disease, an increase in histamine and/or histamine turnover through H3 receptor antagonists may be beneficial for these patients.

The dose-dependent and complete blockade of amphetamine-induced increases in locomotor activity by the H3 antagonists thioperamide and ciproxifan suggest a potential use of H3 antagonists in schizopherina. However, since levels of tele-methylhistamine (a marker to measure the activity of the histamine turnover) have been found to increase in schizophrenic-like symptomatology, increases in histamine levels with H3 antagonists may be contradicted. Yet, it may well have efficacy in the treatment of some symptoms of schizophrenia (e.g. congnition). Two further neurological disorders cataplexy and attention deficit hyperactivity disorder (ADHD) have come into focus as putative candidates for H3 receptor antagonistic treatment. Due to the arousal and sleep-controlling nature of histamine and the relationship of this neurotransmitter to the arousal-inducing orexin/hypocretin system, it is compelling to venture that H3 antagonists may have a therapeutic benefit in disorders of this nature.

Further Reading

Anthonisen, M., Knigge, U., Kjaer, A., Warberg, J. (1997): Histamine and prostaglandin interaction in the central regulation of ACTH secretion. *Neuroendocrinology* **66**: 68–74.

Bekkers, J. M., Vidovic, M., Ymer, S. (1996): Differential effects of histamine on the N-Methyl-D-Aspartate channel in hippocampal slices and culture. *Neuroscience* **72**: 669–677.

Brown, R. E., Federov, N. B., Haas, H. L., Reymann, K. G. (1995): Histaminergic modulation of synaptic plasticity in area CA1 of rat hippocampal slices. *Neuropharmacology* **34**: 181–190.

Chang, R. S. L., Tran, V. T., Snyder, S. H. (1978): Histamine H1-receptors in brain labeled with [^{3}H]metpyramine. *Eur. J. Pharmacol.* **48**: 463–464.

Clark, W. G., Cumby, H. R. (1976): Biphasic changes in body temperature produced by intracerebroventricular injections of histamine in the cat. *J. Physiol.* **261**: 235–253.

Drutel, G., Peitsaro, N., Karlstedt, K., Wieland, K., Smit, M. J., Timmerman, P. P. Leurs, R. (2001): Identification of rat H3 receptor isoforms with different brain expression and signaling properties. *Mol. Pharmacol.* **59**: 1–8.

Lovenberg, T. W., Roland, B. L., Wilson, S. J., Jiang, X., Pyati, J., Huvar, A., Jackson, M. R., Erlander, M. G. (1999): Cloning and functional expression of the human histamine H3 receptor. *Mol. Pharmacol.* **55**: 1101–1107.

Mignot, E., Taheri, S., Nishino, S. (2002): Sleeping with the hypothalamus: therapeutic targets for sleep disorders. *Nat. Neurosci.* **5**[Suppl]: 1071–1075.

Nakamura, T., Itadani, H., Hidaka, Y., Ohta, M., Tanaka, K. (2000): Molecular cloning and characterization of a new human histamine receptor, HH4R. *Biochem. Biophys. Res. Commun.* **279**: 615–620.

Panula, P., Pirvola, U., Auvinen, S., Airaksinen, M. S. (1989): Histamine-immunoreactive nerve fibers in the rat brain. *Neuroscience* **28**: 585–610.

Passani, M. B., Lin, J.-S., Hancock, A., Crochet, S., Blandina, P. (2004): The histamine H3 receptor as a novel therapeutic target for cognitive and sleep disorders. *Trends Pharmacol. Sci.* **25**: 618–625.

Rouleau, M. S., Ligneau, X., Gbhahou, F., Tardivel-Lacombe, J., Stark, H., Schunack, W., Ganellin, C. R., Schwartz, J. C., Arrang, J. M. (2000): High constitutive activity of native receptors regulate histamine neurons in brain. *Nature* **408**: 860–864.

Schwartz, J.-C., Arrang, J. M., Garbarg, M., Pollard, H., Ruat, M. (1991): Histaminergic transmission in the mammalian brain. *Physiol. Rev.* **71**: 1–51.

Schwartz, J.-C., Haas, H. L. (eds) (1992): *The Histamine Receptors*. (*Receptor Biochemistry and Methodology, vol 16*). Wiley Liss, New York.

Taylor, J. E. (1982): Neurochemical and neuropharmacological aspects of histamine receptors. *Neurochem. Int.* **4**: 89–96.

Watanabe, T., Wada, H. (eds) (1991): *Histaminergic Neurons: Morphology and Functions*. CRC Press, Boca Raton.

Witkin, J. M., Nelson, D. L. (2004): Selective histamine H3 receptor antagonists for treatment of cognitive deficiencies and other disorders of the central nervous system. *Pharmacol. Ther.* **103**: 1–20.

3.7 Norepinephrine

3.7.1 General Aspects and History

The catecholamines constitute a family of neurotransmitters (dopamine, epinephrine, norepinephrine) which derive from the same precusor molecule (tyrosine) and thus share an almost identical chemical structure, consisting of a benzene ring with two adjacent hydroxyl groups and an ethylamine side-chain.

In 1921, Loewi found evidence for a substance, which he named "accelerance". In 1936, this substance was identified to be epinephrine (synonym adrenaline). Since norepinephrine (1-β-3,4-dihydroxyphenyl-α-ethanolamine, or synonym noradrenaline) is the direct precursor of epinephrine (Fig. 3.25), it was long thought that norepinephrine was an intermediary substrate in the synthesis of epinephrine.

In the 1940s, evidence accrued that norepinephrine is not a mere intermediary product, but plays a significant role as a neurotransmitter *per se*.

In contrast to epinephrine, which is mainly restricted to the peripheral nervous system, norepinephrine is also a major transmitter in the central nervous system.

The fluorescent histochemical technique which allows the visualization of catecholamines, the so-called glyoxylic acid method developed by Falck and Hillarp (also known as the Falck–Hillarp method), furthered knowledge about the cerebral and extracerebral distribution of this neurotransmitter group.

3.7.2 Localization Within the Central Nervous System

Early work on the localization of norepinephrine using the Falck and Hillarp method, and later immunohistochemical approaches by means of antibodies to the enzymes tyrosine hydroxylase and dopamine-β-hydroxylase, made it possible to screen the noradrenergic systems.

Noradrenergic cell somata were found to be concentrated in the locus coeruleus (LC) and locus subcoeruleus (groups A5, A6, A7), as well as in some areas of the formatio reticularis (groups A1 and C1) and in the nucleus of the tractus solitarius (groups A2 and C2; Fig. 3.26).

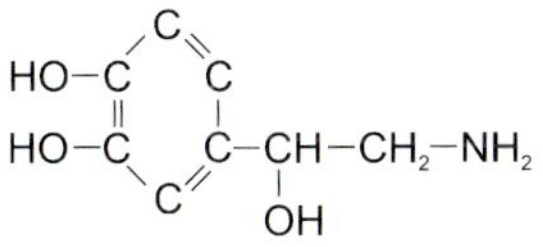

Fig. 3.25 Structure of norepinephrine.

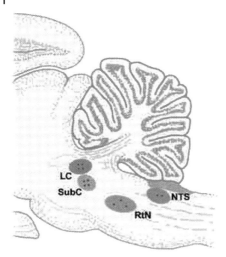

Fig. 3.26 Schematic overview of the distribution of noradrenergic somata in the brain of rats. *Abbreviations*: LC = *locus coeruleus;* NTS = nucleus of the tractus solitarius; SubC = *locus subcoeruleus;* RtN = reticular nuclei.

The locus coeruleus is a source of diffuse noradrenergic projections which innervate the cerebral cortex, the hippocampus, the amygdala, the septum, the thalamus, the hypothalamus and the spinal cord.

The relative small nucleus constitutes the main source of noradrenergic projections in the CNS; however the neurons form abundant collaterals which diverge into several cortical and subcortical areas. Some of the projections from the LC emerge bilaterally, but ipsilateral projections prevail.

Although divergence of the projection pattern from the LC is the dominant feature of the noradrenergic system, a topological organization is obviously related to the internal organization of the nucleus. The medial part of the LC projects to the cortex, whereas neurons which project to the hippocampus are primarily found in more posterior portions of the locus coeruleus.

The noradrenergic neurons of the area A1 and A2 are mixed up to some extent with neurons of group C1 and C2 (which are adrenergic neurons) and their termination fields seem to be locally related. Projections of these areas have been found to terminate in the intermedio-lateral tract of the spinal chord and in some nuclei of the hypothalamus. In addition, some projections of the neurons of group A1 and A2 have been found to terminate in the amygdala and the thalamus.

Some noradrenergic neurons coexpress neuromodulators. For example in the rat, a colocalization of norepinephrine and neuropeptide tyrosine (NPY) has been detected in most of the noradrenergic neurons of the formatio reticularis; and, similarly, about 20% of the neurons in the locus coeruleus coexpress NPY and norepinephrine. A larger fraction of about 70% of the neurons in the locus coeruleus coexpress norepinephrine with galanine.

3.7.3
Biosynthesis and Degradation

The synthesis of all members of the catecholamine family starts from the amino acid tyrosine and runs through several enzymatic steps before it is finalized as the definite end-product (Fig. 3.27).

In sympathetic neurons, the enzymes tyrosine hydroxylase (TH, EC 1.14.16.2), DOPA-decarboxylase (DDC, EC 4.1.1.26), dopamine-β-hydroxylase (DBH, EC 1.14.2.1) and phenylethylamine-N-methyl-transferase (PNMT, EC 2.1.1) are expressed in the cell soma, from which they are transported to the nerve terminals by the anterograde axoplasmic flow. The synthesis of catecholamines begins with a hydroxylation of tyrosine to L-DOPA by tyrosine hydroxylase. The hydroxylation is rate-limiting and takes place in the cytosol of nerve terminals (in peripheral tissue, chromaffin cells carry the same set of enzymes for catecholamine synthesis).

Tyrosine hydroxylase is an oxidase which requires tetrahydrobiopterin and oxygen as cofactors. The activity of this enzyme can be modified by phosphorylation.

The hydroxylation product L-DOPA is then decarboxylated to dopamine by an aromatic acid decarboxylase. In noradrenergic neurons, dopamine is converted to norepinephrine by the copper ion-containing dopamine-β-hydroxylase. This enzyme is highly enriched in noradrenergic neurons and chromaffin cells. Dopamine-β-hydroxylase is constitutively expressed in noradrenergic neurons and represents the key enzymatic difference between dopaminergic and noradrenergic neurons. The cytosolic norepinephrine is subsequently imported through a specific vesicular transport mechanism into synaptic vesicles.

In some neurons of the central nervous system and in chromaffin cells of the medulla of the adrenal gland, norepinephrine is not released, but is further metabolized to generate epinephrine. The formation of epinephrine (which by definition is a transmitter in peripheral tissues) is catalyzed by phenylethanolamine-N-methyltransferase. This enzyme transfers a methyl group from the donor S-adenosylmethionine to the nitrogen of norepinephrine.

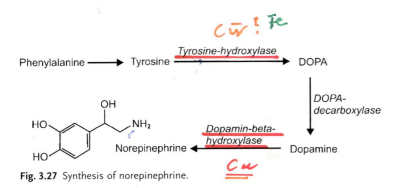

Fig. 3.27 Synthesis of norepinephrine.

The complement of enzymes, which serves for the synthesis of norepinephrine, is subject of neuronal and hormonal control, creating fast and slow adaptive changes in the rate of biosynthesis.

Excess of catecholamines inhibits the activity of tyrosine hydroxylase in a fast negative feedback loop. In contrast, neuromodulators like vasoactive intestinal peptide (VIP) or growth factors like the nerve growth factor (NGF) can upregulate tyrosine hydroxylase activity.

Long-lasting regulation of tyrosine hydroxylase and dopamine-β-hydroxylase take place at the transcriptional level. For instance, repeated stress or administration of reserpine depletes the reservoirs of norepinephrine, which are fueled by up-regulation of the transcription rate of the appropriate genes. The enhancement of transcription leads to an adaptive long-lasting increase in the concentration of the enzymes.

The genes which encode for tyrosine hydroxylase and dopamine-β-hydroxylase are largely homologous in their structure and are located on the same chromosome (chromosome 11).

A selective positive effect on the biosynthesis of phenylethylamine-N-methyl-transferase can be observed by glucocorticoids, while the influence of the latter on dopamine-β-hydroxylase and tyrosine hydroxylase is minor.

The degradation of norepinephrine and its inactivation entails different mechanisms. A prominent mechanism for the inactivation of norepinephrine is its re-uptake from the synaptic cleft.

Norepinephrine transporters (NET) mediate the removal of norepinephrine from the extracellular space, thereby limiting the extent of activation of auto- and heteroadrenoceptors. As with other transporters, the activity of norepinephrine transporters depends on the transmembrane Na^+ gradient.

Some drugs, like cocaine and tricyclic antidepressiva, block the re-uptake of norepinephrine into the nerve terminals and thus lead to an increase in the concentration of extracellular norepinephrine. The consequence is a prolongation of the postsynaptic action of norepinephrine on its receptors. This mechanism is thought to be responsible for some of the psycho-stimulating and euphorizing effects of antidepressives and especially cocaine.

Norepinephrine, like other catecholamines, is metabolized by cytoplasmic catechol-O-methyl-transferase (COMT, EC 2.1.1.6) or by intra-mitochondrial monoamine oxidase (MAO, EC 1.4.3.4).

The monoamine oxidase converts epinephrine and norepinephrine to 3,4-dihydroxymandelic acid. Since MAO is involved in the degradation of norepinephrine, inhibitors of MAO elevate the neuronal level of norepinephrine.

The catechol-O-methyl-transferase transfers the methyl-group of S-adenosyl-methionine to norepinephrine and thus inactivates its substrate.

3.7.4
Receptors and Signal Transduction

When released from the nerve varicosities, norepinephrine interacts with specific receptors. These are collectively designated as adrenoceptors and occur on the plasma membrane of neurons from the central and peripheral nervous system or on peripheral glands or muscle cells (so-called effector cells).

The effects elicited by norepinephrine binding depend on the type of receptor. In 1948, two types of noradrenergic receptors (α and β) were described by Ahlquist to explain the different effects of norepinephrine. To date, the α-adrenoreceptors have been further split into two different subcategories (α1 and α2).

All three adrenoreceptor types have in common the property that they couple to G proteins and show the topology of the seven membrane-spanning domain model.

Each of the receptors can be further divided into several subtypes. Three different subtypes of the α-1 receptors (α-1A, α-1B and α-1D) have been identified, as well as three different subtypes of α-2 receptors (α-2A, α-2B and α-2C). Three further subtypes of β receptors (β1, β2 and β3) complete the catecholamine receptor family.

Table 3.10 lists the different adrenergic receptors, together with some specific agonists and antagonists to which the receptors are susceptible.

The adrenoceptors in the central nervous system are activated by norepinephrine or by specific agonists; and signal transduction involves stimulation of G proteins. The sequence of the seven transmembrane domains is highly conserved among the three adrenoceptor subfamilies; however they share little homology in the third intracellular loop and the carboxy-terminal region.

Table 3.10 The three subfamilies of adrenergic receptors and some pharmacological characteristics.

Subfamily	G protein coupling	Second messenger	Selective agonists	Selective antagonists	Cloned subtypes
α1	G_q	Ca^{2+}	Phenylephrine Methoxamine	Prazosine WB 4101	α_{1A} α_{1B} α_{1D}
α2	G_i	cAMP	Clonidine Dexmedetomidine	Rauwolscin Yohimbine	$\alpha_2 A$ $\alpha_2 B$ $\alpha_2 C$
β	G_s	cAMP	Isoproterenol Terbutalin	Propranolol Metoprolol	β_1 β_2 β_3

Mutagenesis studies exploiting chimeric adrenoceptors have provided evidence that the carboxyl terminus and the third intracellular loop are important for functional binding of the ligand. A displacement of amino acids in the third intracellular loop can disturb the signal transduction downstream of the receptor site. The third intracellular loop and the carboxyl terminus are also essential for desensitization of the adrenoceptors. Desensitization occurs in the continued presence of a ligand and takes the form of a decrease in the response to the agonist over time. The mechanisms involved in desensitization of adrenoceptors include uncoupling from the G protein and removal of the receptor from the plasma membrane (so-called internalization). This internalization stops the ligand-induced effects and allows recycling of the receptor.

The distribution of the different subtypes of adrenoceptors is not homogeneous in the brain and each receptor subtype reveals a characteristic location.

For example, in the cerebral cortex, the adrenoceptors of type α and β are differently distributed among the cortical layers. These topological differences in receptor distribution could account for the local variability of noradrenergic responses in the cortex.

Adrenergic α-1 receptors

Pharmacological differences of the three α1 subtypes have been well established. All subtypes exhibit susceptibility to prazosin.

The α-1A receptors are most sensitive to the α-1-specific antagonist WB4101 and are insensitive to chloroethylclonidine (a site-directed alcylating compound). In contrast, the α-1B receptors exhibit low affinity for WB4101 and they are sensitive to chloroethylclonidine inactivation.

An α-1 clone has been described, which was first thought to code for a new α receptor and was therefore classified as α-1C. However, further studies indicated that the α-1C is identical to the α-1A receptor.

A third subtype is the α-1D receptor, which on the basis of its pharmacology and molecular identity is definitely different from the α-1A and α-1B receptors.

The three subtypes (α-1A, α-1B and α-1D) activate pertussis toxin-insensitive G_q proteins. Linkage of the G_q protein to the receptors results in the activation of phospholipase C, which initiates the generation of 1,4,5-trisphosphate (IP3) and diacylglycerol (DAG).

Adrenergic α-2 receptors

Three different subtypes of the α-2 receptor are known (α-2A, α-2B, α-2C).

A further α-2D receptor has been described in mice and rats. This subtype appears to be a rodent-specific homolog of the human α-2A receptor, with a difference in a single amino acid. A serine present in the sequence of the fifth transmembrane domain of the mouse receptor is a cysteine in the sequence of the human α-2A receptor at the same position. Mutation of this cysteine in the human α-2A receptor to a serine yields a change in the affinity for yohimbine, which is consistent with the affinity of the mouse α-2D receptor.

The early concept was that all members of the α-2 receptor family were coupled to a pertussis toxin-sensitive inhibitory G protein (G_i). However, it turned out that this is true exclusively for the receptor subtype α-2C. While the receptor α-2B is linked to a pertussis toxin-insensitive G protein and is capable of activating adenylate cyclase, the α-2A receptor inhibits adenylate cyclase at low agonist concentrations and shows a reverse effect at high concentrations. Additionally, it has been shown that activation of α-2A receptors involves the stimulation of the mitogen-activated protein kinase (MAPK).

The α-2 receptor family has been found primarily on presynaptic membranes, whereas the α-1 receptors occur at postsynaptic sites.

Adrenergic β receptors

Two subtypes of β adrenoceptors were originally identified as β1 and β2.

The β1 receptors are expressed primarily in neurons, whereas the β2 receptors are found in glia. In addition, a third member of the β adrenoceptors has been identified which shows only a minor affinity for epinephrine or norepinephrine. This receptor subtype was named "atypical" β adrenoceptor or β3 receptor.

A further difference in the β3 receptor as compared to the other β subtypes is its behavior in the presence of CGP12177. This substance is an antagonist of β1 and β2 receptors, but shows a partial agonistic effect when exposed to β3 receptors.

All three subtypes of β adrenoceptors are coupled to G_s proteins, which leads to a stimulation of adenylate cyclase activity.

A prolonged exposure of the β adrenoceptors to their agonists reduces the responsiveness of the receptors (desensitization). Desensitization includes functional uncoupling from the G proteins and removal of the receptor from the membrane. The functional uncoupling from G proteins may occur as a result of phosphorylation of the intracellular domains of the receptor, either by G protein-coupled receptor kinases (GRKs) or by protein kinases.

3.7.5
Biological Effects

The functional consequences of noradrenergic receptor activation can be either inhibitory or excitory. On the one hand, norepinephrine frequently reveals inhibitory effects. Electrical stimulation of the locus coeruleus or the iontophoretic application of norepinephrine induces a decrease in the spontaneous activity of the neurons. On the other hand, norepinephrine seems to potentiate the neuronal responses to visual, auditory or nociceptive stimuli.

As mentioned earlier, norepinephrine has been found to colocalize with some neuropeptides, like NPY or galanine. Under experimental conditions (by increasing the frequency of electrical stimulation of a sympathetic nerve), NPY and norepinephrine are co-released and act synergistically on vasoconstriction. Furthermore, NPY inhibits the release of norepinephrine.

The activation of adrenoceptors of the α- and β-type exhibits almost inverse phsiological effects. In peripheral tissues, for example, activation of α1 adrenoceptors causes vasoconstriction, enhances glycogenolysis and more generally induces the contraction of smooth muscle cells, whereas activation of β adrenoceptors leads to vasodilatation, bronchodilatation and positive ionotropic and chronotropic effects on heart tissue.

Neuronal activation of β adrenoceptors is responsible for hyperpolarization, which depends on the activation of cAMP, accompanied by an increase in membrane resistance.

A further aspect of functional relevance is the potentiating effect of noradrenaline on some neuromodulators and neurotransmitters. For example, it has been found that norepinephrine increases vasoactive intestinal protein (VIP)-induced effects on glycogenolysis and enhances the neuronal responses to excitatory amino acids, like glutamate, in the cerebral cortex.

With respect to the locus coeruleus and its dominant role in the organization of the intracerebral noradrenergic system, the involvement of this nucleus in the regulation of general attention and circadian rhythm is of interest. However, the predominant physiological function of the central noradrenergic system is its response to stress-induced stimuli.

The induction of stress is coupled to an enhanced activity of the locus coeruleus which, because of its widespread projections into cortical and subcortical structures and lower brain stem areas, affects a variety of physiological functions. Evidently, the locus coeruleus seems to work as a relais for the noradrenergic projection, which is regulated by hypothalamic inputs such as the neuropeptide corticotropin-releasing factor (CRF).

Administration of CRF, for example, causes an increase in the plasmatic concentration of norepinephrine and CRF stimulates the synthesis of tyrosine hydroxylase (the enzyme which is necessary for the biosynthesis of norepinephrine) within the locus coeruleus.

3.7.6
Neurological Disorders and Neurodegenerative Diseases

The locus coeruleus and norepinephrinergic neurons in the reticular nuclei of the substantia nigra (A1 and A2) can be targets of degenerative diseases.

In Parkinson's disease, a massive loss of neurons can be observed in both brain areas, particularly if the disease is coupled with dementia, the loss of neurons in the locus coeruleus being significantly higher than in the reticular nuclei. However, the loss of norepinephrinergic neurons is not restricted to the brain stem, but also includes neurons in the cerebral cortex. Although the dopaminergic system plays the dominant role in Parkinson's disease, norepinephrinergic deficits seem to be responsible for the clinical phenotype of this disease, especially in the manifestation of deficits of cognitive function.

Postmortem studies have consistently shown norepinephrinergic system involvement in Alzheimer's disease with decreased norepinephrine levels. The majority of studies indicate significantly decreased cortical and subcortical levels of norepinephrine in the frontal medial gyrus, temporal superior gyrus, cingulated gyrus, hippocampus, amygdala, thalamus, hypothalamus, striatum and the LC.

The loss of norepinephrinergic neurons in the LC has been well established in patients with Alzheimer's disease; and it has not been found in other types of dementia, such as vascular dementia. Concurrent with the loss of NE in the LC is an increase of the norepinephrine metabolite 3-methoxy-4-hydroxyphenylgycol (MHPG), indicating a compensatory mechanism for LC norepinephrine loss. There is a well established link between AD severity and loss of noradrenergic neurons. Norepepinephrine levels in the brain of AD patients have also been found to have an inverse relationship with cognitive impairment.

The norepinephrinergic system seems to be indirectly involved in depression. The compound reserpine can provoke depression and a variety of antidepressive agents prolong the half-life of catecholamines, either by inhibition of their reuptake or by decreasing the metabolic rate (for example, by inhibiting monoamine oxidase; MAO). Modifications to the density of epinephrinergic receptors and phasic variability in the level of norepinephrinergic metabolites have been found in patients suffering from depression. In particular, postsynaptic a_2 receptor down-regulation seems to be prevalent in depression, combined with increased presynaptic receptor sensitivity and increased a_2 receptor density in the LC. Decreased norepinephrinergic receptor sensitivity and increased norepinephrinergic turnover have been noted in patients with anxiety, generalized anxiety disorder and posttraumatic stress disorder.

Some experimental data speak in favor of an involvement of norepinephrine in epilepsy, since norepinephrine inhibits the propagation of seizures or diminishes their extent. Also, vagal nerve stimulation (VNS) has served as a versatile tool in treating patients suffering from refractory epilepsy. Norepinephrine has been suggested to be involved in the prophylactic antiseizure effects of VNS. Most of the data on a link between norepinephrine and epilepsy stem from epidemiological studies and pharmacological effects showing that norepinephrinergic and/or serotonergic transmission are both anticonvulsant and antidepressive. However, most of the the data indicating a direct correlation between norepinephrine and epilepsy were obtained from animal models and need to be substantiated for human epilepsy.

Further Reading

Ben-Menachem, F. (2002): Vagus-nerve stimulation for the treatment of epilepsy. *Lancet Neurol.* **1**: 477–482.

Birnbaumer, L.G. (1990): Proteins in signal transduction. *Annu. Rev. Pharmacol. Toxicol.* **30**: 675–705.

Caron, M.G., Lefkowitz R.J. (1993): Catecholamine receptors: structure, function and regulation. *Recent Prog. Hormone Res.* **48**: 277–290.

Foote, S.L., Bloom, F.E., Aston-Jones, G. (1983): Nucleus locus coeruleus: new evidence of anatomical and physiological specificity. *Physiol. Rev.* **63**: 844–914.

Giorgi, F.S., Pizanelli, C., Biagioni, F., Murri, L., Fornai, F. (2004): The role of norepinephrine in epilepsy: from bench to the bedside. *Neurosci. Biobehav. Rev.* **28**: 507–524.

Hadcock, J. R., Malbon, C. C. (1993): Adrenergic receptors as models for G protein-coupled receptors. *Annu. Rev. Neurosci.* **15**: 87–114.

Herrmann, N., Lanctot, K. L., Khan, L. R. (2004): The role of norepinephrine in the behavioral and psychological symptoms of dementia. *J. Neuropsychiatry Clin. Neurosci.* **16**: 261–276.

Jobe, P. C., Dailey, J. W., Wernicke, J. F. (1999): A noradrenergic and serotonergic hypothesis of the linkage between epilepsy and affective disorders. *Crit. Rev. Neurobiol.* **13**: 317–356.

Kandell, E. R., Schwartz, J. H., Jessell, T. M. (eds) (2002): *Principles of Neural Science, 3rd edn.* Elsevier, Amsterdam.

Maeda, T., Kojima, Y., Arai, R., Fujimiya, M., Kimura, H., Kitahama, K., Geffard, M. (1991): Monoaminergic interaction in the central nervous system: a morphological analysis in the locus coeruleus of the rat. *Comp. Biochem. Physiol.* **98**: 193–202.

Magistretti, P. J., Morrison, J. H. (1988): Noradrenaline- and vasoactive intestinal peptide-containing neuronal systems in neocortex: function convergence with contrasting morphology. *Neuroscience* **24**: 367–378.

Milligan, G., Svoboda, P., Brown, C. M. (1994): Why are there so many adrenoceptor subtypes? *Biochem. Pharmacol.* **48**: 1059–1071.

Moore, R. Y., Bloom, F. E. (1979): Central catecholamine neurone systeme: anatomy and physiology. *Annu. Rev. Neurol.* **2**: 113–168.

Strosberg, A. D. (1995): Structural and functional diversity of β-adrenergic receptors. *Ann. N.Y. Acad. Sci.* **758**: 253–260.

Tota, M. R., Candeloe, M. R., Dixon, R. A. F., Strader, C. D. (1991): Biophysical and genetic analysis of the ligand-binding site of the β-adrenoceptor. *Trends Pharmacol. Sci.* **12**: 4–6.

3.8
Serotonin (5-Hydroxytryptamine)

3.8.1
General Aspects and History

In 1947, serotonin (or 5-hydroxytryptamin; 5-HT) was discovered as a substrate which modulates vasoconstriction. The name serotonin is an acronym composed of *ser*um (because the substance can be found in blood serum) and vaso*ton*ic (because it provides vasotonic properties). In the early 1950s, it became evident that 5-HT and D-lysergic acid diethylamide (LSD), one of the most potent hallucinogenic drugs (Fig. 3.28), possess some common functional features,

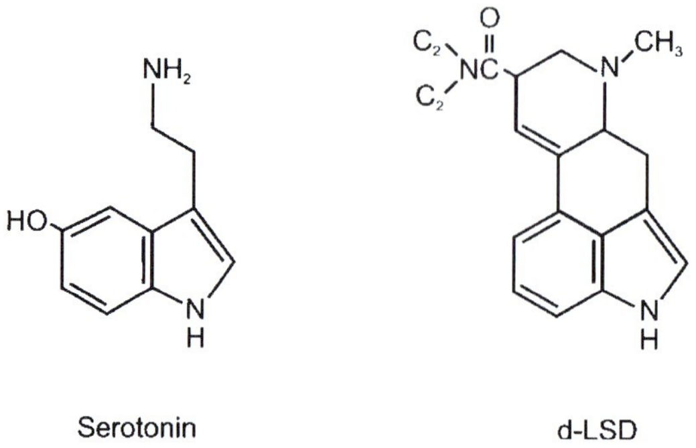

Fig. 3.28 Structure of serotonin and LSD.

3.8 Serotonin (5-Hydroxytryptamine)

but it took nearly another decade before serotonin was identified as a neurotransmitter.

Serotonin belongs to the class of biogenic amines (indolamines or monoamine transmitters) and its biosynthesis resembles that of catecholamines.

3.8.2
Localization Within the Central Nervous System

Fluorescence histochemistry and immunocytochemistry revealed the presence of serotonergic neurons predominantly in different nuclei of the raphe. The general localization of serotonergic neurons in the CNS is depicted in Fig. 3.29. Within the central nervous system, the serotonergic innervation is essentially diffuse. Terminals expressing serotonin can be found in nearly every brain area, but their general density is low. A relatively high density of serotonergic projections occurs in the cerebral cortex, the hippocampus, the amygdala, the basal ganglia, the lateral geniculate nucleus, the suprachiasmatic nucleus, the tectum opticum, the substantia gelatinosa and in the ventral horn of the spinal chord.

3.8.3
Biosynthesis and Degradation

Neurons provide the only source for serotonin in the central nervous system. The precursor of serotonin is the essential amino acid tryptophan and the availability of tryptophan represents the rate-limiting factor in the synthesis of serotonin. Tryptophan crosses the blood–brain barrier by means of an active amino

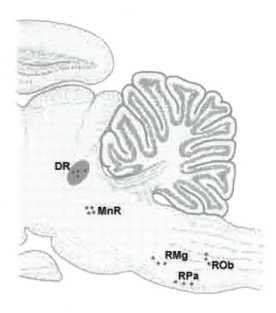

Fig. 3.29 The most important brain areas containing serotonergic neurons. The serotonergic innervation of the brain consists mainly of projections originating in the nuclei of the dorsal raphe (B7) and the medial raphe (B8). *Abbreviations*: DR=dorsal raphe; MnR=medial raphe; RMg=nucleus raphe magnus; ROb=nucleus raphe obscurus; RPa=nucleus raphe pallidus.

acid transport system in competition with leucine, lysine and methionine. The activity of this transporter is facilitated by the presence of glucose and insulin.

The neurogenic uptake of tryptophan involves a membrane transporter. This transporter is not specific for tryptophan but carries other neutral amino acids as well, allowing competition between tryptophan and neutral amino acids. By manipulating the level of extracellular tryptophan, the biosynthesis of serotonin in the brain can be directly influenced.

By the administration of neutral amino acids like valine, leucine, isoleucine or phenylalanine in excess, it is possible to clear tryptophan from the brain.

Under these experminental conditions, the synthesis of serotonin ceases when the intracellular tryptophan stores have been depleted and the initial enzyme in tryptophan synthesis (tryptophan hydroxylase) becomes deprived of its substrate.

Conversely, an excess of tryptophan has the opposite effect, by enhancing the production of neuronal serotonin. In the presence of oxygen molecules and a proton donor (tetrahydropterine), tryptophan hydroxylase converts tryptophan to 5-hydroxytryptophan (5-HTP), which is further processed by decarboxylation to serotonin through an amino acid decarboxylase (AADC). The same decarboxylase is also capable of decarboxylating dihydroxyphenylalanine (DOPA), which is an intermediary product in the biosynthesis of dopamine. Accordingly and in contrast to the tryptophan hydroxylase, AADC is not restricted to serotonergic neurons.

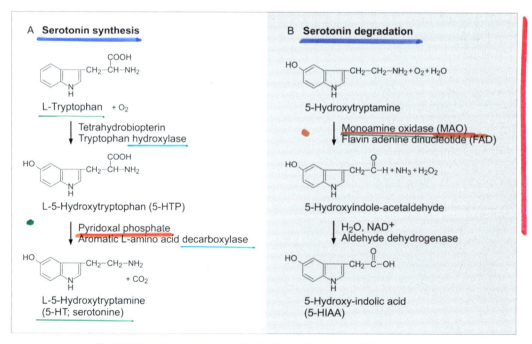

Fig. 3.30 Steps in serotonin synthesis (A) and degradation (B).

Inactivation of serotonin takes place by a specific re-uptake mechanism, which allows the re-uptake of released serotonin from the extracellular space.

The serotonin transporter (Sert) depends on the electrochemical gradient of Na^+ and K^+, like most neurotransmitter transporters. The serotonin transporter is an important pharmacological target for antidepressives. For example, the family of tricyclic antidepressives blocks the recapture of serotonin.

A further mechanism for serotonin inactivation is its enzymatic degradation by monoaminooxidase A (MAO_A) and an aldehyde dehydrogenase. The resulting metabolite, 5-hydroxyindolic acid (5-HIAA), is eliminated via the urinary tract (Fig. 3.30).

3.8.4
Receptors and Signal Transduction

The idea that different serotonin receptors exist in brain tissue was first developed in the 1950s. This hypothesis was proven almost four decades later after molecular biological techniques became available. It is now clear that the quaternary structure of 5-HT receptors consists of at least three different constituents: the transporter, the ligand-gated ion channel and the G protein-coupled receptor (Table 3.11). The largest group of serotonin receptors belongs to the superfamily of G protein-coupled receptors. To date, seven subtypes of 5-HT receptors have been distinguished (5-HT1 to 5-HT7) on the basis of cloning data.

A further subdivision of the seven subclasses has been achieved by classifying them on the basis of their structural and pharmacological properties.

5-HT1 receptors
The subfamily of 5-HT1 receptors consists of seven different members (subtypes 1A, 1B, 1D, 1E, 1F, 1P and 1S). The 5-HT1C receptor was later identified as belonging to the 5-HT2 receptor subfamily. In contrast to the receptors of the 5-HT2 family or the 5HT-4 receptors, the 5-HT1 receptors display a high affinity for serotonin.

Table 3.11 The different subtypes of 5HT receptors and their relationship to the three different molecular structures.

	Subtypes of 5-HT receptor
G protein-coupled receptors	
The family of the 5-HT_1 receptors	5-HT_{1A}, 5-HT_{1B}, 5-HT_{1D}, 5-HT_{1E}, 5-HT_{1F}
The family of the 5-HT_2 receptors	5-HT_{2A}, 5-HT_{2B}, 5-HT_{2C}
Other receptors	5-HT_{4S}, 5-HT_{4L}, 5-HT_{5A}, 5-HT_{5B}, 5-HT_6, 5-HT_7
Ligand-gated ion channels	5-HT_3
Transporters	5-HT "uptake site"

The 5-HT1 receptors are defined on the basis of their putative protein sequence and they are well characterized by their biochemical and pharmacological properties. Selective agonists are, however, sparse; and they exist exclusively for the 5HT-1A receptor (for example, buspirone). The G_i/G_o protein-coupled 5-HT1 receptors generally mediate inhibitory effects either by opening potassium channels or by closing calcium channels.

The 5-HT1A receptors, which constitute autoreceptors, are preferentially expressed on the soma and dendrites of serotonergic neurons. Their activation inhibits the firing rate of serotonergic fibers; and their desensitization after long-term exposure to 5-HT uptake blocker restores the firing rate. Both the human 5-HT1A receptor and the 5-HT1A receptor of rats have been cloned, indicating a predicted length of their amino acid sequence of 421 and 422 amino acids, respectively. The gene of the human 5HT1A receptor subtype is located on human chromosome 5-q1.

The 5-HT1B receptor has a shorter amino acid sequence and consists of 386 amino acids in rats. The human 5-HT1B receptor gene is located on the human chromosome 6q13. This receptor subtype is concentrated in the basal ganglia, striatum and frontal cortex. The receptor is negatively coupled to adenylyl cyclase.

The human 5-HT1D receptors exist in two different isoforms, which have been labeled 5-HT1Dα and 5-HT1Dβ. The 5-HT1Dα receptor has a putative length of 377 amino acids, while the 5-HT1Dβ receptor consists of 390 amino acids. The receptor gene is located on human gene 1p36.3–p34.3. A general functional feature of autoreceptors is their capacity for self-regulating the release of their neurotransmitter. In rats, this behavior has been described for the 5-HT1B receptors, while the receptor subtype 5-HT1D shows autoreceptive function in humans and guinea pigs.

Using [^{3}H] 5-HT as a radioligand, another 5HT binding site was discovered, which was named 5-HT1E. Human brain binding studies have reported that 5-HT1E receptors (representing up to 60% of 5HT1 receptor binding) are concentrated in the caudate putamen, with lower levels in the amygdala, frontal cortex and globus pallidus. The receptor has been mapped to human chromosome 6q14–q15. The 5HT1E receptor is negatively linked to adenylyl cyclase and consists of a 365-amino-acid protein with seven transmembrane domains.

The 5-HT1F receptor subtype is most closely related to the 5-HT1E receptor with a sequence homology of 70% across the seven transmembrane domains. In addition, this receptor exhibits intermediate transmembrane homology with several other 5-HT1 receptors: the 5-HT1Dα (63%), 5-HT1Dβ (60%) and 5-HT1A (53%) receptor subtypes. *In situ* hybridizations have demonstrated that mRNA coding for the receptor is concentrated in the dorsal raphe, hippocampus and cortex and is also present in the striatum, thalamus and hypothalamus. The receptor is negatively linked to adenylyl cyclase.

5-HT1P sites are labeled by [^{3}H]5-HT and display a pharmacology distinct from other 5-HT receptors. 5-HT1P receptors were mainly located in the gut but, because they have not been identified in the central nervous system until now, they will not be discussed here in detail and are mentioned only for completeness.

The 5-HT1S receptors are mainly expressed in the spinal chord. Although these receptors appear to be the predominant 5-HT1 receptor population in the spinal cord, no significant density of 5-HT1S receptors has been found in the brain stem or frontal cortex.

5-HT2 receptors

In contrast to 5-HT1 receptors, 5-HT2 receptors mediate slow excitatory effects through a decrease in potassium conductance or, alternatively, an increase in non-selective cation conductance. The 5-HT2 receptor subfamily consists of three members: 5-HT2A, 5-HT2B and 5-HT2C.

The 5-HT2A receptor is composed of 471 amino acids, whereas the 5-HT2B receptor in rats is composed of 479 amino acids (or 504 amino acids in mice). The 5-HT2C receptors also differ in their amino acid sequence, depending on the species from which they were cloned. In rats, this receptor is composed of 460 amino acids, while in mice the receptor is composed of 459 amino acids; and the human 5-HT2C receptor is 458 amino acids in length. The 5-HT2 receptors are coupled to phospholipase C and binding of the ligand leads to activation of phosphoinositol metabolism.

Different selective 5-HT2 receptor antagonists have been developed, including cyroheptadine, mertergoline, lisurid, methysergid, and pizotifen. The compound amitriptylin shows dual properties, working as an antagonist of the 5-HT2 receptors and as a serotonin-uptake blocker.

5-HT3 receptors

The unique feature of the 5-HT3 receptors, as compared to the other members of the 5-HT receptor family, is that they are not coupled to G proteins. This receptor type exhibits ionotropic features and belongs to the superfamily of ligand-gated ion channels. Not surprisingly, the 5-HT receptor family is structurally and functionally different from the other serotonergic receptors. The 5-HT3 receptor has a putative length of 487 amino acids and forms four transmembrane-spanning segments.

Activation of 5-HT3 receptors elicits an increase in Na^+ and K^+ permeability and makes the membrane impermeable for divalent cations. Consequently, activation of 5-HT3 receptors depolarizes the neuron, which therefore shows an excitatory response.

In contrast to the 5-HT1 and 5-HT2 receptors, which are located on neurons, glia and muscle cells, the 5-HT3 receptors are exclusively expressed in neurons. 5-HT3 receptors have a much lower affinity for their specific ligand than 5-HT1 receptors and have been found at high density in the area postrema of the hindbrain, the entorhinal cortex and the amygdala.

Some selective antagonists, like graniseton, ondanseton and tropiseton, act specifically upon 5-HT3 receptors.

5-HT4 receptors

Two different 5HT4 receptors have been cloned which are generated by alternative splicing. Signal transduction of the 5-HT4 receptors involves cAMP, which distinguishes them from the other 5-HT receptors.

Activation of 5-HT4 receptors in the hippocampus elicits an increase in intracellular cAMP levels and a decrease in K^+ conductance.

Two potent pharmacological compounds are known to possess a high affinity for 5-HT4 receptors, namely renzaprid and zacoprid.

5-HT5, 5-HT6 and 5-HT7 receptors

Two 5-HT receptors identified from rat cDNA and cloned were found to have 88% overall sequence homology, yet were not closely related to any other 5-HT receptor family. These receptors have been termed 5-HT5A and 5HT5B. The 5-HT5A receptor is composed of 357 amino acids and the 5-HT5B receptor is composed of 370 amino acids.

Functional expression in COS7 cells indicates that both subtypes are G protein-coupled receptors. Their topology fulfills the criteria of the seven transmembrane-spanning model. The mRNA coding for the 5-HT5 receptors is located in the cortex, the olfactory bulb, the hippocampus, the habenula and the cerebellum. However, it is assumed that this receptor subtype is mainly expressed on glial cells.

By screening cDNA libraries, a 5-HT6 receptor gene has been found and cloned. This receptor shares only low amino acid homology (about 40%) with the other 5-HT receptors. The rat receptor consists of 438 amino acids with seven transmembrane domains and is positively coupled to adenylyl cyclase via G-proteins. 5-HT6 mRNA has been detected in the cortex, the hippocampus, the striatum, the amygdala, the nucleus accumbens and in the olfactory tubercle.

5-HT7 is the most recent 5-HT receptor identified by molecular cloning. The 5-HT7 receptor, expressed mainly in the central nervous system, has been cloned from several species, including guinea pigs, rats and humans. Alternative splicing probably accounts for two splice variants in rats. Both forms are positively coupled to adenylate cyclase. The short form of the rat 5-HT7 receptor contains 435 amino acids, whereas the long form contains 448 amino acids. The human 5-HT7 receptor, which does not exist in two isoforms, falls in between, with 445 amino acids. The human 5-HT7 receptor gene has been localized to chromosome 10.

Receptor autoradiography and *in situ* hybridization have been exploited to delineate the distribution of the 5-HT7 receptor and its mRNA in rat brain. The location data indicate that 5-HT7 receptors are concentrated in layers 1–3 of the cortex, as well as in neurons of the septum, the hippocampus, the thalamus, the hypothalamus, the centromedial amygdala, the periaquaductal gray and in neurons of the superior colliculus.

3.8.5
Biological Effects

Because of its receptor diversity and divergence in transduction pathways, serotonin modulates several biological functions in the central nervous system. The serotonin system exhibits sexual dimorphism, since differences in the expression of mRNA and in the binding of serotonin receptors (subtypes 1A and 2A) between sexes have been described by *in situ* hybridization and autoradiography. For instance, mRNA of the 5-HT1A receptor shows distinctly different expression patterns in female and male rats.

Serotonin influences processes related to memory and learning, sexual behavior and feeding behavior. The latter becomes apparent in transgenic mice, which overexpress 5-HT2C receptors. The transgenics reveal a significantly higher body weight than wild types. In addition, the 5-HT2 antagonist ketanserin inhibits salt appetite induced by sodium depletion. Serotonin seems also be involved in regulating agressive behavior. For example, it has been shown that, after a certain period of isolation, 5-HT1B-deficient mice become more aggressive than wild-type littermates. Since these findings are based on animal models, one must be cautious about assuming that they are directly relevant to human beings.

Within the central nervous system, serotonin plays a significant role in nociception. Experimental data obtained by electrical stimulation of the raphe nuclei speak in favor of this assumption, since the electrical stimulus induces powerful analgesia. In addition, the selective destruction of serotonergic cells (by the neurotoxin 5,7-dihydroxytryptophan) directly affects nociception.

There is ample pharmacologic evidence that serotonin receptors in the brain can activate the hypothalamic–pituitary–adrenocortical (HPA) axis. Direct-acting serotonin agonists, serotonin-uptake inhibitors, serotonin releasers and the serotonin precursor L-5-hydroxytryptophan all increase release of adrenocorticotrophin (ACTH) and corticosterone. Serotonin-containing nerve terminals make synaptic contact with corticotrophin-releasing factor (CRF)-containing cells in the hypothalamus; and serotonin, as well as serotonin agonists, stimulate corticotropin release from the hypothalamus. Current evidence suggests that both 5-HT1A and 5-HT2 receptor subtypes are involved in the regulation of the secretion of corticotrophin-releasing factors. However, the physiology of the serotonergic regulation of the HPA axis is not well understood. Serotonin-containing neurons also appear to influence the secretion of other pituitary hormones, especially prolactin and gonadotropins.

3.8.6
Neurological Disorders and Neurodegenerative Diseases

Alterations in serotonin function have been linked to anxiety, affective, eating and sleep disorders.

In clinical trials, the 5-HT1A agonist buspirone has been found to be useful in the treatment of anxiety. Buspirone has a side-effect profile that differs from the anxiolytic effect of benzodiazepines insofar as it has no sedating effect. 5-HT1A agonists thus offer a useful tool in the treatment of anxiety.

A role for serotonin in the origin of migraine has been supported by changes in circulating levels of serotonin and its metabolites during the cycle of migraine attacks, along with the ability of serotonin-releasing agents to induce migraine-like symptoms. An involvement of serotonin in migraine is further supported by the efficacy of serotonin receptor ligands. Sumatriptan is an agonist on 5-HT1D and 5-HT1B receptor subtypes and is effective in treating migraine pain and associated symptoms. Recently, selective 5-HT1F agonists have been proposed for the treatment of migraine, without the side-effects associated with the 5-HT1D and 5-HT1B receptor agonists. A triggering role has also been suggested for 5-HT2B receptors in the initiation of migraine, suggesting that the application of selective 5-HT2B receptor antagonists might be effective in migraine treatment. Thus, compounds that modulate 5-HT1B, 5-HT1D, 5-HT1F and 5-HT2B receptors either have or may have clinical relevance in the therapy of migraine headache.

The serotonergic system plays a significant role in the generation of depression. Functional deficiencies in serotonin and norepinephrine have been implicated in the pathophysiology of depressive syndromes; and restoration of the normal function of the 5-HT- and NE-associated signaling pathway has been the target of antidepressants. This strategy is based on the monoamine theory of depression. In this context, it is essential to mention the observation that the monoamine hypothesis of 5-HT and NE deficiencies fails to explain the whole mechanims of antidepressants; and account must be taken of the additional hypotheses, including the cytokine hypothesis of depression, the hypothalamic–pituitary–thyroid hypothesis of depression, as well as the role of brain-derived neurotrophic factor (BDNF) and cyclic AMP response elements.

Nevertheless, the monoamine hypthesis has been fruitful in conceptualizing and developing potent antidepressants. Hypoactivity of the central serotonergic system has been demonstrated in neuronal subpopulations in depressed patients, leading to the therapeutical concept of 5-HT uptake blocker application in the treatment of depression.

Selective serotonin re-uptake inhibitors (SSRI), e.g. citalopram, fluoxetine, paroxetine and sertraline, are as effective as tricyclic antidepressants. An advantage of the serotonin re-uptake inhibitors over the tricyclic antidepressants is that they induce a lower tolerance. Drugs that inhibit serotonin and norepinephrin re-uptake (SNRIs), e.g. venlafaxine, milnacipram, and duloxetine, have become a further potent class of antidepressiva. There is no evidence for major differences between the SSRIs and SNRIs in their efficacy in treating anxiety disorders. In contrast to SSRIs, which are generally ineffective in treating chronic pain, all SNRIs seem to be helpful in relieving chronic pain associated with and independent of depression.

5-HT1A agonists seem to have antidepressant properties in some animal models of depression. A noteworthy issue in this context (and somewhat contradictory to the anxiolytic effect of 5-HT receptor activation) is the finding that some forms of phobia seem to be associated with an increase in serotonin levels, indicating that the serotinergic system is only one player in the generation of psychotic syndromes (see above).

In the treatment of psychosis, antagonists of both dopamine D2 and 5HT-2 receptors seem to be efficacious; and they reveal fewer extrapyramidal symptoms than neuroleptics that block only dopamine receptors.

A partial degeneration of serotonergic neurons occurs in some pathological states, like Alzheimer's and Chorea–Huntington's disease. In addition, in patients suffering from Alzheimer's disease, the density of 5-HT1A and 5-HT2A receptors in the cortex decreases. These deficits in serotonergic transmission might be relevant to the frequently observed depressive mood in Alzheimer's patients.

Concerning Chorea–Huntington, it has been shown that this inherited degenerative disease is coupled with a significant reduction in 5-HT1D receptors in the substantia nigra. However, the functional aspects of this reduction are essentially unknown.

Further Reading

Aghajanian, G. K. (1994): Serotonin and the action of LSD in the brain. *Psychiatr. Ann.* **24**: 137–141.

Boess, F. G., Martin, I. L. (1994): Molecular biology of 5-HT receptors. *Neuropharmacology* **33**: 275–317.

Fuller, R. W. (1990): Serotonin receptors and neuroendocrine responses. *Neuropsychopharmacology* **3**: 495–502.

Goodnick, P. J., Goldstein, B. J. (1998): Selective serotonin re-uptake inhibitors in affective disorders – I. Basic pharmacology. *J. Psychopharmacol.* **12**[Suppl B]: 5–20.

Gustafson, E. L., Durkin, M. M., Bard, J. A., Zgombick, J., Branchek, T. A. (1996): A receptor autoradiographic and *in situ* hybridization analysis of the distribution of the 5-ht7 receptor in rat brain. *Br. J. Pharmacol.* **117**: 657–666.

Iversen, I. (2000): Neurotransmitter transporter: fruitful targets for CNS drug discovery. *Mol. Psychiatry* **5**: 357–362.

Jayanthi, L. D., Ramamoorthy S. (2005): Regulation of monoamine transporters: influence of psychostimulants and therapeutic antidepressants. *AAPS J.* **7**: E728–E738.

Johnson, K. W., Phebus, L. A., Cohen. M. L. (1998): Serotonin in migraine: theories, animal models and emerging therapies. *Prog. Drug Res.* **51**: 219–244.

Leonard, B. E. (1994): Serotonin receptors – where are they going? *Int. Clin. Psychopharmacol.* **9**[Suppl 1]: 7–17.

Maeda, T., Fujimiya, M., Kitahama, K., Imai, H., Kimura, H. (1989): Serotonin neurons and their physiological roles. *Arch. Histol. Cytol.* **52**[Suppl]: 113–120.

Nair, G. V., Gurbel, P. A., O'Conner, C. M., Gattis, W. A., Murugesan, S. R., Serebruany, V. L. (1999): Depression, coronary events, platelet inhibition, and serotonin re-uptake inhibitors. *Am. J. Cardiol.* **84**: 321–323.

Pazos, A., Palacios, J. M. (1985): Quantitative autoradiographic mapping of serotonin receptors in the rat brain. I. Serotonin-1 receptors. *Brain Res.* **346**: 205–230.

Peroutka, S. J. (1994): Molecular biology of 5-hydroxytryptamine receptor subtypes. *Synapse* **18**: 241–260.

Stahl, S. M., Grady, M. M., Briley, M. (2005): SNRIs: their pharamocology, clinical efficacy, and tolerability in comparison with other classes of antidepressants. *CNS Spectr.* **10**: 732–747.

Sussman, S. (1994): The potential benefits of serotonin receptor-specific agents. *J. Clin. Psychiatr.* **55**[Suppl 2]: 45–51.

Vizi, E. S. (2000): Role of high-affinity receptors and membrane transporters in nonsynaptic communication and drug action in the central nervous system. *Pharmacol. Rev.* **52**: 63–89.

Whitaker-Azmitia, P. M., Peroutka, S. J. (1990): *The Neuropharmacology of Serotonin*. Academic Sciences, New York.

White, K. J., Walline, C. C., Barker, E. L. (2005): Serotonin transporters: implications for antidepressant drug development. *AAPS J.* **7**: E421–E433.

Zhang, L., Ma, W., Barker, J. L., Rubinow, D. R. (1999): Sex differences in expression of serotonin receptors (subtypes 1A and 2A) in rat brain: a possible role of testosterone. *Neuroscience* **94**: 251–259.

Zifa, E., Fillion, G. (1992): 5-Hydroxytryptamine receptors. *Pharmacol. Rev.* **44**: 401–458.

4
Neuromodulators

In recent years, the recognition that neuroactive substances others than the classic neurotransmitters are widespread in the CNS has gained considerable attention. These neuroactive substances have been termed neuromodulators (see also Section 1.1: *Neuroactive Substances*). A main feature of neuromodulators is that they are capable of modulating synaptic transmission. Neuromodulators cover a broad spectrum of neuroactive substances, ranging from neuropeptides to steroids.

Some of them reveal neurotransmitter-like characteristics and have been termed "putative" neurotransmitters.

Neuromodulators can be distinguished from the classic neurotransmitters by several criteria:

- The amount of expression is generally lower than that of neurotransmitters.
- In contrast to neurotransmitters, neuromodulators are effective at low concentrations.
- Normally, neuromodulators are incapable of inducing the rapid changes in signal transmission which may be seen with neurotransmitters. They exert slow and long-lasting actions, mainly by affecting G protein-coupled receptors and subsequent second messenger systems.
- Neuromodulators act indirectly by interacting with neurotransmitters (in such cases, they coexist with neurotransmitters in the nerve terminals). Their direct effects on synaptic transmission are weaker than those of neurotransmitters.
- Specific rapid inactivation mechanisms, as e.g. re-uptake mechanisms are rare.

Depending on their chemical origin, the family of neuromodulators is composed of three members: neuropeptides, the derivatives of arachidonic acid and gaseous molecules, including nitric oxide and carbon monoxide.

Neuropeptide-secreting neurons were among the first specialized cells to appear in the evolution of primitive nervous systems; and neuropeptides are strongly conserved in phylogeny. A likely interpretation of the biological conservatism of neuropeptides is that they perform some essential feature in brain function.

Compared with neurotransmitters, the synthesis of neuropeptides requires complex metabolic pathways, including proteolytic processing from pre-neuro-

---

Neurotransmitters and Neuromodulators. Handbook of Receptors and Biological Effects. 2nd Ed.
Oliver von Bohlen und Halbach and Rolf Dermietzel
Copyright © 2006 WILEY-VCH Verlag GmbH & Co. KGaA, Weinheim
ISBN: 3-527-31307-9

peptides to smaller pro-neuropeptides and finally mature neuropeptides in the Golgi apparatus.

The rate of synthesis of neuropeptides in the soma is low and they are transported anterogradually into the nerve terminals. Neuropeptides are commonly stored in large, dense-cored vesicles and can colocalize with additional neuromodulators or neurotransmitters.

Derivatives of arachidonic acid act similarly to neuropeptides. They are synthesized by complex metabolic processing from precursors and the neuroactive products act through specific receptors.

Some gases are produced in brain tissues and exhibit physiological effects on neuronal signaling. Among these gases is nitric oxide which *per definitionem* constitutes a neuromodulator. This gaseous neuromodulator has an extremely brief half-life and is transmitted by diffusion. In contrast to neurotransmitters and most of the neuromodulators, the gaseous substance nitric oxide (NO) is not stored in vesicles, but is synthesized on demand and does not bind to membrane-associated receptors.

4.1
Adrenocorticotropic Hormone

4.1.1
General Aspects and History

The adrenocorticotropic hormone (ACTH) was the first pituitary peptide isolated from hypophyseal extracts (Li et al. 1955).

Adrenocorticotropic hormone (ACTH), also known as corticotropin, is secreted by the anterior lobe of the pituitary gland and is responsible for the stimulation of steroid secretion from the cortex of the adrenal glands.

Corticotropin-releasing hormone (CRH) is the most prominent inducer of ACTH secretion. Furthermore, vasopressin, angiotensin II, vasoactive intestinal peptide (VIP), growth hormone-releasing hormone (GHRH), norepinephrine and epinephrine also influence the secretion of ACTH. However, gene expression of proopiomelanocortin (the precursor of ACTH) in corticotropic cells is exclusively stimulated by CRH. Adrenal steroids via a negative feedback mechanism inhibit the secretion of ACTH.

ACTH is synthesized in several forms. ACTH 1–39 (Fig. 4.1) represents the complete biologically active form of ACTH and consists of 39 amino acids. Some additional biologically active fragments of ACTH have also been demonstrated, including the fragments ACTH 11–24 and ACTH 4–10.

> Ser-Tyr-Ser-Met-Glu-His-Phe-Arg-Trp-Gly-Lys-Pro-Val-Gly-Lys-
> Lys-Arg-Arg-Pro-Val-Lys-Val-Tyr-Pro-Ala-Gly-Glu-Asp-Asp-Glu-
> Ala-Ser-Glu-Ala-Phe-Pro-Leu-Glu-Phe

Fig. 4.1 The amino acid sequence of ACTH.

4.1.2
Localization Within the Central Nervous System

ACTH 1–39 and ACTH 11–24 have been demonstrated by immunohistochemistry in the hypothalamus, the thalamus, the amygdala and in the periaqueductal gray.

ACTH 4–10 has a more restricted distribution and is expressed in the ventral horn of the spinal chord, the eminentia mediana, the septal area and the subventricular zone of the third ventricle.

ACTH 1–39 is cleaved from the precursor proopiomelanocortin and can be further processed into shorter forms which also show neurotropic and behavioral effects. The shorter form ACTH 11–24 exhibits biological activity similar to that of ACTH 1–39. It shows corticotropic and melanotropic properties.

4.1.3
Biosynthesis and Degradation

ACTH, alpha melanocyte-stimulating hormone (a-MSH) and the endorphins are processed from a common precursor, proopiomelanocortin. Biosynthesis of ACTH from this precursor is described in detail in the section on proopiomelanocortin (Section 4.28).

ACTH is secreted from the anterior pituitary in response to the hypothalamic corticotropin-releasing hormone. There is a diurnal variation in the release of ACTH and a subsequent variation in the concentration of endogenous corticosteroids. The levels of both are higher during the early morning hours.

4.1.4
Receptors and Signal Transduction

Melanotropic receptors are responsible for the signal transduction of ACTH and a-MSH. They reveal the typical seven membrane-spanning domains of G protein-coupled receptors. Recognition sites within the extracellular domains are important for ligand binding. The intracytoplasmic loops, in particular the third loop (TM3) and the intracellular carboxyl terminus, are essential for transduction of the signal to the G proteins. The transduction pathway involves the activation of adenylate cyclase and the activation of phosphokinase A (PKA).

Receptors of the MC1 type bind specifically with a-MSH, whereas the MC2 receptor binds selectively ACTH. Both receptors share a homology of about 39%.

The receptor types MC3, MC4, MC5 act as neutral receptors, because they bind a-MSH or ACTH with nearly the same affinity.

Binding sites for ACTH in brain tissue have been identified in the bed nucleus of the stria terminalis, in the eminentia mediana and in hypothalamic nuclei, including the ventromedial nucleus, the dorsomedial nucleus, the paraventricular nucleus and the arcuate nucleus.

Binding sites for the short fragment ACTH 4–10 and the neurotrophic analog Org 2766 have been identified in the hypothalamus, the preoptic area, the septum and the hippocampus.

4.1.5
Biological Effects

The main biological function of ACTH is the stimulation of the adrenal cortex. ACTH elicits an increase in the production of glucocorticoids, mineralocorticoids and androgenic steroids, collectively labeled as corticosteroids. Glucocorticoids are so named because they promote the mobilization of energy by carbohydrates; and mineralocorticoids because they influence water and salt homeostasis. Aldosterone is the most potent mineralocorticoid as it accounts for 95% of activity in this class of corticosteroids. They control primarily electrolyte homeostasis by targeting to the distal tubules of the kidney to increase sodium resorption and reduce potassium elimination, thereby maintaining osmotic balance in the urine and preventing serum acidosis.

Glucocorticoids are secreted by the zona fasciculata (and to a lesser extent by the zona reticularis). The predominant glucocorticoid is hydrocortisone (cortisol) which accounts for 95% of glucocorticosteroid activity.

ACTH 1–39 and ACTH 1–24 reveal both corticotropic and neurotropic properties. The corticotropic properties are dependent on the sequence of amino acids 11–24 of ACTH. The corticotropic effects are initiated by the binding of ACTH to the MC2 receptors in the adrenal cortex. Via its corticotropic effect, ACTH influences the serotoninergic and the catecholaminergic systems.

ACTH and its fragments are involved in neurotransmission of the central nervous system. It enhances the synthesis of acetylcholine, stimulates the uptake of choline into synaptosomes and enhances its calcium-dependent release. The stimulation of the re-uptake of choline seems to be restricted to some brain areas, including the hippocampus, the hypothalamus, the olfactory tract, the pons and the medulla oblongata. In the cerebellum and in the anterior thalamus ACTH exerts an opposite effect.

In vitro studies have shown that ACTH is capable of activating neurogenesis since it reveals neurotrophic effects upon cultivated cerebral neurons. It stimulates the outgrowth of neurites in a dose-dependent manner and enhances their metabolic activity. Strand and Kung (1980) were the first to demonstrate that

ACTH promotes the regeneration of crushed sciatic nerves, assessed by an accelerated return of motoric and sensoric functions in adrenalectomized animals.

ACTH 4–10 and ACTH 1–24 have been demonstrated to increase the density of neuronal networks in dissociated cerebral neurons. ACTH 1–24 has the full biological activity of the 39-amino-acid ACTH, thus possessing both neurotrophic and corticotropic characteristics.

ACTH 4–10 is a neurotrophic fragment of ACTH 1–39. Neonatal administration of ACTH 4–10 accelerates the maturation of the developing neuromuscular system (Smith and Strand 1981).

Adrenocorticotropic hormones seem also to influence cognitive functions like memorizing and learning, since administration of ACTH facilitates the acquisition and recall of learned tasks which are coupled with fear or anxiety.

Furthermore, it has been demonstrated that ACTH plays a role in sexual behavior and the determination of sexual dimorphism. Prenatal administration of ACTH elicits changes in the serotoninergic system, (e.g. by increasing 5-HT uptake) and induces changes in the dopaminergic innervation of the brain which persists until adulthood. These anatomical changes are coupled to severe dysfunctions of sexual activity in males, while sexual activity is not significantly altered in females.

4.1.6
Neurological Disorders and Neurodegenerative Diseases

ACTH is a prominent mediator in stress-related effects. Acutely, glucocorticoids act to inhibit stress-induced corticotrophin-releasing factor (CRF) and ACTH secretion through their actions in brain and anterior pituitary. With chronic stress, glucocorticoid feedback inhibition of ACTH secretion changes markedly. Chronically stressed rats characteristically exhibit facilitated ACTH responses to acute, novel stressors. Stress-enhanced release of ACTH and prolonged exposure to ACTH-evoked increase in glucocorticoids can induce hypertension and immuno-suppression.

Not only chronic stress, but also adverse early life experiences can have a negative impact later in life. For example, maternal separation leads to anxiety-like behavior in rats. These separated animals also have enhanced basal ACTH levels and display a blunted ACTH response and altered neurotransmitter levels in response to subsequent stressors.

ACTH-dependent hypersecretion of glucocorticoids induces Cushing's syndrome. Cushing's syndrome is a hormonal disorder caused by prolonged exposure of the body's tissues to high levels of the hormone cortisol. The symptoms of Cushing's syndrome vary, but most people show upper body obesity, a rounded face, an increase in the neck region and thinning arms and legs. Children tend to be obese with slowed growth rates. Additional symptoms afflict the skin, which becomes fragile and thin. Most patients suffer from severe fatigue, weak muscles, high blood pressure and high blood sugar. Irritability, anxiety and depression are common.

Cushing's syndrome can develop due to an overproduction of cortisol, because of over-stimulation by ACTH.

The following factors may contribute to increased ACTH production and consequently to Cushing's syndrome:
- Pituitary adenomas. These are benign, or non-cancerous, tumors of the pituitary gland which secrete increased amounts of ACTH.
- Some benign or malignant (cancerous) tumors that arise outside the pituitary can produce ACTH. This condition is known as ectopic ACTH syndrome. In addition, peripheral tumors of the adrenal gland can induce Cushing's syndrome.

Cushing's disease is a form of Cushing's syndrome, with persistent inappropriate hypercortisolism that results from pituitary ACTH hypersecretion. The episodic secretion of ACTH is similar to the normal; however, the frequency and amplitude of the secretory episodes lack the normal circadian rhythm.

ACTH secretion in Cushing's disease is under hypothalamic control; however, there are observations that suggest that brain centers superior to the hypophysiotropic area of the hypothalamus are involved in the pathophysiology of Cushing's disease. Thus, there is a dispute whether Crushing's disease is primarily a CNS or a pituitary disorder (Fehm and Voigt 1979).

Further Reading

Aron, D.C., Findling, J.W., Tyrrell, J.B. (1987): Cushing's disease. *Endocrinol. Metab. Clin. N. Am.* **16**:705–730.

Dallman, M.F., Akana, S.F., Strack, A.M., Scribner, K.S., Pecoraro, N., La Fleur, S.E., Houshyar, H., Gomez, F. (2004): Chronic stress-induced effects of corticosterone on brain: direct and indirect. *Ann. N.Y. Acad. Sci.* **1018**:141–150.

Daniels, W.M., Pietersen, C.Y., Carstens, M.E., Stein, D.J. (2004): Maternal separation in rats leads to anxiety-like behavior and a blunted ACTH response and altered neurotransmitter levels in response to a subsequent stressor. *Metab. Brain Dis.* **19**:3–14.

Darlington, C.L., Gilchrist, D.P.D., Smith, P.F. (1996): Melanocortins and lesion-induced plasticity in the CNS: a review. *Brain Res. Rev.* **22**: 245–257.

Fehm, H.L., Voigt. K.H. (1979): Pathophysiology of Cushing's disease. *Pathobiol. Annu.* **9**:225–255.

File, S.E., Vellucci, S. (1978): Studies on the role of ACTH and 5-HT in anxiety, using an animal mode. *J. Pharmacol. Exp. Ther.* **30**: 105–110.

Hannigan, J., Isaacson, R. (1985): The effects of ORG 2766 on the performance of sham, neocortical and hippocampal-lesioned rats in a food search task. *Pharmacol. Biochem. Behav.* **23**: 1019–1027.

Hol, E.M., Gispen, W.H., Bär, P.R. (1995): ACTH related peptides: receptors and signal transduction systems involved in their neurotropic and neuroprotective actions. *Peptides* **16**: 979–993.

Li, C.H., Geschwind, I.I., Dixon, J.S., et al. (1955): Corticotropins (ACTH): isolation of a-corticotropin from sheep pituitary glands. *J. Biol. Chem.* **213**: 171–185.

Newell-Price, J. (2003): Proopiomelanocortin gene expression and DNA methylation: implications for Cushing's syndrome and beyond. *J. Endocrinol.* **177**:365–372.

Richter-Landsberg, C., Bruns, I., Flohr, H. (1987): ACTH neuropeptides influence development and differentiation of embryonic rat cerebral cells in culture. *Neurosci. Res. Commun.* **1**: 153–162.

Schwyzer, R. (1977): ACTH: a short introductory review. *Ann. N.Y. Acad. Sci.* **247**: 3–26.
Segarra, A.C., Luine, V.N., Strand, F.L. (1991): Sexual behavior of male rats is differentially affected by timing of perinatal ACTH administration. *Physiol. Behav.* **50**: 689–697.
Smith, C.M., Strand, F.L. (1981): Neuromuscular response of the immature rat to ACTH/MSH 4–10. *Peptides* **2**: 197–206.
Strand, F.L., Kung, T.T. (1980): ACTH accelerates recovery of neuromuscular functions following crushing of peripheral nerve. *Peptides* **1**: 135–138.
Van der Neut, R., Bär, P.R., Sodaar, P., Gispen, W.H. (1988): Tropic influences of α-MSH and ACTH (4–10) on neuronal outgrowth *in vitro*. *Peptides* **9**: 1015–1020.
Vrezas, I., Willenberg, H.S., Mansmann, G., Hiroi, N., Fritzen, R., Bornstein, S.R. (2003): Ectopic adrenocorticotropin (ACTH) and corticotropin-releasing hormone (CRH) production in the adrenal gland: basic and clinical aspects. *Microsc. Res. Tech.* **61**: 308–314.

4.2
Anandamide (Endocannabinoids)

4.2.1
General Aspects and History

The Indian plant *Cannabis sativa* shows psychotropic properties as a result of its activating a specific endogenous receptor in the brain. The most potent psychoactive substance of the *Cannabis* plant is cannabinoid Δ^9-tetrahydrocannabinol (THC), although other substances can modify the action of THC or have pharmacological effects of their own. The psychotropic substance THC was isolated in 1964 by Mechoulam and coworkers (Mechoulam et al. 1994).

The psychotropic action of THC is mediated through a plasmalemmal receptor, designated as CB1 receptor. This receptor was characterized by Matsuda and coworkers (1990). The discovery of this receptor indicated that endogenous cannabinoids may occur in the brain, which acts as physiological ligands for CB1.

In 1993, Munro and coworkers characterized a second cannabinoid receptor, which is primarily expressed in macrophages. This was accordingly named CB2 receptor.

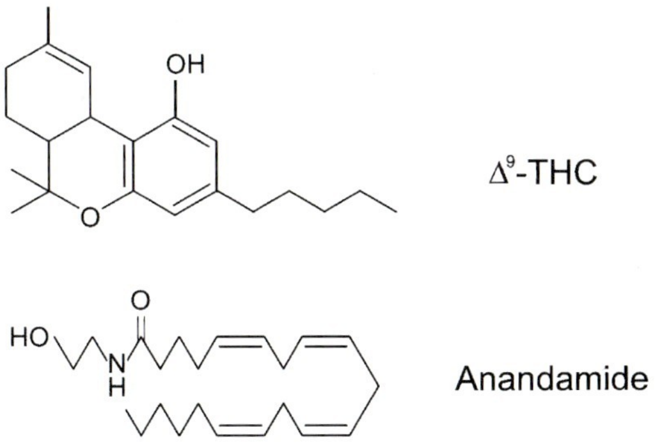

Fig. 4.2 Structure of Δ^9-tetrahydrocannabinol (THC) and anandamide. Anandamide is the endogenous ligand of the CB receptor. Tetrahydrocannabinol is a potent exogenous psychoactive substance.

At the same time, Mechoulam and coworkers (1994) isolated an endogenous substance, which proved capable of activating the cannabinoid receptors in very low concentrations.

This substance was identified as a derivate of arachidonic acid and the term anandamide was coined to reflect this fact. Anandamide (Fig. 4.2) is about 10-fold less potent and has a shorter duration of action than THC.

Anandamide is found in the central nervous system, where it is synthesized and degraded. It is believed that other endogenous ligands exist, with structural relationship to anandamides.

4.2.2
Localization Within the Central Nervous System

The exact localization of anandamides in brain tissues is unknown. However, *in vitro* evidence exists that cultivated cerebral neurons of rats are able to synthesize anandamides.

The presence of an enzyme (anandamide amidohydrolase) which degrades anandamides and the presence of CB1 receptors which specifically bind this arachidonic derivative have strengthened the hypothesis that anandamides are produced within the central nervous system and that these substances play a role in neurotransmission.

The molecular machinery responsible for anandamide production as well as its degradation can be found in the CNS.

4.2.3
Biosynthesis and Degradation

The synthesis of anandamide was first evaluated in primary cell cultures. The initiation of anandamide production is induced by depolarizing stimuli and is calcium-dependent.

The formation of anandamide from its precursor N-arachidonyl-phosphatidyl-ethanolamine (NAPE) is catalyzed by the enzyme phospholipase D.

In cultured primary neurons, depolarization induces the release of anandamide and elicits the biosynthesis of its precursor NAPE.

Extracellular anandamide is rapidly recaptured by neuronal and non-neuronal cells through a mechanism that meets four key criteria of carrier-mediated transport: fast rate, temperature dependence, saturability and substrate selectivity. In contrast with transport systems for classic neurotransmitters, [^{3}H]anandamide re-uptake is neither dependent on external Na$^+$ ions nor affected by metabolic inhibitors, suggesting that it may be mediated by a process of carrier-facilitated diffusion. However, the mechanism has not been molecularly characterized yet.

Because of this rapid deactivation process, the endocannabinoids may primarily act near their sites of synthesis by binding to and activating cannabinoid receptors on the surface of neighboring cells.

Endocannabinoid re-uptake has been demonstrated in discrete brain regions and in various tissues and cells throughout the body. Inhibitors of endocannabinoid re-uptake include N-(4-hydroxyphenyl)-arachidonylamide (AM404), which blocks transport with IC_{50} values in the low micromolar range.

Anandamide can be transformed by the activity of two different lipoxygenases (12-LOX and 15-LOX) into 12-hydroperoxy-anandamide or 15-hydroperoxy-anandamide (Fig. 4.3). The resulting anandamides are further metabolized by the activity of a hydrolase into arachidonic acid.

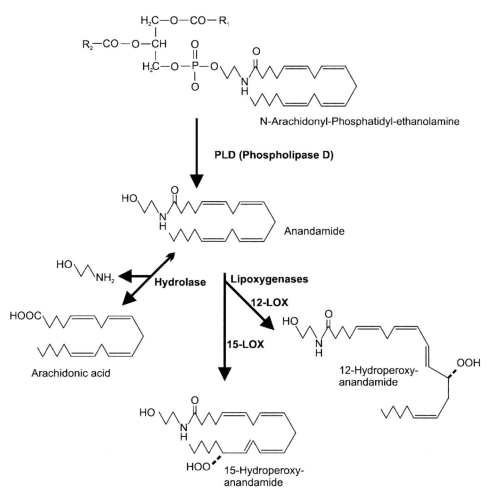

Fig. 4.3 Biosynthesis and degradation of anandamide. The formation of anandamide from the specific precursor is catalyzed by phospholipase D. The degradation of anandamide involves either hydrolases or lipoxygenases (12-LOX and 15-LOX).

The endocannabinoids are hydrolyzed by an intracellular membrane-bound enzyme, termed anandamide amidohydrolase (AAH). AAH may act as a general hydrolytic enzyme not only for anandamide and other fatty acid ethanolamides, but also for primary amides. Moreover, like other hydrolase enzymes, AAH may act in reverse, catalyzing the synthesis of anandamide from free arachidonate and ethanolamine.

4.2.4
Receptors and Signal Transduction

Both endocannabinoid receptors (CB1 and CB2) belong to the superfamily of G protein-coupled receptors with seven transmembrane-spanning domains. Additionally to the CB1 receptors, a splice variant, CB1A, has been described. Activation of both receptors is achieved by endocannabinoids, but they differ in their selectivity. While the CB1 receptor is activated by anandamide, the CB2 receptor is selectively activated by N-palmitoyl-ethanolamine.

Activation of CB receptors initiates a signal transduction, which involves the mobilization of arachidonic acid and thus induces an increase in the intracellular production of eicosanoids.

Cannabinoid receptors are not restricted to the central nervous system. Thus, they also occur in many peripheral tissues, including the immune system, reproductive and gastrointestinal tracts, sympathetic ganglia, endocrine glands, arteries, lung and heart.

CB1 receptors

Genes encoding orthologs of the mammalian CB1 receptor have been identified in fishes, amphibians and birds, indicating that CB1 receptors may occur throughout the vertebrates. The genomes of the invertebrates *Drosophila melanogaster* and *Caenorhabditis elegans*, however, do not contain CB1 orthologs, indicating that CB1-like cannabinoid receptors may have evolved after the divergence of deuterostomes and protostomes.

The cDNA of the CB1-receptor is 5.7 kb in length and encodes for a putative protein which contains seven hydrophobic domains. The CB1 receptor (Fig. 4.4) is coupled to G proteins of the G_i/G_o type. Anandamide and endocannabinoids appear to bind to CB1 receptors in transmembrane alpha helices 2, 3, 4 and 5, leading to the inhibition of the activity of adenylate cyclase, which gives rise to a subsequent inhibition of cAMP production.

Electrophysiological studies reveal that anandamide, via CB1 receptor binding, inhibits high-voltage-activated calcium channels (so-called N-type channels). CB1 receptors are primarily expressed at presynaptic sites and thus allow modulation of neurotransmitter release through retrograde signaling by endocannabinoids. Such a modulation was described in the cerebellum and the hippocampus, where the activation of CB1 receptors elicits a depolarization-induced suppression of inhibition (DSI). DSI represents a short inhibition of neurotransmitter release which is initiated by the postsynaptic release of endocannabinoids.

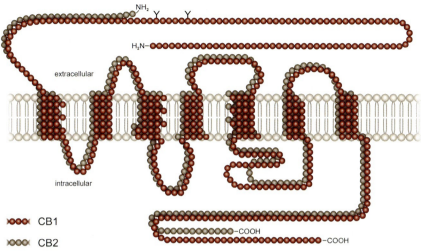

Fig. 4.4 Scheme of the structure of the CB1 and CB2 receptors. Both receptors belong to the GPCR family. CB1 shows extended amino and carboxyl termini.

The secretion of endocannabinoids is stimulated through postsynaptic increase of intracellular calcium by:
1. Activation of voltage-sensitive calcium channels which stimulates the production of endocannabinoids via phospolipase D.
2. Receptor-mediated release where activation of metabotropic glutamate or nicotinergic acetylcholine receptors seem to be involved. The latter pathway is G protein-dependent and requires phospholipase C.

This receptor-driven pathway is responsible for endocannoabinoid-mediated long-term depression (LTD). Also, the facilitation of LTP production through endocannabinoid release has been described for CA1 neurons, indicating that these neuromodulators play a substantial role in modulating synaptic efficacy in different cell types and brain regions (Fig. 4.5).

CB1 receptors exhibit binding specificity to anandamides, but they can also bind several other endogenous ligands which are present in the central nervous system. These substances include homo-γ-linolenyl-ethanolamide and docosatetraenyl-ethanolamide, as well as 2-arachidonylglycerol, all of which bind to both CB1 receptors and CB2 receptors.

The human CB1 receptor and the CB1 receptor of rats share a homology of 97.3% in their amino acid sequence. The distribution of CB1 receptors in the human central nervous system and in the central nervous system of rats is quite similar.

CB1 receptors have been identified within the cortex, the olfactory bulb, the hippocampal formation, the caudate-putamen, the globus pallidus, the substantia nigra, the nucleus accumbens, the entopeduncular nucleus and the molecular layer of the cerebellum (Fig. 4.6).

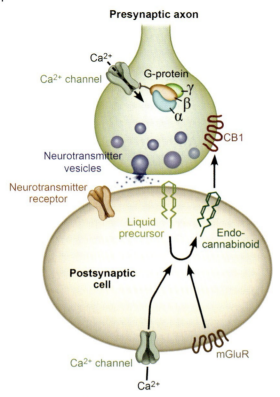

Fig. 4.5 Retrograde signaling by endocannabinoids. Postsynaptic depolarization opens voltage-dependent Ca^{2+} channels. An increase in postsynaptic Ca^{2+} elicits an activation of phospholipase D, which leads to endocannabinoid synthesis from lipid precursors. Activation of postsynaptic mGluRs can also generate endocannabinoids. A pathway which seems to involve phospholipase C and the generation of diacylglycerol is further cleaved by diacylglycerol lipase to yield 2-arachidonyl-glycerol. Endocannabinoids then leave the postsynaptic cell and work as retrograde messengers by activating presynaptic CB1 receptors. Postsynaptic G protein activation liberates $G_{\beta\gamma}$, which then directly inhibits presynaptic Ca^{2+} influx. This decreases the probability of release of a vesicle of neurotransmitter (adapted from Wilson and Nicoll 2002).

CB2 receptors

The topology of the CB2 receptor shows the same seven transmembrane domain-spanning motif as the CB1 receptor. The amino acid sequence of the human CB2 receptor exhibits a homology of 44% with the rat CB1 receptor. The mouse and rat CB2 genes have also been cloned and encode proteins of 347 amino acids (mouse) and 361 amino acids (rat), with only about 80% identity to the human CB2 receptor (Fig. 4.4). The CB2 receptor is found in peripheral organs and has not been detected in the CNS thus far.

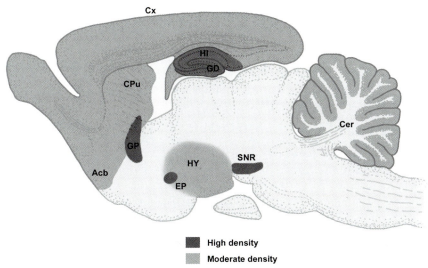

Fig. 4.6 Schematic overview of the distribution of CB1 receptors in the central nervous system. *Abbreviations*: Acb = *nucleus accumbens*; Cer = Cerebellum; CPu = caudate-putamen; Cx = Cortex; EP = entopeduncular nucleus; GD = *gyrus dentatus*; GP = *globus pallidus*; HI = hippocampus; HY = hypothalamus; SNR = *substantia nigra pars reticularis*.

The receptor is expressed in mononuclear macrophages and in mast cells, suggesting that anandamides may play a role in immunomodulation through binding to CB2 receptors.

The signal transduction mechanisms of the CB2 receptor are largely unknown; however some evidence suggests that the CB2 receptor is also G protein-coupled and inhibits adenylate cyclase activity.

Other binding sites

In vitro studies indicate that anandamides can inhibit gap junction-mediated signal transmission in striatal astrocytes. The inhibition is sensitive to pertussis toxin and to some extent to glytheric acid, but can neither be mimicked by cannabinoid agonists nor blocked by the CB1 receptor antagonist SR-14176A.

These data suggest different binding sites for anandamide in gap junction inhibition and in CB1- or CB2-mediated effects.

4.2.5
Biological Effects

The cannabinoids are involved in a variety of biological functions, such as antinociception, hypothermia, catalepsy, depression of spontaneous locomotor activity and perhaps mechanisms underlying learning and memory storage.

The injection of anandamide triggers effects which are analogous to those of THC. They differ in their kinetics, since they are somewhat faster than those induced by THC and also somewhat weaker.

The antinociceptive effects of THC can be blocked by opioidergic antagonists of the κ receptors. Curiously, this effect is not blocked by anandamide, indicating that THC and anandamide activate different pathways in nociception. An alternative explanation is that THC not only activates receptors of the CB1 and CB2 types, but also represents a ligand for additional receptors involved in nociception.

THC is also capable of modulating the metabolic and secretory activity of lymphocytes and macrophages. Among the known effects of THC on these cells is the suppression of interferon production, the reduction in secretion of tumor-necrosis factor alpha (TNF-α) and the enhanced production of interleukin1 (IL-1). In addition, THC seems to inhibit or stimulate the proliferation of T- or B-lymphocytes, according to the dose. At low concentration, it stimulates the proliferation of lymphocytes, while it decreases their proliferation at high concentration.

The anandamid system is also involved in mechanisms related to anxiety. Data from animal tests provide evidence of dose-dependent bidirectional modulation of anxiety by the cannabinoid system that appears to involve CB1 and non-CB1 cannabinoid receptors. CB1 receptor knockout mice show anxiogenic-like and depressive-like phenotypes in several behavioral paradigms. Along this line, pharmacological blockade of CB1 receptors induces anxiety in rats and inhibition of anandamide metabolism produces anxiolytic-like effects.

4.2.6
Neurological Disorders and Neurodegenerative Diseases

Cannabis has been historically used to relieve some of the symptoms associated with central nervous system disorders. Nowadays, there are anecdotal evidences for the use of cannabis in many patients suffering from multiple sclerosis or chronic pain. Considerable evidence indicates that the endocannabinoid system plays an essential role in pain regulation. For example, *in vivo* microdialysis experiments have shown that peripheral injections of the chemical irritant formalin are accompanied by increases in anandamide outflow within the periaqueductal gray (PAG), a brain region involved in pain-processing. Since activation of CB1 receptors in the PAG causes profound analgesia, it is speculated that inhibitors of anandamide inactivation may be useful in the pharmacotherapy of pain, particularly in instances where opiates are ineffective.

Recent research in animal models of multiple sclerosis has demonstrated the efficacy of cannabinoids in ameliorating symptoms such as spasticity and tremor. Although cannabinoid treatment of multiple sclerosis symptoms has been shown to be both well tolerated and effective in a number of subjective tests in several small-scale clinical trials, objective measures demonstrating the efficacy of cannabinoids are still lacking (Croxford and Miller 2004).

Further Reading

Aceto, M. D., Scates, S. M., Lowe, J., Martin, B. J. (1995): Cannabinoid precipitated withdrawal: introduction by a purported antagonist SR 141716A. *Eur. J. Pharmacol.* **282**: 1–3.

Bayewitch, M., Avidor-Reiss, T., Levy, R., Barg, J., Mechoulam, R., Vogel, Z. (1995): The peripheral cannabinoid receptor: adenylate cyclase inhibition and G protein-coupling. *FEBS Lett.* **375**: 143–147.

Belue, R. C., Howlett, A. C., Westlake, T. M., Hutchings, D. E. (1994): The ontogeny of cannabinoid receptors in the brain of postnatal and aging rats. *Neurotoxicol. Teratol.* **17**: 25–30.

Croxford, J. L., Miller, S. D. (2004): Towards cannabis and cannabinoid treatment of multiple sclerosis. *Drugs Today* **40**: 663–676.

Deutsch, D. G., Chin, S. A. (1993): Enzymatic synthesis and degeneration of anandamide, a cannabinoid receptor antagonist. *Biochem. Pharmacol.* **46**: 791–796.

Devane, W. A., Dysarz, F. A., Johnson, M. R., Melvin, L. S., Howlett, A. C. (1988): Determination and characterization of a cannabinoid receptor in the rat brain. *Mol. Pharmacol.* **34**: 605–613.

De Vries, T. J., Schoffelmeer, A. N. (2005): Cannabinoid CB1 receptors control conditioned drug seeking. *Trends Pharmacol. Sci.* **26**: 420–426.

Dewey, W. L. (1986): Cannabinoid pharmacology. *Pharmacol. Rev.* **38**: 151–178.

DiMarzo, V., Fontana, A., Cades, H., Schinelli, S., Cimino, G., Schwartz, J.-C., Piomelli, D. (1994): Formation and inactivation of endogenous cannabinoid anandamide in central neurons. *Nature* **372**: 686–691.

Elphick, M. R., Egertova, M. (2001): The neurobiology and evolution of cannabinoid signalling. *Philos. Trans. R. Soc. Lond. B* **356**: 381–408.

Fride, E. (2005): Endocannabinoids in the central nervous system: from neuronal networks to behavior. *Curr. Drug Target CNS Neurol. Disord.* **4**: 633–642.

Fride, E., Mechoulam, R. (1993): Pharmacological activity of the cannabinoid agonist anandamide, a brain constitute. *Eur. J. Pharmacol.* **231**: 313–314.

Giuffrida A, Beltramo M, Piomelli D. (2001): Mechanisms of endocannabinoid inactivation: biochemistry and pharmacology. *J. Pharmacol. Exp. Ther.* **298**: 7–14.

Grotenhermen F. (2005): Cannabinoids. *Curr. Drug Targets CNS Neurol. Disord.* **4**: 507–530.

Harris, L. S., Carchman, R. A., Martin, B. R. (1978): Evidence for the exsistence of specific cannabinoid binding sites. *Life Sci.* **22**: 1131–1138.

Herkenham, M., Lynn, A. B., Johnson, M. R., Melvin, L. S., deCosta, H. R., Rice, K. C. (1991): Characterization and localization of cannabinoid receptors in rat brain: a quantitative *in vitro* autoradiographic study. *J. Neurosci.* **11**: 563–583.

Howlett, A. C. (1995): Pharmacology of cannabinoid receptors. *Annu. Rev. Pharmacol. Toxicol.* **35**: 607–634.

Howlett, A. C., Johnson, M. R., Melvin, L. S., Milne, G. M. (1988): Nonclassical cannabinoid analgetics inhibit adenylate cyclase: development of a cannabinoid receptor model. *Mol. Pharmacol.* **33**: 297–302.

Hunter, S. A., Burstein, S. H. (1997): Receptor mediation in cannabinoid stimulated arachidonic acid mobilization and anandamide synthesis. *Life Sci.* **60**: 1563–1573.

Klein, T. W., Lane, B., Newton, C. C., Friedman, H. (2000): The cannabinoid system and cytokine network. *Proc. Soc. Exp. Biol. Med.* **225**: 1–8.

Kreitzer, A. C. (2005): Neurotransmission: emerging roles of endocannabinoids. *Curr. Biol.* **15**: R549–R551.

Lambert, D. M., Fowler, C. J. (2005): The endocannabinoid system: drug targets, lead compounds, and potential therapeutic applications. *J. Med. Chem.* **48**: 5059–5087.

Manzanares, J., Uriguen, L., Rubio, G., Palomo, T. (2004): Role of endocannabinoid system in mental diseases. *Neurotox. Res.* **6**: 213–224.

Martin, B. R. (1986): Cellular effects of cannabinoids. *Pharmacol. Rev.* **38**: 45–74.

Martin, B. R., Welch, S. P., Abood, M. (1994): Progress toward understanding the cannabinoid receptor and its second messenger systems. *Adv. Pharmacol.* **25**: 341–397.

Matsuda, L. A., Lolait, S. J., Brownstein, M. J., Young, A. C., Bonner, T. L. (1990): Structure of a cannabinoid receptor and functional expression of the cloned cDNA. *Nature* **346**: 561–564.

Mechoulam, R., Hanus, L., Martin, B. R. (1994): Search for endogenous ligands of the cannabinoid receptor. *Biochem. Pharmacol.* **48**: 1537–1544.

Safo, P. K., Regehr, W. G. (2005): Endocannabinoids control the induction of cerebellar LTD. *Neuron* **48**: 647–659.

Smith, P. B., Compton, D. R., Welch, S. P., Rajdan, R. K., Mechoulam, R., Martin, B. R. (1994): The pharmacological activity of anandamide, a putative endogenous cannabinoid, in mice. *J. Pharmacol. Exp. Ther.* **270**: 219–227.

Thakur, G. A., Duclos, R. I., Makriyannis, A. (2005): Natural cannabinoids: templates for drug discovery. *Life Sci.* **78**: 454–466.

Thomas, B. F., Wie, X., Martin, B. R. (1992): Characterization and autoradiographic localization of the cannabinoid binding site in rat brain using [^{3}H]11-OH-Δ^9-THC-DMH. *J. Pharmacol. Exp. Ther.* **263**: 1383–1390.

Ueda, N., Kurahashi, Y., Yamamoto, K., Yamamoto, S., Tokunaga, T. (1996): Enzymes for anandamide biosynthesis and metabolism. *J. Lipid Med. Cell Signal.* **14**: 57–61.

Valaverde, O. (2005): Participation of the cannabinoid system in the regulation of emotional-like behaviour. *Curr. Pharm. Des.* **11**: 3421–3429.

Viveros, M. P., Marco, E. M., File, S. E. (2005): Endocannabinoid system and stress and anxiety responses. *Pharmacol. Biochem. Behav.* **81**: 331–342.

Wilson, R. I., Nicoll R. A. (2001): Endogenous cannabinoids mediate retrograde signaling at hippocampal synapses. *Nature* **410**: 588–592.

Wilson, R. I., Nicoll, R. A. (2002) Endocannabionoid signaling in the brain. *Science* **269**: 678–682.

4.3
Angiotensin

4.3.1
General Aspects and History

In 1898, more then a century ago, Tigerstedt and Bergman discovered that the injection of renal extracts exerts profound effects upon blood pressure. Accordingly, this extract was called renin. Several years later, it was found that the active substrate, renin, is a specific protease which represents the key enzyme for angiotensin production. Two groups working in the 1930s–1940s, Braun-Menendez and coworkers as well as Page and Helmer, isolated a hypertensive substance from the blood, named angiotensin (ANG), which is produced through the catalytic action of renin.

The renin–angiotensin system is well known for its regulation of blood pressure and fluid homeostasis. In recent decades, it has become apparent that local autocrine or paracrine renin–angiotensin systems may exist in different tissues. All components of the renin–angiotensin system can be found, for example, in the gonads, placenta, pancreas, adipose tissue, heart, retina and pituitary.

Bickerton and Buckley (1961) found that intraventricular injection of angiotensin II induces a centrally mediated pressor response. Since angiotensin II does not readily cross the blood–brain barrier, it is unlikely that brain angiotensin II originates in the peripheral renin–angiotensin system. Based on this experiment and on subsequent detailed analysis of the distribution and functions of the renin–angiotensin system, it is now firmly established that a complete renin–angiotensin system exists in the brain, which is distinctly separate from the peripheral renin–angiotensin system and comprises all necessary precursors and enzymes required for the formation and metabolism of the biologically active forms of angiotensin.

Several traditional roles have been attributed to the brain renin–angiotensin system, including the regulation of blood pressure and the regulation of salt and water homeostasis. However, central actions of angiotensins are not exclusively associated with these traditional roles. Thus, a variety of studies have shown that central angiotensins are also involved in sexual behavior, stress, learning and memory.

4.3.2
Localization Within the Central Nervous System

The biologically inactive form angiotensin I is intracellularly converted to the active form, angiotensin II. Angiotensin II can be metabolized to a shorter fragment, angiotensin IV, which is also biologically active. In addition, angiotensin II can be metabolized to a further biologically active fragment, angiotensin II (1–7).

Immunohistochemical and autoradiographic studies have been used to determine the distribution of angiotensin II in the central nervous system.

Large numbers of neurons expressing angiotensin II have been identified in the circumventricular organs (the subfornical organ, the organum vasculosum of the terminal lamina and the area postrema). In addition, cells expressing angiotensin II have been encountered, including the hypothalamus, the thalamus and the amygdala. A schematic overview of the distribution of angiotensin II-expressing cells of the central nervous system is shown in Fig. 4.7.

A further biologically active form of the angiotensins is a fragment of angiotensin II, termed angiotensin IV. Although specific binding sides for angiotensin IV (namely AT4 receptors) exist within the brain, data concerning the distribution of angiotensin IV within the central nervous system is not yet available.

Angiotensin II (1–7) has been identified in the amygdala, the hypothalamus, the bed nucleus of the stria terminalis, the substantia innominata, the median eminence and in the medulla oblongata.

4.3.3
Biosynthesis and Degradation

The precursor of angiotensin I is angiotensinogen, which is synthesized in both the liver and in the central nervous system. Renin (EC 3.4.23.15) is the enzyme which catalyzes the formation of angiotensin I from the precursor angiotensinogen. Renin, which is an endopeptidase aspartyl-protease, clips off a fragment of

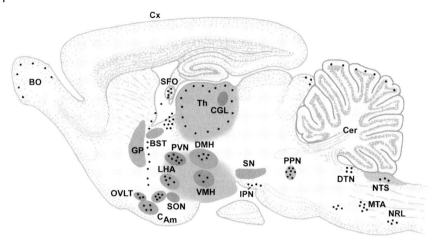

Fig. 4.7 Schematic overview of the distribution of angiotensinergic cells in the brain. *Abbreviations*: BO = *bulbus olfactorius*; BST = bed nucleus of the stria terminalis; C$_{Am}$ = central nucleus of the amygdala; Cer = cerebellum; CGL = corpus geniculatum, lateral; Cx = Cortex; DMH = dorso-medial nucleus of the hypothalamus; DTN = dorsal tegmental nucleus; GP = *globus pallidus*; IPN = interpeduncular nucleus; LHA = lateral ypothalamic area; MTA = magnocellular tegmental area; NRL = lateral nucleus reticularis; NTS = nucleus of the *tractus solitarius*; OVLT = *organum vasculosum* of the *lamina terminalis*; PPN = pedunculopontine nucleus; PVN = paraventricular nucleus of the hypothalamus; SFO = subfornical organ; SN = *substantia nigra*; SON = supraoptic nucleus; Th = thalamus; VMH = ventromedial nucleus of the hypothalamus.

four amino acids from the amino-terminal region of angiotensinogen and thereby catalyzes the synthesis of angiotensin I.

By the activity of the angiotensin-converting enzyme (ACE, EC 3.4.15.1), a dipeptidyl carboxypeptidase, angiotensin I is hydrolyzed at its carboxyl terminus, leading to the generation of the octapeptide angiotensin II. Angiotensin II seem to represent the first biologically active form of the angiotensins. Angiotensin II is not only generated in the brain via this classic pathway, using renin and angiotensin-converting enzyme, it can also be generated directly from angiotensinogen by cathepsin G or tonin. In conformity with tachykinins and opioid peptides, a limited hydrolysis of angiotensin II results in the formation of fragments with retained but modified biological activity. Two different fragments have been identified so far which possess biological activity: angiotensin IV and angiotensin (1–7).

Aminopeptidase A (EC 3.4.11.7) converts angiotensin II to the 2–8 fragment of angiotensin II, named angiotensin III (Fig. 4.8A). This peptide is further cleaved by aminopeptidase N (EC 3.4.11.2) to form the 3–8 hexapeptide fragment of angiotensin II, with the amino acid sequence Val-Tyr-Ile-His-Pro-Phe. This biologically active 3–8 fragment of angiotensin II is called angiotensin IV. In addition, angiotensin IV may also be formed by aminopeptidases acting on angiotensin I prior to conversion to angiotensin II by ACE or may be generated by the activity of aminopeptidase A and N directly from angiotensin II.

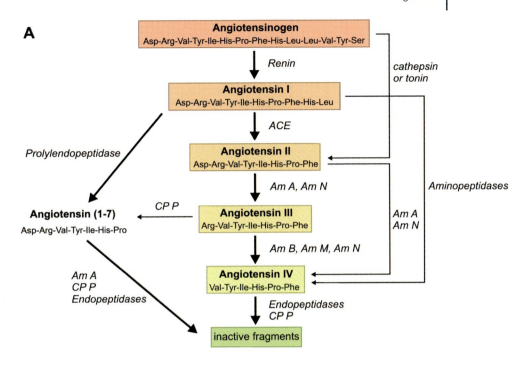

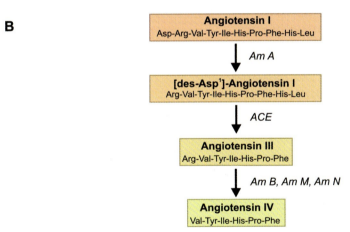

Fig. 4.8 Biosynthesis and structures of angiotensins. (A) Biosynthesis of the bioactive angiotensins angiotensin II, angiotensin (1–7) and angiotensin IV. (B) Alternative pathway of the biosynthesis of angiotensin IV from angiotensin I, involving [des-Asp1]-angiotensin I. *Abbreviations*: ACE = angiotensin converting enzyme; Am = aminopeptidase; CP P = carboxypeptidase P.

Angiotensin (1–7) is a primary product of angiotensin I (by the activity of specific endopeptidases, like EC 3.4.24.11, EC 3.4.24.15 and EC 3.4.21.26), although it may also be cleaved from angiotensin II by a post-proline carboxypeptidase.

Moreover, a further pathway exists which involves the nonapeptide [des-Asp1]-angiotensin I (Fig. 4.8 B). This angiotensin is converted directly to angiotensin III via ACE. At least in the rat hypothalamus, angiotensin I is mainly degraded to [des-Asp1]-angiotensin I instead of angiotensin II.

Angiotensin II can be metabolized by aminopeptidase A to angiotensin III and further converted to angiotensin IV by aminopeptidase B. Additional endopeptidases and carboxypeptidases are involved in downstream degradation of angiotensin IV, producing biologically inactive fragments. In addition, angiotensin (1–7) can be degraded to angiotensin (2–7) by aminopeptidase A activity at the Asp–Arg bond.

4.3.4
Receptors and Signal Transduction

Angiotensin II binds to two different receptors, which are referred to as AT1 and AT2 receptors. The AT1 receptor accounts for most of the peripheral and cerebral effects of angiotensin II, including cardiovascular regulation, hormone secretion and fluid balancing. Blockade of the AT1 receptor inhibits angiotensin II-induced changes in blood pressure and prevents the induction of drinking behavior. The AT1 receptor consists of a monomer of 359 amino acids. The receptor possesses seven transmembrane domains and is coupled to G proteins. Two subtypes (AT1a and AT1b) have been characterized.

Binding of angiotensin II to the AT1 receptor results in an activation of G proteins and involves the activation of tyrosine kinases and phospholipase C (PLCγ), which cleaves phosphatodylinositol-4,5-bisphosphate to diacylglycerol (DAG) and inositol-1,4,5-trisphospate (IP3). The increase in IP3 triggers an increase in intracellular calcium. DAG activates proteinkinase C which, in turn, phosphorylates a variety of intracellular proteins (Fig. 4.9).

The AT2 receptor is also capable of binding angiotensin II. The functional significance of this receptor is not fully understood, but it might be involved in angiogenesis and the regulation of cerebral blood flow. AT2 receptors are highly expressed in neonatal tissue. AT2 levels decrease sharply after birth, but persist into adulthood in the brain and in some other tissues in rats. Additional findings have led to the proposal that the AT2 receptor might be involved in differentiation, development and/or apoptosis.

Recently it was demonstrated that, in neurons cultured from newborn hypothalamus and brain stem of rats, angiotensin II contributes to apoptosis via AT2 receptors and activation of a serine/threonine phosphatase.

As with the AT1 receptor, the AT2 receptor shows a monomeric structure. The AT2 receptor is composed of 363 amino acids and its sequence displays a homology of about 34% with the amino acid sequence of the AT1 receptor.

The AT2 receptor seems to be coupled to G proteins and exhibits the same topology.

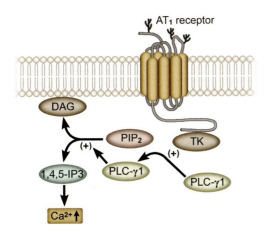

Fig. 4.9 Schematic drawing of an AT1 receptor and the intracellular signal cascade which is activated after the binding of angiotensin II to the receptor.

Binding sites for angiotensin IV are also widely distributed in peripheral tissues as well as in the brain. Specific binding sites for angiotensin IV are found in both neurons and astrocytes. These binding sites are referred to as AT4 receptors.

By protein purification and peptide sequencing, the AT4 receptor was identified in 2001 to be identical with insulin-regulated aminopeptidase and, thus, the known angiotensin AT4 ligands are potent, competitive inhibitors of insulin-regulated aminopeptidase. The signaling cascade, activated after the binding of angiotensin IV to the AT4 receptor, is largely unknown. AT4 activation has been shown to stimulate expression and release of plasminogen activator inhibitor-1 (PAI-1), an effect which can be blocked by AT4 antagonists. In addition to angiotensin IV, hemorphins, a class of endogenous peptides obtained by hydro-

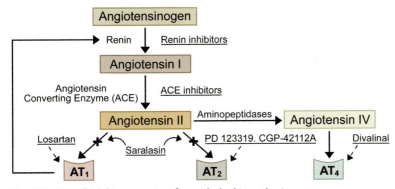

Fig. 4.10 Several inhibitors can interfere with the biosynthesis of angiotensin. In addition, several antagonists are available to block selectively one receptor type (losartan, PD 123319, CGP-42112A or divalinal) or both angiotensin II receptors (saralasin).

Table 4.1 Localization of AT1, AT2 and AT4 receptors in the brain, as determined by immunohistochemistry and autoradiography. *Abbreviations*: NTS: nucleus of the *tractus solitarius*; OVLT: *organum vasculosum* of the *laminae terminalis*.

Area/receptor		AT1	AT2	AT4
Cortex	Frontal and parietal cortex			+
	Piriform cortex	+		+
	Entorhinal cortex	+		+
	Insular cortex			+
	Cingulate cortex	+	+	
Hippocampus	CA1–CA3; dentate gyrus	+	+	+
	Subiculum	+		
Amygdala	Lateral and basomedial nucleus	+		+
	Basolateral nucleus	+	+	+
	Central and medial nucleus	+	+	+
Globus pallidus				+
Caudate-putamen		+		+
Subfornical organ, OVLT		+		
Lateral septal area			+	+
Medial septal area		+		+
Thalamus	Habenula		+	+
	Anterior thalamus	+	+	+
	Mediodorsal, ventroposterior and centromedial thalamus		+	+
	Ventrolateral thalamus			+
	Lateral geniculate nucleus			+
	Medial geniculate nucleus		+	
	Reticular nucleus		+	
	Zona incerta			+
Hypothalamus	Median eminence	+		
	Paraventricular nucleus	+	+	+
	Periventricular nucleus	+		
	Supraoptic nucleus	+	+	+
	Suprachiasmatic nucleus	+		+
	Preoptic nucleus	+		+
	Dorsomedial hypothalamus	+		+
	Ventromedial hypothalamus			+
	Arcuate nucleus	+		+
Mesencephalon	Superior colliculus	+	+	+
	Inferior colliculus		+	+
	Periaqueductal gray	+		+
	Ventral tegmental area		+	+
	Substantia nigra	+		+
	Locus coeruleus	+	+	+
	Pontine reticular nucleus	+		
Cerebellum			+	+
Area postrema		+		
NTS		+		
Inferior olive		+	+	+

lysis of the beta chain of hemoglobin, can also interact with the AT4 receptor. Therefore, it might be possible that hemorphins acts as endogenous competitors of angiotensin IV.

Pharmacologically active antagonists directed to the three angiotensin receptors (AT1, AT2 and AT4) have been developed (Fig. 4.10).

In the brain, the presence of the three receptor types (AT1, AT2 and AT4, differing considerably in their location) has been described (Table 4.1).

Several studies have shown that angiotensin (1–7) does not act through angiotensin II receptors (AT1 and AT2); and thus it is speculated that a specific receptor for angiotensin (1–7) might exist. A specific receptor for angiotensin (1–7) has not been identified or cloned so far. However, there are reports that the *Mas* proto-oncogene product may encode an angiotensin (1–7) receptor.

The existence of an AT3 receptor, as described in cultured neuroblastoma, is still questionable since neither specific agonists nor specific functions have been attributed to this putative receptor.

4.3.5
Biological Effects

The brain renin–angiotensin system is independent of the peripheral renin–angiotensin system. However, circulating angiotensin can cross the blood–brain barrier at specific sites and interact with angiotensin receptors of the brain. The sites of entry are the so-called circumventricular organs: the area postrema, the subfornical organ and the organum vasculosum of the terminal lamina (OVLT). These hemoneuronal regions exhibit reduced barrier properties because of the presence of fenestrated capillaries, which allow a passage of small molecules from the circulating blood into the brain parenchyma or vice versa.

Angiotensins are involved in water and electrolyte balance. Intraventricular injection of angiotensin, for example, induces drinking behavior as well as salt appetite. A further prominent property of angiotensin is its vasoactive effect on cerebral blood vessels. Application of angiotensin II results in vasoconstriction of cerebral arterioles, while angiotensin IV results in vasodilatation.

In the brain, angiotensin acts through its receptors on the regulation of sexual hormones and thereby influences sexual behavior. Intraventricular application of angiotensin increases the release of vasopressin and leads to an elevation of circulating LH.

Angiotensin interacts with a number of neurotransmitters and neuromodulators. Angiotensin II enhances the release, but not the production, of norepinephrine. The effects of angiotensin on serotonin are dose-dependent: at high doses angiotensin enhances serotonin synthesis and at low doses it inhibits synthesis. In addition, it has been shown that angiotensin is involved in the regulation of corticoid hormones, since it modulates the release of ACTH as well as the level of glucocorticoids and enhances the release of corticotropin-releasing hormone.

The manipulation of angiotensin levels in the brain appears to influence cognitive processing, especially the acquisition and recall of a newly learned task.

There is also some evidence that angiotensin-converting enzyme inhibitors (ACE-inhibitors) have positive effects upon memory and cognition by reducing the level of active angiotensin.

These data are supported by a number of experiments. For example, intraventricular administration of angiotensin II reduces the learning capacity in rodents. In addition, animals which received ACE-inhibitors prior to a learning task showed a better learning behavior as controls, since they need less time to solve newly learned tasks. Almost identical results have been achieved by the application of an AT1 receptor antagonist. These findings corroborate the idea that via AT1 receptors angiotensin II influences learning and memory in a negative way. Supportive data come from electrophysiological experiments which indicate that activation of AT1 receptors can inhibit long-term potentiation, a standard model of synaptic plasticity suggested as being involved in learning and memory storage.

Angiotensin IV is also a biologically active peptide, which derives from angiotensinogen. It is speculated that angiotensin IV binds to a specific receptor, which has been named the AT4 receptor.

The effects of angiotensin IV are in some respects the opposite to those of angiotensin II. While angiotensin II is a potent vasoconstrictor, angiotensin IV is a vasodilator. Opposing effects have also been described in the amygdala, where angiotensin II enhances field potentials, while angiotensin IV reduces them. Some data indicate that the role of angiotensin IV is different from that of angiotensin II in learning and memory storage, since intraventricularly injected angiotensin IV seems to enhance learning in animal experiments. In addition, angiotensin IV binding sites have been identified in brain regions (which are known to be involved in cognitive functions) where angiotensin II receptors are abundant.

Angiotensin II (1–7) is a biologically active fragment of angiotensin II. It is unclear whether it binds to a specific receptor. Angiotensin (1–7) has an blood pressure-lowering effect, acting as a vasodilatator and an antihypertensive substance. Angiotensin (1–7) binding sites have been demonstrated in the central nervous system (amygdala, hypothalamus, medulla oblongata); however, little is known about the action of angiotensin (1–7) in the nervous system. It has been demonstrated that angiotensin (1–7) modulates the release of vasopressin and increased substance P levels in the central nervous system. Furthermore, angiotensin (1–7) could mediate antihypertensive effects by stimulating the synthesis and release of prostaglandines and nitric oxide.

4.3.6
Neurological Disorders and Neurodegenerative Diseases

Cerebral blood flow and ischemia
Angiotensin II regulates cerebral blood flow by stimulating cerebral vasoconstriction via AT1 receptors. Treatment with ACE inhibitors, like captopril, appears to improve cerebral blood flow and protects against damage induced by cerebral ischemia in spontaneously hypertensive rats. With regard to the regulation of cerebral blood flow, it appears that angiotensin II and IV can induce op-

posite effects. Thus, angiotensin II induces reductions in cerebral blood flow, while angiotensin IV increases cerebral blood flow. Thus, it is not the biosynthesis of biological active forms of angiotensins which should be blocked, but the signaling via AT1 receptors. Indeed, it has been shown that only an effective blockade of brain AT1 receptors has neuroprotective effects in ischemic neuronal tissue and improves the neurological outcome of focal brain ischemia.

The requirements for an effective AT1 antagonist include sufficient penetration to cross the blood–brain barrier when administered orally, effective and selective inhibition of AT1 receptors in the brain and long-lasting receptor blockade. One of the AT1 receptor antagonists which meet these criteria is candesartan, which is highly effective in crossing the blood–brain barrier. Indeed, in rats, chronic treatment with the AT1 receptor antagonist candesartan markedly decreased cerebral ischemia as a consequence of MCAO.

Angiotensins and neurodegenerative disorders

Recent reports support the hypothesis of tight links between vascular and neurodegenerative diseases, e.g. increased risk of Alzheimer's dementia for hypertensive subjects and decreased risk of dementia for elderly treated with hypotensive drugs. As a major player of vascular homeostasis, the renin–angiotensin system constitutes an interesting source of candidate genes which might be involved in neurodegenerative disorders.

Despite that, no association between Alzheimer's disease and polymorphisms in the angiotensinogen and renin genes have been found. Therefore, investigators focussed on a possible association between polymorphisms in the ACE gene and Alzheimer's disease, but controversial results were obtained. Thus, no association between polymorphism in the ACE gene was found in a Russian population and in a population from North America or Italy, while associations were found in a Japanese population.

Concerning Parkinson's disease and a possible linkage with polymorphisms in genes of the renin–angiotensin system, the results are as controversial as in the case of Alzheimer's disease. Thus, an association between genetic polymorphism of the ACE gene and Parkinson's disease has been found in Taiwan but not in Australia. However, decreased angiotensin-binding sites in the substantia nigra of postmortem brains from patients with Parkinson's disease have been found.

Additional research efforts are required to resolve the differences in the findings concerning the association of ACE gene polymorphisms and neurodegenerative disorders, and to clarify the role of the brain renin–angiotensin system in Alzheimer's or Parkinson's disease.

Further Reading

Albiston, A. L., McDowall, S. G., Matsacos, D., Sim, P., Clune, E., Mustafa, T., Lee, J., Mendelsohn, F. A., Simpson, R. J., Connolly, L. M., Chai, S. Y. (2001): Evidence that the angiotensin IV (AT4) receptor is the enzyme insulin regulated aminopeptidase. *J. Biol. Chem.* **276**: 48623–48626.

Allen, A. M., MacGregor, D. P., McKinley, M. J., Mendelsohn, F. A. O. (1999): Angiotensin II receptors in the human brain. *Reg. Peptides* **79**: 1–7.

Amouyel, P., Richard, F., Berr, C., David-Fromentin, I., Helbecque, N. (2000): The renin angiotensin system and Alzheimer's disease. *Ann. N.Y. Acad. Sci.* **903**: 437–441.

Bickerton, R. K., Buckley, J. P. (1961): Evidence for a central mechanism in angiotensin induced hypertension. *Proc. Soc. Exp. Biol. Med.* **106**: 834–839.

Braszko, J. J., Wisniewski, K., Kupryszewski, G., Witczuk, B. (1987): Psychotropic effects of angiotensin II and III in rats: locomotor and exploratory vs cognitive behaviour. *Behav. Brain Res.* **25**: 195–203.

Braszko, J. J., Kupryszewski, G., Witczuk, B., Wisniewski, K. (1988): Angiotensin II-(3-8)-hexapeptide affects motor activity, performance of passive avoidance and a conditioned avoidance response in rats. *Neuroscience* **27**: 777–783.

Chappell, M. C., Brosnihan, K. B., Diz, D. I., Ferrario, C. M. (1989): Identification of angiotensin-(1-7) in rat brain. Evidence for differential processing of angiotensin peptides. *J. Biol. Chem.* **264**: 16518–16523.

De Gasparo, M., Catt, K. J., Inagami, T., Wright, J. W., Unger, T. (2000) International union of pharmacology. XXIII. The angiotensin II receptors. *Pharmacol. Rev.* **52**: 415–472.

Ganten, D., Minnich, J. L., Granger, P., Hayduk, K., Brecht, H. M., Barbeau, A., Boucher, R., Genest, J. (1971): Angiotensin-forming enzyme in brain tissue. *Science* **173**: 64–65.

Gard, P. R. (2002): The role of angiotensin II in cognition and behaviour. *Eur. J. Pharmacol.* **438**: 1–14.

Gehlert, D. R., Speth, R. C., Wamsley, J. K. (1986): Quantitative autoradiography of angiotensin II receptors in the spontaneously hypertensive rat brain. *Peptides* **7**: 1021–1027.

Groth, W., Blume, A., Gohlke, P., Unger, T., Culman, J. (2003): Chronic pretreatment with candesartan improves recovery from focal cerebral ischaemia in rats. *J. Hypertens.* **21**: 2175–2182.

Jöhren, O., Inagami, T., Saavedra, J. M. (1995): AT1A, AT1B and AT2 angiotensin receptor subtype gene expression in rat brain. *NeuroReport* **6**: 2549–2552.

Lin, J. J., Yueh, K. C., Chang, D. C., Lin, S. Z. (2002): Association between genetic polymorphism of angiotensin-converting enzyme gene and Parkinson's disease. *J. Neurol. Sci.* **199**: 25–29.

Mendelsohn, F. A. O. (1985): Localization and properties of angiotensin receptors. *J. Hypertens.* **3**: 307–316.

Phillips, M. I. (1987): Functions of brain angiotensin. *Annu. Rev. Physiol.* **49**: 413–435.

Roberts, K. A., Krebs, L. T., Kramar, E. A., Shaffer, M. J., Harding, J. W., Wright, J. W. (1995): Autoradiographic identification of brain angiotensin IV binding sites and differential c-Fos expression following intracerebroventricular injection of angiotensin II and IV in rats. *Brain Res.* **682**: 13–21.

Rowe, B. P., Saylor, D. L., Speth, R. C., Absher, D. R. (1995): Angiotensin (1–7) binding at angiotensin II receptors in the rat brain. *Regul. Peptides* **56**: 139–146.

Saavedra, J. M. (1992): Brain and pituitary angiotensin. *Endocrine Rev.* **13**: 329–380.

von Bohlen und Halbach, O. (2003): Angiotensin IV in the central nervous system. *Cell Tiss. Res.* **311**: 1–9.

von Bohlen und Halbach, O. (2005): The renin–angiotensin system in the mammalian central nervous system. *Curr. Protein Peptide Sci.* **6**:355–371.

von Bohlen und Halbach, O., Albrecht, D. (1998): Angiotensin II inhibits long-term potentiation in the lateral nucleus of the amygdala through AT1-receptors. *Peptides* **19**: 1031–1036.

Wayner, M. J., Armstrong, D. L., Polan Curtain, J. L., Denny, J. B. (1993): Role of angiotensin II and AT1 receptors in hippocampal LTP. *Pharmacol. Biochem. Behav.* **45**: 455–464.

Wright, J. W., Harding, J. W. (1994): Brain angiotensin receptor subtypes in the control of physiological and behavioral responses. *Neurosci. Biobehav. Rev.* **18**: 21–53.

Wright, J. W., Harding, J. W. (1995): Brain angiotensin receptor subtypes AT1, AT2, and AT4 and their functions. *Regul. Peptides* **59**: 269–295.

Wright, J. W., Harding, J. W. (2004): The brain angiotensin system and extracellular matrix molecules in neural plasticity, learning, and memory. *Prog. Neurobiol.* **72**: 263–293.

4.4
Atrial Natriuretic Factor

4.4.1
General Aspects and History

The natriuretic peptides are widely distributed throughout the whole organism and are present to some extent in the central nervous system.

In the 1950s, Kirsch (1956) observed that the atrium contains granules. In 1980, De Bold and coworkers (1981) found that these granules seemed to be secretory. Injection of extracts into rats revealed natriuretic as well as diuretic properties. In 1984, Kangawa and Matsuo purified a peptide of 28 amino acids. They labeled this peptide atrial naturetic factor (ANF) because of its abundance in cardiomyocytes of the right atrium and because of the naturetic effect provoked by the peptide. Frequently used synonyms include arterio-natriuretic peptide (ANP), atriopeptin, cardiodilatin and atrial natriuretic hormone (ANH).

Within the central nervous system, three different forms of atrial natriuretic factor are distinguished: the atrial natriuretic factor (ANF) itself, the brain natriuretic peptide (BNP), which represents a homolog of 26–32 amino acids, and the C-type natriuretic peptide (CNP). CNP is composed of 22 amino acids.

4.4.2
Localization Within the Central Nervous System

The natriuretic peptide is found preferentially in brain tissues, but it is also present in some peripheral organs. Most immunohistochemical studies do not distinguish between ANF and BNP, because of a lack of differentiating antibodies, however it is likely that ANF and BNP factors are differentially distributed.

CNP is more widely distributed than either ANF or BNP. The neuropeptides ANF and BNP are located predominantly in brain areas, which are involved in cardiovascular regulation and in the control of natriuresis and diuresis. Thus, it comes as no surprise that high densities of ANF-containing somata are found in the preoptic paraventricular and arcuate nuclei of the hypothalamus as well as in the central nucleus of the amygdala and the nucleus of the tractus solitarius (Fig. 4.11). ANF-immunoreactive fibers and terminals coexist in the same brain nuclei. Further immunoreactivity is found in the olfactory bulb, the median eminence, the locus coeruleus, the area postrema, the subfornical organ and the dorsal motor nucleus of the vagus nerve.

In mouse CNS, gene transcripts for CNP were present at the onset of neurogenesis, embryonic day 10.5 (E10.5), primarily in the dorsal part of the ventricular zone throughout the hindbrain and spinal cord. On E14.5, new CNP signals were observed in the ventrolateral spinal cord. In contrast, ANP and BNP gene transcripts were not detected in embryonic brain of mice.

Immunohistochemical studies indicate that CNP is widely distributed in the central nervous system, with a pronounced expression in the cells of various hy-

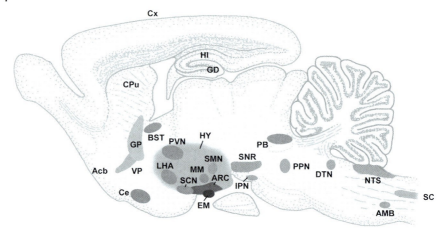

Fig. 4.11 Schematic representation of the distribution of ANF-producing cells in the central nervous system. *Abbreviations*: Acb = *nucleus accumbens*; AMB = *nucleus ambiguus*; ARC = arcuate nucleus; BST = bed nucleus of the *stria terminalis*; Ce = central nucleus of the amygdala; CPu = caudate-putamen; Cx = Cortex; DTN = dorsal tegmental nucleus; EM = *eminentia mediana*; GD = gyrus *dentatus*; GP = *globus pallidus*; HI = hippocampus; HY = hypothalamus; IPN = interpeduncular nucleus; LHA = lateral hypothalamic area; MM = medial mammillary nucleus; NTS = nucleus of the tractus solitarius; PB = parabrachial nucleus; PPN = pedunculopontine nucleus; PVN = paraventricular nucleus of the hypothalamus; SC = spinal cord; SCN = suprachiasmatic nucleus; SMN = supramamillary nucleus; SNR = Substantia nigra pars reticularis; VP = ventral pallidum.

pothalamic and thalamic nuclei. Furthermore, the existence of CNP-expressing cells has been demonstrated in different nuclei of the brain stem. CNP has also been detected in small quantities in the spinal cord and in the pituitary. As compared to the wealth of knowledge concerning the expression of ANF and BNF, a thorough analysis of CNP expression has yet to be performed.

4.4.3
Biosynthesis and Degradation

Natriuretic peptides (ANF, BNP and CNP) comprise a family of structurally related peptides which are derived from three different genes and which share a 17-amino-acid internal ring. All three peptides have in common the fact that they are processed from precursors of high molecular weight. The gene which encodes for the atrial natriuretic factor consists of three exons and two introns and is located on human chromosome 1.

Transcription gives rise to the primary RNA of the precursor prepro-ANF with a putative length of 152 amino acids. Within the endoplasmic reticulum the prepro-ANF is processed to pro-ANF, a precursor of 126 amino acids.

The pro-ANF is stored in vesicular form. Upon a stimulus which elicits the secretion of ANF, the pro-ANF is fragmented by proteolytic cleavage. The carboxy-terminal fragment of pro-ANF, which consists of 28 amino acids, represents the biologically active native form of ANF. Through the cleavage process, three additional peptides are generated: the long-acting natriuretic peptide (LANP) and two additional fragments, all with biological activity. These two fragments are known as "vessel dilator" and as "kaliuretic peptide".

The half-life of ANF in the blood is short (about 5 min), whereas LANP survives for more than 24 h in the circulation. The duration of ANF activity is limited insofar as it is rapidly inactivated by hydrolases and different peptidases. Enkephalinase (EC 24.11) is a prominent peptidase involved in the degradation of ANF and is abundantly expressed in several tissues, including the central nervous system. Moreover, it is known that neutral endopeptidases rapidly degrade ANF, without having an influence upon BNP metabolism.

As with the ANF-coding gene, the BNP gene is located on human chromosome 1. This gene shows a structural similarity to the ANF gene and gives rise to a BNP precursor protein. In contrast to ANF, different species-specific isoforms of BNF are known. Human BNF, for example, consists of 36 amino acids, while rat BNF is composed of 45 amino acids.

The biosynthesis of CNP is somewhat different from that of the ANF and BNP. The prepro-CNP, a peptide of 126 amino acids, is cleaved to pro-CNP, a peptide of 103 amino acids. This precursor is stored in vesicular form. The precursor is further processed to a fragment of 22 amino acids (CNP-22) and an additional fragment of 53 amino acids (CNP-53).

In vitro analysis has shown that cytokines – interleukin-1 and the transforming growth factor (TGF-β) – influence the release of CNP and that CNP in turn is able to modulate the effects of the cytokines.

The degradation mechanisms of CNP are largely unknown.

4.4.4
Receptors and Signal Transduction

In the central nervous system, the existence of specific binding sites for the natriuretic peptides has been shown. The receptors are confined to certain brain areas, predominantly the circumventricular organs, i.e. the area postrema, the organum vasculosum of the lamina terminalis, the subfornical organ and the ependyma of the ventricles. However, a moderate density of these receptors is known to occur in the olfactory bulb, the habenula and some nuclei of the hypothalamus, in the hippocampus and in various thalamic nuclei and the cerebellum.

The receptors are not located exclusively on neurons, but are also found on asterocytes. The receptors family consists of the following three members: NPRa (or R1), NPRb and NPRc (or R2).

ANF preferentially binds to the receptors NPRa and NPRc, whereas BNP binds to NPRc receptors and CNP to the receptors NPRb and NPRc.

A common feature of the NPRa and NPRb receptors is that they couple to guanylate cyclase. Both metabotropic receptor types reveal identical topologies consisting of an extracellular domain, a transmembrane-spanning domain and two intracellular domains. One of the intracellular domains possesses a protein kinase-binding site, while the other activates guanylate cyclase and thereby could increase the formation of cGMP.

Agonist binding results in the activation of guanylate cyclase, which in turn elevates the level of intracellular cyclic guanosine monophosphate (cGMP).

NPRa and NPRb, although identical in their structure, reveal different affinities for their specific ligands. The third member of the natriuretic peptide receptor family is the NPRc receptor. The NPRc subtype is completely different from the NPRa and NPRb subtypes, both in respect of its structure and in its signal transduction.

The NPRc receptor is composed of one extracellular domain, one transmembrane-spanning domain and a single intracellular domain. Unlike the NPRa and NPRb subtypes, the NPRc receptor is not linked to guanylate cyclase, but is internalized as a ligand–receptor complex which dissociates intracellularly.

The NPRc receptor reveals nearly equal affinities for ANF, BNP and CNP. The internalization of the NPRc receptor efficiently limits the steady-state concentration of the extracellular natriuretic peptides. For this reason, NPRc belongs to the class of so-called "clearance receptors", i.e. receptors which regulate the extracellular concentrations of their ligands through receptor-mediated uptake.

However, some data indicate that the NPRc receptor, at least in some brain areas, is also directly involved in modulating the effects of natiuretic peptides.

In the CNS of man and rodents, NPRa is found mainly in cortex and hippocampus, whereas NPRb is present in the amygdala and several brainstem regulatory sites. NPRc is found widely within the CNS, e.g. in the neocortex, the limbic cortex, the hippocampal area and the amygdala.

4.4.5
Biological Effects

The central effects of the natriuretic peptides are comparable to those in the periphery, i.e. augmentation of natriuresis and diuresis and modulation of salt and water homeostasis. Natriuretic peptides are specifically involved in the regulation of the hypothalamo–pituitary–adrenocortical (HPA) system: in man and rodents ANF inhibits the HPA system at all regulatory levels, while CNP stimulates the release of cortisol.

Angiotensin is the antagonistic counterpart of the natriuretic peptides with respect to central water and salt homeostasis. Intraventricular application of angiotensin, for example, induces thirst, whereas intraventricular application of natriuretic peptides suppresses thirst.

The central cardiovascular effects of the natiuretic peptides depend on their control of vasopressin secretion from the paraventricular and supraoptic nucleus of the hypothalamus.

In the periphery, angiotensin II and the natriuretic peptides also exhibit opposite effects. Most of the peripheral effects are related to fluid homeostasis of the organism. ANF enhances natriuresis and diuresis by regulating the filtration rate in the glomeruli and by increasing renal perfusion. In addition, ANF blocks the vasopressin-dependent reabsorption of Na^+, resulting in an enhanced diuresis and natriuresis.

Besides its direct vasodilator action, ANF exerts indirect effects on the cardiovascular system by inhibiting the secretion of aldosteron and renin.

Since BNP shows co-secretion with ANF, it is believed that BNP cooperates with ANF in the regulation of blood pressure and in salt homeostasis. Consistent with this idea is the finding that hypertensive patients show an increase of both peptides under a low-Na^+ diet.

In contrast to BNP and ANF, the C-type natriuretic peptide (CNP) has minor effects upon natriuresis and diuresis. CNP seems to possess an anti-proliferative action on smooth muscle cells *in vitro* and induces vasodilatation as well as relaxation of smooth muscle cells in the bronchi.

The natriuretic peptides are also thought to play a role in higher brain functions. In rodents, ANF was found to reduce anxiety levels, whereas CNP induced the opposite effect. In patients with panic disorder, basal ANF plasma levels are lower in comparison to healthy volunteers, but ANF secretion is faster and more pronounced during experimentally induced panic attacks. Interestingly, panic anxiety and concomitant ACTH and cortisol secretion elicited by stimulation with cholecystokinin tetrapeptide were also attenuated by ANF infusions in patients as well as in healthy volunteers, indicating a role for ANF in anxiety.

The central corticotropin-releasing factor (CRF)-ergic system is known to play a critical role in anxiety and other behavioral stress responses. It has been shown that ANF, but also BNP and CNP, exert anxiolytic-like effects in behavioral studies.

4.4.6
Neurological Disorders and Neurodegenerative Diseases

Under certain pathological conditions, such as sepsis, trauma or major surgery, systemic hypotension and an intrinsic myocardial dysfunctions occur. Natriuretic peptides have emerged as valuable markers to detect left ventricular dysfunction in congestive heart failure of different origins. Increased plasma levels of circulating natriuretic peptides, ANF, amino-terminal pro-atrial natriuretic peptide, BNP and its amino-terminal moiety amino-terminal pro-brain natriuretic peptide have also been found under critical cardiovascular conditions; and all of these peptides have been reported to reflect left ventricular dysfunction in these patients (Witthaut 2004).

Apart from that, anti-ischemic properties have been reported for ANF. An association between the ANF/TC2238 polymorphic site and stroke occurrence has been reported. In cerebral ischemia, peri-infarct cortical spreading depression (CSD)-like depolarization potentiates infarct growth, whereas preconditioning with a CSD episode protects against subsequent ischemic insult. Acute, unilateral CSD in rats increases ANP mRNA up to 80% in layers II and VI of ipsilateral cortex, indicating that ANF may convey neuroprotective effects.

Further Reading

Barr, C. S., Rhodes, P., Struthers, A. D. (1996): C-type natriuretic peptide. *Peptides* **17**: 1243–1251.

Cantin, M., Genest, J. (1985): The heart and the artrial natriuretic factors. *Endocrinol. Rev.* **6**: 107–127.

Chabot, J. G., Morel, G., Quirion, R. (1990): Artrial natriuretic factors and the central nervous system: receptors and functions. *Curr. Aspects Neurosci.* **1**: 65–95.

De Bold, A. J., Borenstein, H. B., Veress, A. T., et al. (1981): A rapid and potent natriuretic response to intravenous injection of atrial myocardial extracts in rats. *Life Sci.* **28**: 89–94.

Deschepper, C. F. (1998): Peptide receptors on astrocytes. *Front. Neuroendocrinol.* **19** 20–46.

DiCicco-Bloom, E., Lelievre, V., Zhou, X., Rodriguez, W., Tam, J., Waschek, J. A. (2004): Embryonic expression and multifunctional actions of the natriuretic peptides and receptors in the developing nervous system. *Dev. Biol.* **271**: 161–175.

Ganguly, A. (1992): Atrial natriuretic peptide-induced inhibition of aldosterone secretion: a quest for mediator(s) [editorial]. *Am. J. Physiol.* **263**: E181–E194.

Jacobs, J. W., Vlasuk, G. P. (1987): Atrial natriuretic factor receptors. *Endocrinol. Metab. Clin.* **16**: 63–77.

Kirsch, B. (1956): Electron microscopy of the atrium of the heart. I. Guinea pig. *Exp. Med. Surg.* **14**: 99–112.

Levin, E. R. (1993): Natriuretic peptide C-receptor: more than a clearance receptor. *Am. J. Physiol.* **264**: E483–E489.

Nakao, K., Ogawa, Y., Suga, S., Imura, H. (1992): Molecular biology and biochemistry of the natriuretic peptide receptors. *J. Hypertens.* **10**: 1111–1114.

Oliveira, M. H., Antunes Rodrigues, J., Gutkowska, J., Leal, A. M., Elias, L. L., Moreira, A. C. (1997): Atrial natriuretic peptide and feeding activity patterns in rats. *Braz. J. Med. Biol. Res.* **30**: 465–469.

Rubattu, S., Stanzione, R., Di Angelantonio, E., Zanda, B., Evangelista, A., Tarasi, D., Gigante, B., Pirisi, A., Brunetti, E., Volpe, M. (2004): Atrial natriuretic peptide gene polymorphisms and risk of ischemic stroke in humans. *Stroke* **35**: 814–818.

Samson, W. K., Quirion, R. (eds) (1990): *Atrial Natriuretic Peptides*. CRC Press, Boca Raton, Fla.

Walther, T., Stepan, H., Pankow, K., Becker, M., Schultheiss, H. P., Siems, W. E. (2004): Biochemical analysis of neutral endopeptidase activity reveals independent catabolism of atrial and brain natriuretic peptide. *J. Biol. Chem.* **385**: 179–184.

Waschek, J. A. (2004): Developmental actions of natriuretic peptides in the brain and skeleton. *Cell Mol. Life Sci.* **61**: 2332–2342.

Wiedemann, K., Jahn, H., Kellne, M. (2000): Effects of natriuretic peptides upon hypothalamo–pituitary–adrenocortical system activity and anxiety behaviour. *Exp. Clin. Endocrinol. Diabetes* **108**: 5–13.

Wiggins, A. K., Shen, P. J., Gundlach, A. L. (2003): Atrial natriuretic peptide expression is increased in rat cerebral cortex following spreading depression: possible contribution to sd-induced neuroprotection. *Neuroscience* **118**: 715–726.

Witthaut, R. (2004): Science review: natriuretic peptides in critical illness. *Crit. Care* **8**: 342–349.

Yeung, V. T., Lai, C. K., Cockram C. S., Teoh, R., Young J. D., Yandle, T. G., Nicholls, M. G. (1991): Atrial natriuretic peptide in the central nervous system. *Neuroendocrinology* **53** [Suppl 1]: 18–24.

4.5
Bombesin and Related Neuropeptides

4.5.1
General Aspects and History

Bombesin was first isolated from skin of the amphibian *Bombina bombina*. Later, additional structurally related peptides were discovered and divided into three groups:
1. The bombesin family, which includes bombesin and alytesin.
2. The ranatensin family, which includes ranatensin, litorin and their derivatives.
3. The phyllolitorin family (Erspamer et al. 1984).

The first mammalian bombesin-like peptide was isolated from porcine gastric tissue and was named gastrin-releasing peptide (GRP) because of its potent gastrin-releasing competence (McDonald et al. 1979). GRP shows high homology with bombesin and alytesin. Similarly, neuromedin B (NmB) was identified from porcine spinal cord as a novel decapaptide which belong to the ranatensin family (Minamino et al. 1983).

4.5.2
Localization Within the Central Nervous System

Bombesin-like peptides are present in the central nervous system as well as in the gastrointestinal tract of mammals, where they modulate smooth-muscle contraction, exocrine and endocrine processes, metabolism and behavior. By using *in situ* hybridization, the distribution of the GRP and NmB mRNA expressing cells in the brain was analyzed and it was found that the distribution of cells expressing either GRP or NmB mRNA is quite distinct. In general, there is a higher GRP mRNA expression than NmB mRNA expression in the brain.

Strong GRP mRNA expression was observed in the hippocampal formation and in several nuclei of the amygdala. Moderate GRP mRNA expression levels were found in the anterior olfactory nucleus, the isocortex, the medial geniculate nucleus, the suprachiasmatic nucleus, the medial preoptic nucleus and the parabrachial nucleus.

A prominent NmB mRNA expression has been detected in the olfactory bulb as well as in the polymorph layer of the dentate gyrus. Moderate levels of NmB mRNA expression have been found in the substantia nigra and in the ventral tegmental area. Moreover, NmB mRNA expression is found in the central nucleus of the amygdala and in neurons located in the raphe nuclei and in somatosensory and motor nuclei of the brain stem.

4.5.3
Biosynthesis and Degradation

Human mature GRP is processed from a precursor form which consists of a 148-amino-acid signal sequence, a GRP sequence of 27 amino acids and a carboxy-terminal extension peptide.

The cDNA encoding human NmB was isolated in the 1980s by Krane and coworkers (1988). NmB is embedded in a 76-amino-acid precursor consisting of a signal peptide, a 32-amino-acid sequence for the large form of NmB and a carboxy-terminal extension peptide.

GRP and NmB share only 48% identity at the nucleotide level and localize to different chromosomes. Thus, the human GRP gene is localized on chromosome 18, whereas the human NmB gene is located on chromosome 15.

4.5.4
Receptors and Signal Transduction

In the 1990s, Spindel and coworkers (1990) cloned a bombesin/GRP receptor from mouse cells (Swiss 3T3). Analysis of the putative amino acid sequence of this GRP receptor revealed that the receptor has seven hydrophobic transmembrane domains. Thus, it is likely that this receptor belongs to the family of the G protein-coupled receptors (GPCRs).

The human GRP receptor gene is localized on chromosome X (at X p22).

GRP receptors have a high affinity for GRP and bombesin, but a low affinity for NmB. The GRP receptors are mainly expressed in hypothalamic regions, but are also found in the dentate gyrus. Moreover, GRP receptor expression is found in the nucleus ambiguus.

In 1991, the NmB receptor was cloned from a rat cDNA library. The NmB receptor shows a higher affinity to NmB than to GRP or bombesin. The NmB receptor gene is localized to human chromosome 6 and to mouse chromosome 10.

Prominent NmB receptor expression has been described in the olfactory region and thalamic areas, the nucleus ambiguus and the dentate gyrus.

Gorbulev and coworkers (1992) cloned a further subtype of a bombesin-like peptide receptor. The receptor showed a high similarity to GRP receptors (52%) and NmB receptors (47%) and was therefore designated as bombesin-like peptide receptor subtype 3 (BRS-3). The affinity of the BRS-3 receptor for bombesin is lower than the affinity of GRP receptors or NmB receptors. A high-affinity ligand for the BRS-3 receptor has not been identified so far.

The human BRS-3 gene is mapped to chromosome X (X q26–q28); and the mouse BRS-3 gene is also mapped to chromosome X.

The expression of this receptor in the brain is restricted. BRS-3 receptors can be found in the medial habenula, some hypothalamic areas and in several areas of the hindbrain.

It is important to note that, beside the mentioned receptors, some further bombesin-binding receptors have been cloned in amphibians and chicken. However, these receptors have not been detected in mammalian brain yet.

4.5.5
Biological Effects

Bombesin, GRP and NMB show an inhibitory effect on food intake. Central administration of bombesin, for example, inhibits food intake in mice. Infusion of bombesin in combination with loxiglumide, a cholecystokinin (CCK) receptor blocker, results in suppression of food intake, suggesting that the appetite-suppressing effect of bombesin is independent of the presence of CCK.

Since both GRP and NmB receptors are expressed in hyopthalamic areas which are known to be involved in feeding behavior, it is likely that the action of bombesin and bombesin-related peptides is mediated by both receptors.

Interestingly, BRS-3-deficient mice are hypertensive and show mild obesity, associated with an impairment of glucose metabolism.

Signaling via GRP receptors seem to play a role in amygdala-dependent memory of fear. Recent data show that GRP receptor-deficient mice exhibit enhanced long-term potentiation (LTP) and a greater and more persistent long-term fear memory. This role of GRP receptors in memory storage seems to be specific for the amygdala, since these mice performed normally in hippocampus-dependent Morris maze tests.

There are also indications that bombesin can exert sexual-dimorphism effects. For example, in male rats bombesin inhibits GH secretion, whereas in female rat bombesin stimulates GH release.

Further Reading

de Graaf, C., Blom, W.A.M., Smeets, P.A.M., Stafleu, A., Hendriks, H.F.J. (2004): Biomarkers of satiation and satiety. *Am. J. Clin. Nutr.* **79**: 946–961.

Erspamer, V., Erspamer, G.F., Mazzanti, G., Endean, R. (1984): Active peptides in the skins of one hundred amphibian species from Australia and Papua New Guinea. *Comp. Biochem. Physiol.* **77**: 99–108.

Gorbulev, V., Akhundoca, A., Buchner, H., Fahrenholz, F. (1992): Molecular cloning of a new bombesin receptor subtype expressed in uterus during pregnancy. *Eur. J. Biochem.* **208**: 405–410.

Giustina, A., Veldhuis, J.D. (1998): Pathophysiology of the neuroregulation of growth hormone secretion in experimental animals and the human. *Endocrine Rev.* **19**: 717–797.

Krane, I.M., Naylor, S.L., Helin-Davis, D., Chin, W.W., Spindler, E.R. (1988): Molecular cloning of cDNAs encoding the human bombesin-like peptide neuromedin B. Chromosomal localization and comparison to cDNAs encoding its amphibian homolog ranatensin. *J. Biol. Chem.* **263**: 13317–13323.

Lebarq-Verheyden, A.M., Krystal, G., Sartor, O., Way, J., Battey, J.F. (1988): The rat prepro-astrin releasing peptide gene is transcribed from two initiation sites in the brain. *Mol. Endocrinol.* **2**: 556–563.

Merali, Z., McIntosh, J., Anisman, H. (1999): Role of bombesin-related peptides in the control of food intake. *Neuropeptides* **33**: 376386

McDonald, T.J., Jornvall, H., Nilsson, G., Vagne, M., Ghatei, M., Bloom, S.R., Mutt, V. (1979): Characterization of a gastrin releasing peptide from porcine non-antral gastric tissue. *Biochem. Biophys. Res. Commun.* **90**: 227–233.

Minamino, N., Kangawa, K., Matsuo, H. (1983): Neuromedin B: a novel bombesin-like peptide identified in porcine spinal cord. *Biochem. Biophys. Res. Commun.* **114**: 541–548.

Ohki-Hamazaki, H., Watase, K., Yamamoto, K., Ogura, H., Yamano, M., Yamada, K., Maeno, H., Imaki, J., Kikuyama, S., Wada, E., Wada, K. (1997): Mice lacking bombesin receptor subtype-3 develop metabolic defects and obesity. *Nature* **1390**:165–169.

Shumyatsky, G.P., Tsvetkov, E., Malleret, G., Vronskaya, S., Hatton, M., Hampton, L., Battey, J.F., Dulac, C., Kandel, E.R., Bolshakov, V.Y. (2002): Identification of a signaling network in lateral nucleus of amygdala important for inhibiting memory specifically related to learned fear. *Cell* **11**: 905–918.

Spindel, E.R., Gibson, B.W., Reeve, J.R., Kelly, M. (1990): Cloning of cDNAs encoding amphibian bombesin: evidence for the relationship between bombesin and gastrin-releasing peptide. *Proc. Natl Acad. Sci.* **87**: 9813–9817.

Spindel, E.R., Giladi, E., Brehm, P., Goodman, R.H., Segerson, T.P. (1990): Cloning and functional characterization of a complementary DNA encoding the murine fibroblast bombesin/gastrin-releasing peptide receptor. *Mol. Endocrinol.* **4**: 1956–1963.

Wada, E., Way, J., Lebarq-Verheyden, A.M., Battey, J.F. (1990): Neuromedin B and gastrin-releasing peptide mRNAs are differentially distributed in the rat nervous system. *J. Neurosci.* **10**: 2917–2930.

Wada, E., Way.J., Shapira, H., Kusano, K., Lebacq-Verheyden, A.M., Coy, D., Jensen, R., Battery, J. (1991): cDNA cloning, characterization, and brain region-specific expression of a neuromedin-B-preferring bombesin receptor. *Neuron* **6**: 421–430.

Woodruff, G.N., Hall, M.D., Reynolds, T., Pinnock, R.D. (1996): Bombesin receptors in the brain. *Ann. N.Y. Acad. Sci.* **780**:223–243.

4.6
Calcitonin and Calcitonin Gene-related Protein

4.6.1
General Aspects and History

Copp and Cameron (1961) discovered calcitonin (CT) as a substance which is released by high calcium perfusion of the thyroid and which induces a lowering of plasma calcium. Calcitonin acts primarily on bone, but has also a direct action on the gastrointestinal secretion as well as on the kidney. In addition, calcitonin has direct and indirect effects on the CNS.

Calcitonin is a phylogenetically ancient peptide, found in mammals (including humans), reptiles, amphibians, fishes and even in *Escherichia coli*. Calcitonin is a peptide which consists of 32 amino acids. A structural analysis of the calcitonin-encoding gene revealed that this gene encodes also for a second biologically active peptide. This peptide was termed the calcitonin gene-related protein (CGRP). Hence, calcitonin and the calcitonin gene-related peptide are two products expressed by the same gene (Fig. 4.12). This gene is referred to as Calc I.

The main targets of calcitonin are peripheral organs, where it is involved primarily in the regulation of Ca^{2+} homeostasis. However, receptors which bind calcitonin in a specific manner have also been discovered in the central nervous system.

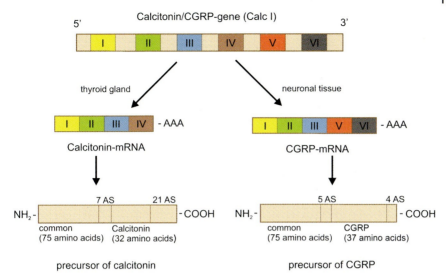

Fig. 4.12 The gene Calc I is composed of six exons. The primary transcripts of this gene give rise to two different RNAs. One of these transcripts codes for the precursor of calcitonin; and the other transcript codes for the precursor of the CGRP.

CRGP is a peptide which consists of 37 amino acids. Its distribution and biological activity are broadly consistent with the notion that CGRP represents a neuromodulator *per se*.

4.6.2
Localization Within the Central Nervous System

In higher vertebrates, calcitonin and related proteins are produced by peripheral endocrine glands. These hormones are, however, also found in the neural tissues of lower vertebrates and invertebrates that lack these endocrine organs, suggesting that neural tissue may be an ancestral site of this peptide synthesis. Indeed, the demonstration of CNS receptors for calcitonin and related peptides and their induction of neurological actions suggest that these hormones arose as neuropeptides.

Peripheral calcitonin is produced, stored and secreted by C-cells of the thyroid gland.

CGRP has a more widespread occurrence. It is expressed in the peripheral nervous system as well as in several areas of the central nervous system, including the limbic system and the hypothalamus. In addition, CGRP is found in high concentrations in the spinal cord and in the trigeminal ganglia. Additional peripheral sources are blood vessels, heart and the gastrointestinal tract. CGRP occurs in small quantities in the blood circulation. In humans, the plasma concentration of CGRP shows a circadian rhythm, ranging from 0.4 pmol to 100 pmol, with the highest concentration at nighttime.

Colocalization of CGRP with other neuromodulators such as cholecystokinin (CCK), substance P, somatostatin or NPY has frequently been described.

4.6.3
Biosynthesis and Degradation

The gene which encodes for calcitonin and CGRP is located on chromosome 11 in humans or on chromosome 6 in mice. The gene structure reveals six exons separated by five introns. The differential expression of both biologically active peptides occurs at the posttranscriptional level through alternative splicing.

Two different mRNAs, encoding for either calcitonin or CGRP, are spliced from the primary transcript. A further gene has been identified and named Calc II. This second gene encodes for a second form of CGRP, designated as CGRP-β or CGRP II.

Human CGRP-β differs from CGRP at three positions but, in rats, at only one position. The gene which encodes for CGRP-β is also located on human chromosome 11. A disulfide bridge at position 2–7 and a phenylamide at the carboxy-terminal region are two peculiar features of the primary structure of CGRP.

The amino acid sequence of CGRP is strongly conserved among mammals. Calcitonin-like CGRP is provided with a disulfide-bridge at position 1–7 and a prolinamide at the carboxy-terminal region. These features seem to be essential for the biological activity of the polypeptides. The amino acid sequence of calcitonin is less well conserved and this is true also for CGRP.

4.6.4
Receptors and Signal Transduction

The calcitonin receptor belongs to the class of heptaspan transmembrane proteins, which are G protein-coupled. The calcitonin receptor exists in two isoforms: the $C1_a$ and $C1_b$ form. The exact mechanism involved in signal transduction of either receptor is largely unknown. Some evidence indicates that the activation of calcitonin receptors induces an increase in the cAMP concentration.

The highest receptor concentrations have been reported in the peripheral target tissues: bones, lymphocytes as well as some tumor cells.

Central localization includes the pituitary gland, the nucleus accumbens, the preoptic area, the paraventricular nucleus of the hypothalamus and the arcuate nucleus.

CGRP receptors are more abundant than calcitonin receptors in the central nervous system, with maximal concentrations in the cerebellum and the spinal cord. Lower concentrations are present in the cortex, the hippocampus, the caudate-putamen, the substantia nigra, the hypothalamus, the pons, the mesencephalon and the anterior pituitary (Fig. 4.13). Blood vessels in the leptomeninges also express CGRP receptors.

The central CGRP receptor is not coupled to the adenylate cyclase pathway and can be coactivated by calcitonin.

Additional types of CGRP receptor have been identified in cardiomyocytes, in the pancreas and in blood vessels, including coronal, femoral and superior mesenteric arteries. The cardiac CGRP receptor cannot be co-activated by calcitonin; and activation of this receptor impinges on the adenylate cyclase pathway. Recent discoveries indicate that receptor diversity of the calcitonin peptide family is created by heterodimeric complexes, involving a new class of proteins which seem to be essential for the phenotyping of GPRCs in general. These receptor activity-modifying proteins (RAMP) have first been identified by expression cloning of the calcitonin gene-related peptide. Three of these proteins have been identified: RAMP1, 2 and 3. RAMP1 interacts with the calcitonin receptor-like receptor (CL-R) to yield a high-affinity CGRP receptor, whereas interaction with RAMP2 or RAMP3 generates receptors that are preferentially selective for adrenomedullin, a peptide that is primarily involved in angiogenesis. In the central nervous system, adrenomedullin acts to regulate blood pressure: RAMP2 mRNA has been found in numerous brain areas, including autonomic nuclei such as the paraventricular nucleus, supraoptic nucleus, arcuate nucleus and ventromedial nucleus of the hippocampus, as well as the nucleus of the solitary tract, area postrema and dorsal motor nucleus of the vagus. Interaction of calcitonin receptor with RAMPs generates different amylin receptor phenotypes. Amylin is a pancreatic hormone peptide involved in satiation and works through the acti-

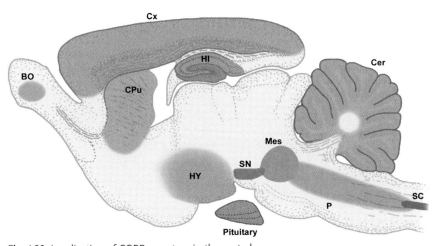

Fig. 4.13 Localization of CGRP receptors in the central nervous system. *Abbreviations*: BO=bulbus olfactorius; Cer=cerebellum; CPu=caudate putamen; HI=hippocampus; HY=hypothalamus; Mes=mesencephalon; P=pons; SC=spinal cord (dorsal horn); SN=*substantia nigra*.

vation of neurons in the area postrema. The intriguing finding of the heteromerisation of the calcitonin peptide receptor family with RAMPs is that the functional phenotype of GPCRs depends largely on the interaction with accessory proteins of the RAMP-type.

4.6.5
Biological Effects

The preferred peripheral substrate of calcitonin activity is the cellular component of the skeleton. Calcitonin exerts an inhibitory effect on osteoclasts and reduces the synthesis of osteoporins, proteins which are synthesized by osteoclasts and are responsible for attaching osteoclasts to the bone matrix. Through the inactivation of osteoclasts, the release of free calcium from the bone matrix is suppressed, which makes calcitonin in the regulation of calcium homeostasis the most potent endocrine antagonist to the parathyroid parathormone.

Calcitonin is sometimes employed therapeutically to relieve the symptoms of osteoporosis, although details of its mechanism of action remain unclear and clacitonin treatment has been replaced by bisphosphonate.

Furthermore, calcitonin plays a role in the processing of renal vitamin D3.

It has also been demonstrated that calcitonin induces analgesia, apparently a centrally mediated effect.

Behavioral studies indicate that the administration of calcitonin suppresses feeding behavior and exerts stress-reducing properties. Central administration of CGRP induces tachycardia and hypertension while – paradoxically – systemic peripheral administration induces hypotension. CGRP is a potent vasodilator with long-lasting effects, causing prolonged hypotension. A CGRP concentration of about 56 pmol l^{-1} is sufficient to induce hypotony in humans. Thus, CGRP can be considered to be one of the most potent endogenous vasodilators. The vasodilatatory effect of CGRP is due to a direct effect on blood vessels and does not require the release of neurotransmitters. CGRP-dependent vasodilatation occurs in a number of blood vessels, including coronal arteries and cerebral vessels. Subcutaneous injection elicits a direct local vasodilatatory response at the site of injection.

The effect of CGRP on cerebral vessels is more pronounced than that induced by substance P.

The presence of CGRP and its receptors in brain tissue argues strongly in favor of its having a neuromodulatory function.

It is assumed that CGRP is involved in the central regulation of visceral and limbic functions. Furthermore, CGRP seems to be involved in the differentiation of dopaminergic neurons in the midbrain, where it acts as a differentiation-promoting factor during neurogenesis when dopaminergic neurons assemble in the ventral mesencephalon and project their neurites towards the neostriatum.

The phenotypic expression of the calcitonin peptide family, as indicated above, seems to depend largely on heterodimerization with the RAMP class of proteins.

Several novel members of the CGRP family have been identified. In 2003, a novel biologically active peptide, designated calcitonin receptor-stimulating peptide (CRSP), has been isolated from the acid extract of pig brain by Katafuchi and coworkers (2003). Today, three different members, CRSP-1, CRSP-2 and CRSP-3 have been identified. These peptides are specific ligands for the calcitonin receptor. Interestingly, these three CRSPs are mainly expressed in the CNS and within the thyroid. However, their physiological functions within the brain are largely unknown.

4.6.6
Neurological Disorders and Neurodegenerative Diseases

It is generally thought that local vasodilatation of intracranial extracerebral blood vessels and a consequent stimulation of surrounding trigeminal sensory nervous pain pathways is a key mechanism underlying the generation of headache pain associated with migraine. This activation of the "trigeminovascular system" is thought to cause the release of vasoactive sensory neuropeptides, especially CGRP, that increase the pain response. The activated trigeminal nerves convey nociceptive information to central neurons in the brain stem trigeminal sensory nuclei that in turn relay the pain signals to higher centers where headache pain is perceived. Indeed, enriched localization of CGRP in trigeminal sensory ganglia has been detected as well as CGRP has been detected in increased amounts in external jugular venous blood during migraine attacks and intravenous administration of CGRP causes headache and migraine in migraineurs. In addition, studies in patients have revealed a clear association between headache and the release of CGRP. These findings suggest that the increase in CGRP observed during spontaneous migraine attacks may play a causative role and, thus, CGRP antagonists may represent a novel therapeutic approach to the treatment of migraine.

Further Reading
Brain, S. D. (1997): Sensory neuropeptides: their role in inflammation and wound healing. *Immunopharmacology* **37**: 133–152.
Brian, J. E. Jr, Faraci, F. M., Heistad, D. D. (1996): Recent insights into the regulation of cerebral circulation. *Clin. Exp. Pharmacol. Physiol.* **23**: 449–457.
Bürvenich, S., Unsicker, K., Kriegelstein, K. (1998): Calcitonin-gene related peptide promotes differentiation but not survival of rat mesencephalic dopaminergic neurons in vitro. *Neuroscience* **86**: 1165–1172.
Copp, D. H., Cameron, E. C. (1961): Demonstration of a hypercalcemic factor (calcitonin) in commercial parathyroid extract. *Science* **134**: 2038–2039.
Edvinsson, L. (2001): Aspects on the pathophysiology of migraine and cluster headache. *Pharmacol. Toxicol.* **89**: 65–73.
Feuerstein, G., Willette, R., Aiyar N. (1995): Clinical perspectives of calcitonin gene related peptide pharmacology. *Can. J. Physiol. Pharmacol.* **73**: 1070–1074.
Fischer, J. A., Tobler, P. H., Kufmann, M., Born, W., Henke, H., Cooper, P. E., Sagar, S. M., Martin, J. B. (1981): Calcitonin: regional distribution of the hormone and its binding sites in the human brain and pituitary. *Proc. Natl Acad. Sci. USA* **78**: 7801–7805.

Geppetti, P. (1993): Sensory neuropeptide release by bradykinin: mechanisms and pathophysiological implications. *Regul. Peptides* **47**: 1–23.

Hargreaves, R. J., Shepheard. S. L.(1999): Pathophysiology of migraine – new insights. *Can. J. Neurol. Sci. Suppl.* **3**: S12–S19.

Hilton, J. M., Chai, S. Y., Sexton, P. M. (1995): *In vitro* autoradiographical localization of the calcitonin isoforms C_{1a} and C_{1b} in rat brain. *Neuroscience* **69**: 1223–1237.

Hull, K. L., Fathimani, K., Sharma, P., Harvey, S. (1998): Calcitropic peptides: neural perspectives. *Comp. Biochem. Physiol. C. Pharmacol. Toxicol. Endocrinol.* **119**: 389–410.

Katafuchi, T., Minamino, N. (2004): Structure and biological properties of three calcitonin receptor-stimulating peptides, novel members of the calcitonin gene-related peptide family. *Peptides* **25**: 2039–2045.

Katafuchi, T., Kikumoto, K., Hamano, K., Kangawa, K., Matsuo, H., Minamino, N. (2003): Calcitonin receptor-stimulating peptide, a new member of the calcitonin gene-related peptide family. Its isolation from porcine brain, structure, tissue distribution, and biological activity. *J. Biol. Chem.* **278**:12046–12054.

Lassen, L. H., Haderslev, P. A., Jacobsen, V. B., Iversen, H. K., Sperling, B., Olesen, J. (2002): CGRP may play a causative role in migraine. *Cephalalgia* **22**: 54–61.

Moore, M. R., Black, P. M. (1991): Neuropeptides. *Neurosurg. Rev.* **14**: 97–110.

Rosenfeld, M. G., Mermod, J. J., Amara, S. G., Swanson, L. W., Sawchenko, P. E., Rivier, J., Vale, W. W., Evans, R. M. (1983): Production of a novel neuropeptide encoded by the calcitonin gene via tissue-specific RNA processing. *Nature* **304**: 129–135.

Tschopp, F. A., Henke, H., Petermann, J. B., Tobler, P. H., Janzer, R., Höckfelt, T., Lundberg, J. M., Cuello, C., Fischer, J. A. (1985): Calcitonin gene-related peptide and its binding sites in human central nervous system and pituitary. *Proc. Natl Acad. Sci. USA* **82**: 248–252.

Udawela, M., Hay, D. L., Sexton, P. M. (2004): The receptor activity modifying protein family of G protein coupled receptor accessory protein. *Sem. Cell Develop. Biol.* **15**: 299–308.

Zaidi, M., Moonga, B. S., Bevis, P. J., Bascal, Z. A., Breimer, L. H. (1990): The calcitonin gene peptides: biology and clinical relevance. *Crit. Rev. Clin. Lab. Sci.* **28**: 109–174.

4.7
Cholecystokinin

4.7.1
General Aspects and History

In 1928, cholecystokinin (CCK) was first described by Ivy and Oldberg as a substance that causes gallbladder contractions. In the 1940s, Harper and Raper recognized that intestinal extracts also stimulated secretion of pancreatic enzymes and coined the term pancreozymin. Sequence analysis of the peptide indicated that CCK and pancreozymin were identical with a length of 33 amino acids; and the term cholecystokinin for both peptides is now generally accepted. In 1975, van der Haagen and coworkers discovered cholecystokinin as one of the first gastrointestinal peptides in mammalian brain.

CCK carries different biologically active fragments, which constitute the CCK family. The best known are: caerulein (CCK-10), octapeptide (CCK-8), pentagastrin (CCK-5) and tetrapeptide (CCK-4). In brain tissue, CCK-8 seems to be the most dominant member of the cholecystokinin family.

CCK possesses an amidated carboxy-terminal pentapeptide (Gly-Trp-Asp-Met-Phe-NH_2) which is identical with gastrin. The carboxyl terminus of CCK is the biologically active portion of the hormone. Because of strong sequence homology at the carboxyl terminus between CCK and gastrin, each peptide is able to interact with the corresponding receptor of the other.

4.7.2
Localization Within the Central Nervous System

Cholecystokinin shows a heterogeneous distribution in both the peripheral and central nervous systems, with specific binding sites and clearly defined projections.

CCK-8 [Asp-Tyr(SO$_3$H)-Met-Gly-Trp-Met-Asp-Phe-NH$_2$] is widely distributed throughout the central nervous system of mammals. This neuropeptide has been detected in cortical areas and in limbic structures such as the hippocampus and the amygdala. Furthermore, CCK-8 has been found in neurons of the olfactory tubercle, the substantia nigra, the ventromedial thalamus, the septum, the nucleus accumbens, the ventral tegmental area, the interpeduncular nucleus, the hypothalamus, the posterior lobe of the pituitary and the spinal cord.

Some of these CCK-containing neurons give rise to widespread projections. The main projections arising from different CCK-containing neurons are summarized in Fig. 4.14.

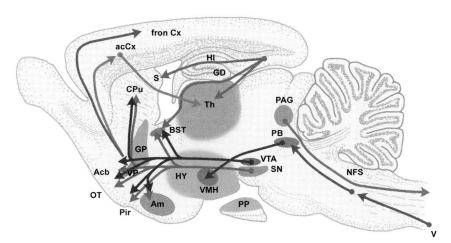

Fig. 4.14 Schematic overview of the main projections (arrows) of CCK-containing neurons in the rat brain. The areas of CCK-containing neurons are symbolized by gray circles. *Abbreviations*: Acb = *nucleus accumbens*; AcCx = anterior cingular cortex; Am = amygdala; BST = bed nucleus of the *stria terminalis*; CPu = caudate-putamen; fron Cx = frontal cortex; GD = *gyrus dentatus*; GP = *globus pallidus*; HI = hippocampus; HY = hypothalamus; NFS = nucleus of the *fasciculus solitarius*; OT = olfactory tubercle; PAG = periaqueductal gray; PB = parabrachial nucleus; Pir = piriform cortex; PP = posterior pituitary; S = subiculum; SN = *substantia nigra*; Th = thalamus; V = vagal nerve; VMH = ventromedial nucleus of the hypothalamus; VP = ventral pallidum; VTA = ventral tegmental area.

CCK-8 is often colocalized with neurotransmitters or with other neuromodulators. In the cerebral cortex and the hippocampus, CCK-8 colocalizes with GABA. In the substantia nigra or the ventral tegmental area, colocalization of CCK with dopamine has been detected, while in the periaqueductal gray matter colocalization of CCK with substance P or enkephalins has been encountered.

In addition, it is known that CCK colocalizes with norepinephrine, serotonin and vasopressin.

4.7.3
Biosynthesis and Degradation

The gene which encodes cholecystokinin, is located on human chromosome 3. It has a length of seven kilobases with three exons, the second and third of which encode the prepropeptide. Regulation of CCK gene expression depends on food intake. This elicites promotor activation during fasting; conversely, somatostatin supresses transcription.

The mRNA (with a length of 345 nucleotides) encodes for a precursor (procholecystokinin or proCCK) with a length of 115 amino acids. This precursor consists of a putative 20-amino-acid signal peptide, a 25-amino-acid spacer peptide, the bio-

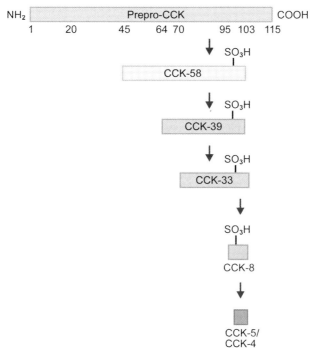

Fig. 4.15 Scheme of the maturation of some forms of cholecystokinin from its precursor molecule procholecystokinin.

logically active CCK-58 (the major processed form in most tissues), and a 12-amino-acid extension at the carboxyl terminus. Cholecystokinin has been isolated in different isoforms which vary in length but reveal identical carboxy-terminal regions. CCK is stored in synaptic vesicles in neuronal somata and terminals, from which it is released upon depolarization in a calcium-dependent manner.

Posttranslational modifications, including proteolytic cleavage, sulfation of thyrosins and aminidation of phenylalanine, result in the formation of different fragments of cholecystokinin. Among these fragments are CCK-58, CCK-39, CCK-33, CCK-25, CCK-8, CCK-5 and CCK-4.

Besides CCK-8, a second peptide CCK-4 displays a physiological role in the brain. In peripheral organs, some active forms with higher molecular masses have been encountered. Different enzymes, in particular aminopeptidases, like neutral endopeptidases, thiol- and serine proteases are involved in the degradation of CCK-8.

The latter peptidases also catalyze the generation of CCK-5 and CCK-4 (Fig. 4.15). Cholecystokinin shares a common carboxy-terminal fragment of five peptides with gastrin and both peptides are acknowledged as phylogenetically "old" peptides.

4.7.4
Receptors and Signal Transduction

CCK is one of the most common neuropeptides in the mammalian brain. An important functional distinction must be made between the CCK tetrapeptide and octapeptide. CCK-4 and CCK-8 possess different affinities towards CCK receptors, vary in their distribution in the periphery and in the brain and differ in their effects on behavior. CCK receptors are widely distributed in the central nervous system. High concentrations of CCK receptors are found in the cerebral cortex, limbic system, nucleus accumbens, olfactory tubercle and basal ganglia. Lower concentrations have been reported in the hippocampus, the hypothalamus, the lower medullary regions and the spinal cord.

Two subtypes of CCK receptors have been identified by means of *in vitro* pharmacology, autoradiography and electrophysiological experiments: CCK-A and CCK-B receptors. The two receptors can be distinguished by their different affinity to sulfated and non-sulfated forms of cholecystokinin and by their affinity to long and short forms of cholecystokinin.

The sulfated form of CCK8 shows high affinity for the CCK-A receptor while the non-sulfated form of CCK8 and CCK4, gastrin and pentagastrin reveal a 10 000-fold lower affinity to this receptor. The CCK-B receptor exhibits high affinity and selectivity for CCK4, gastrin, pentagastrin (CCK5) and the non-sulfated form of CCK8. Sulfated CCK8 has a slightly lower affinity to this receptor.

CCK-B receptors have been encountered mainly in the brain ("b" stands for brain), whereas CCK-A receptors are constitutively expressed in tissues of the alimentary tract ("a" stands for alimentary) and are closely involved in the endocrine regulation of digestion and feeding behavior.

CCK-A receptors are not restricted to peripheral organs, but are also found in some regions of the central nervous system which enclose the area postrema, the nucleus of the tractus solitarius, the substantia nigra, the hypothalamus and the medulla oblongata.

CCK-A receptors that occur in the dorsal horn of the spinal cord of primates, particularly in the substantia gelatinosa, have been associated with nociception.

The CCK-B receptor is more prevalent in the brain than CCK-A receptors and is present to only a minor extent in the periphery, where it has been described in the stomach and pancreas.

In the central nervous system, the CCK-B receptor is predominantly expressed in those limbic regions which are associated with the generation of emotions, namely the amygdala and the hippocampus. Furthermore, CCK-B receptors have been localized in the cortex, the olfactory bulb, the nucleus accumbens, the ventromedial hypothalamus and the area postrema, as well as in the nucleus of the tractus solitarius and the dorsal nucleus of the vagus.

An interesting morphological and functional relationship exists between the two types of CCK receptor. They are found not only in the same brain regions, albeit with different densities, but also at identical postsynaptic sites. It has been shown that CCK-B receptors are preferentially excitatory with a prolonged excitatory action whereas activation of CCK-A receptors elicits an inhibitory postsynaptic response. However, the effects of CCK agonists on CCK-A receptors are quite variable, ranging from short-term excitation, prolonged excitation to inhibition.

The behavioral effects of different agonists vary to a large extent and it is suggested that the responses depend on the differential somato-dendritic and presynaptic distribution of both types of receptors.

Both receptors have been cloned. They exhibit the heptaspan motif of metabotropic G protein-coupled receptors and utilize the phospholipase C/DAG pathway for signal transmission. CCK-A and CCK-B receptors showed a homology of about 48%. It is suggested that both receptors occur in different subtypes.

The CCK-B receptors are not only recognized by CCK but also by gastrin. Moreover, a CCK-type C receptor has recently been described, which recognizes gastrin. The CCK-C is not a purely new receptor subtype, since CCK-C represents a splice variant of the CCK-B receptor. Nevertheless, the term CCK-C or "cancer" receptor has been proposed to signify the relationship of this receptor to neoplasia. Interestingly, CCK-C receptors are not only found in peripheral cancers, but have also been identified in human glioma and rat gliosarcoma.

4.7.5
Biological Effects

CCK shows different effects in the mesocorticolimbic and nigrostriatal pathways. It either excites neurons or potentiates the inhibitory effects of dopamine when dopamine and CCK are co-applied. It is thought that the first excitatory effect is mediated by CCK-A receptors, while the second effect is believed to depend on CCK-B receptors.

Within the central nervous system, cholecystokinin is involved in different biological processes, including:
- *Satiety*: It has been suggested that cholecystokinin is the neurotransmitter responsible for the satiety signals to the brain. Cholecystokinin reduces food intake. It seems likely that the CCK release could indirectly involve the activation of hypothalamic areas, but a direct central effect can not be excluded.
- *Nociception*: Several studies indicate that CCK could induce analgesic effects. In the physiological control of nociception, there is also an involvement of opioids. Several endogenous neuropeptides are able to reduce the effects of opioids. These peptides are therefore also known as "anti-opioids". Both CCK and the neuropeptide FF (NPFF) have been identified as members of this group. Thus, by increasing CCK levels, a decrease in opioid-derived anti-nociception can be induced. In addition, systemic or intrathecal administered morphine increases extracellular level of CCK in the spinal cord dorsal horn of rats (de Araujo et al. 1998).
- *Fear*: Anxiogenic properties of CCK have also been suggested, since it was observed that microinjection of CCK-8 into the nucleus accumbens facilitated the extinction of active avoidance behavior and attenuated the retention of passive avoidance behavior. However, microinjection of these peptides into the central amygdaloid nucleus caused opposite effects in these behavioral tests (Fekete et al. 1984).

    Some years later, it was shown that intravenous injections of the CCK-4 in humans could induce a short-lasting (1–4 min) panic-like attack at doses between 20 µg and 80 µg, while doses of 80–100 µg induced severe anxiety, but no panic-like attack (De Montigny 1989).

    CCK-8, applied by microiontophoresis to hippocampal pyramidal neurons, has a powerful excitatory effect, whereas the non-sulfated CCK octapeptide has no such effect. Low doses of benzodiazepines depress the spontaneous activity of hippocampal pyramidal neurons. Furthermore, benzodiazepines at very low doses antagonize selectively the CCK-8-induced activation of rat hippocampal pyramidal neurons. This antagonistic action might be involved in the anxiolytic effect of these drugs (Bradwejn and de Montigny 1984).

    An important issue concerning the implications of CCK peptides in the activation of panic attacks is the question whether antagonists of CCK receptors show anxiolytic effects in CCK-induced and in naturally occurring panic attacks. Several non-peptide CCK receptor antagonists with reliable specificity and the ability to penetrate the blood–brain barrier have been synthesized in the past few years. Accumulating data from studies exploiting these antagonists suggest that CCK-B receptors are in fact involved in the anxiety-inducing properties of CCK-4.
- *Regulation of body temperature*: The administration of cholecystokinin into the hypothalamus reduces body temperature. Microiontophoretically injected CCK into the preoptic area of the hypothalamus enhances the firing rate of heat-sensitive neurons and reduces the firing rate of neurons which have been shown to be sensitive to cold. It is therefore believed that cholecystokinin

counteracts neuronal mechanisms involved in the up-regulation of body temperature.

Central microinjection of CCK in rats induces a thermogenic response that can be attenuated by CCK-B receptor antagonists, but some authors also observe hypothermia. By contrast to its central fever-inducing effect, in rodents exposed to cold, CCK-8 elicits a dose-dependent hypothermia on peripheral injection, probably acting on CCK-A receptors. It is suggested that neuronal CCK may have a specific role in the development of hyperthermia and endogenous CCKergic mechanisms could contribute to the mediation of fever.

- *Learning and memory*: Experimental data suggest that cholecystokinin participates in processes related to memorizing and learning. Thus, it has been shown that CCK-8S facilitates LTP in the hippocampus through CCK-B receptor activation. Moreover, in behavioral tests, it has been shown that administration of CCK-B agonists improve learning tasks, whereas CCK-B antagonists injected into the hippocampus abolish the improving effect of CCK-B agonists. Moreover, CCK-B receptor-deficient mice have an impairment of performance in the spatial memory task, indicating that the CCK-ergic system through its action upon CCK-B receptors can improve the performance in specific learning tasks.

4.7.6
Neurological Disorders and Neurodegenerative Diseases

A strong reduction of cholecystokinin-containing neurons in the striatum has been observed in Huntington's chorea disease. It has also been suggested that CCK-containing dopaminergic neurons, in particular in the mesolimbic pathway, are implicated in disorders such as schizophrenia and Parkinson's disease.

Schizophrenia also leads to a reduction in levels of cholecystokinin in the cerebrospinal fluid and in certain brain areas, such as the cortical and subcortical regions of the temporal lobe. A reduction in the density of CCK receptors in the frontal cortex and in the hippocampus has also been demonstrated in these patients. There are several reports showing associations of polymorphisms of the human cholecystokinin receptor A and B genes with schizophrenia; and there is evidence that an association between polymorphisms of the cholecystokinin gene with schizophrenia may exist.

Dopamine is colocalized with CCK in the ventro-tegmental area and in the substantia nigra, where CCK increases the firing rate of the dopaminergic neurons. When simultaneously administered, CCK and dopamine facilitate the inhibitory effects of dopamine on nigro- and ventrotegmental neurons. Patients with Parkinson's disease show a reduction in CCK levels of about 40% in the cerebrospinal fluid and within the substantia nigra. Interestingly, an association between polymorphism of the cholecystokinin gene and idiopathic Parkinson's disease has been reported.

Further Reading

Balschun, D., Reymann, K.G. (1994): Cholecystokinin (CCK-8S) prolongs 'unsaturated' theta-pulse induced long-term potentiation in rat hippocampal CA1 *in vitro*. *Neuropeptides* **26**: 421–417.

Beinfeld, M.C., Palkovits, M. (1982): Distribution of cholecystokinin (CCK) in rat lower brain stem nuclei. *Brain Res.* **238**: 260–265.

Bradwejn, J., De Montigny, C. (1984): Benzodiazepines antagonize cholecystokinin-induced activation of rat hippocampal neurones. *Nature* **312**: 363–364.

Chen, D.-Y., Deutsch, J.A., Gonzalez, M.F., Gu, Y. (1993): The induction and supression of c-fos expression in the rat brain by cholecystokinin and its antagonist L364,718. *Neurosci. Lett.* **149**: 91–94.

Cosi, C., Altar, A., Wood, P.L. (1989): Effect of cholecystokinin on acetylcholine turnover and dopamine release in the rat striatum and cortex. *Eur. J. Pharmacol.* **165**: 209–214.

Crawley, J.N. (1991): Cholecystokinin–dopamine interactions. *Trends Pharmacol. Sci.* **12**: 232–236.

Crawley, J.N., Corwin, R.L. (1994): Biological actions of cholecystokinin. *Peptides* **15**: 731–755.

De Araujo, L.G., Alster, P., Brodin, E., Wiesenfeld-Hallin, Z. (1998): Differential release of cholecystokinin by morphine in rat spinal cord. *Neurosci. Lett.* **245**: 13–16.

De Montigny, C. (1989): Cholecystokinin tetrapeptide induces panic-like attacks in healthy volunteers. Preliminary findings. *Arch. Gen. Psychiatry* **46**: 511–517.

Dourish, C.T., Hill, D.R. (1987): Classification and function of CCK receptors. *Trends Pharmacol. Sci.* **8**: 207–208.

Fekete, M., Lengyel, A., Hegedus, B., Penke, B., Zarandy, M., Toth, G., Telegdy G. (1984): Further analysis of the effects of cholecystokinin octapeptides on avoidance behaviour in rats. *Eur. J. Pharmacol.* **98**: 79–91.

Fink, H., Morgenstern, R., Ott, T. (1991): CCK-8 modulates D2 receptor agonist-induced hypermotility in the nucleus accumbens. *Brain Res. Bull.* **26**: 437–440.

Fujii, C., Harada, S., Ohkoshi, N., Hayashi, A., Yoshizawa, K., Ishizuka, C., Nakamura, T. (1999): Association between polymorphism of the cholecystokinin gene and idiopathic Parkinson's disease. *Clin. Genet.* **56**: 394–399.

Gibbs, J., Young, R.C., Smith, G.P. (1973): Cholecystokinin decreases food intake in rats. *J. Comp. Physiol. Psychol.* **84**: 488–495.

Hadjiivanova C., Belcheva S., Belcheva I. (2003): Cholecystokinin and learning and memory processes. *Acta Physiol. Pharmacol. Bulg.* **27**: 83–88.

Harro, J., Oreland, L. (1993): Cholecystokinin receptors and memory: a radial maze study. *Pharmacol. Biochem. Behav.* **44**: 509–517.

Hommer, D.W., Skirboll, L.R. (1983): Cholecystikinin-like peptides potentiate apomorphine-induced inhibition of dopamine neurons. *Eur. J. Pharmacol.* **91**: 151–152.

Lefranc, F., Chaboteaux, C., Belot, N., Brotchi, J., Salmon, I., Kiss, R. (2003): Determination of RNA expression for cholecystokinin/gastrin receptors (CCKA, CCKB and CCKC) in human tumors of the central and peripheral nervous system. *Int. J. Oncol.* **22**: 213–219.

Lotstra, F., Vanderhaeghen, J.J. (1987): Distribution of immunreactive cholecystokinin in the human hippocampus. *Peptides* **8**: 911–920.

Markstein, R., Hökfeldt, T. (1984): Effect of cholecystokinin-octapeptide on dopamine release from slices of cat caudate nucleus. *J. Neurosci.* **4**: 570–575.

Mollereau, C., Roumy, M., Zajac, J.M. (2005): Opioid-modulating peptides: mechanisms of action. *Curr. Top. Med. Chem.* **5**: 341–355.

Moran, T.H., Robinson, P., Goldrich, M.S., McHuggh, P.R. (1986): Two brain cholecystokinin receptors: implications for behavioral actions. *Brain Res.* **362**: 175–179.

Morley, J.E. (1982): Minireview: The ascent of cholecystokinin (CCK) from gut to brain. *Life Sci.* **30**: 479–493.

Palacois, J.M., Savasta, M., Mengod, G. (1989): Does cholecystokinin colocalize with dopamine in the human substantia nigra? *Brain Res.* **488**: 369–375.

Saito, A., Sankaran, H., Goldfine, I.D., Williams, J.A. (1980): Cholecystokinin receptors in the brain: characterization and distribution. *Science* **208**: 1155–1156.

Wang, Z., Wassink, T., Andreasen, N.C., Crowe, R.R. (2002): Possible association of a cholecystokinin promoter variant to schizophrenia. *Am.J. Med. Genet.* **114**: 479–482.

Zhang, D.M., Bula, W., Stellar, E. (1986): Brain cholecystokinin as a satiety peptide. *Physiol. Behav.* **36**: 1183–1186.

Zoltan, S., Szekely, M., Hummel, Z., Balasko, M., Romanovsky, A.A., Petervari, E. (2004): Cholecystokinin: possible mediator of fever and hypothermia. *Front. Biosci.* **9**: 301–308.

4.8
Corticotropin-releasing Factor

4.8.1
General Aspects and History

In 1955, Guillemin and Schally were able to prove the existence of a hypothalamic factor which was able to stimulate the secretion of adrenocorticotropic hormone (ACTH). Since then, several hypothalamic substrates prone to stimulating the secretion of ACTH have been discovered. The identity of the primary secretagogue of ACTH, however, remained a matter of controversy for a long time. First, arginine–vasopressin (AVP) was considered to represent the prime activator of corticotropes. Several years after the initial description by Guillemin and Schally, Vale and coworkers (1981) succeeded to isolate a peptide of 41 amino acids, which became known as corticotropin-releasing factor [CRF; synonyms are: corticotropin-releasing hormone (CRH) or corticoliberin]. This peptide stimulates the secretion of ACTH in a highly specific manner and is considered to represent the original regulator of ACTH release.

Analysis of the sequence of CRF revealed that it belongs to a family of structurally and functionally closely related proteins, like glucagon, vasoactive intestinal polypeptide (VIP), growth hormone-releasing hormone (GHRH) and pituitary adenylate cyclase-activating peptide (PACAP).

4.8.2
Localization Within the Central Nervous System

By means of immunohistochemical and radioimmunoassay techniques, CRF was found to be heterogeneously distributed throughout the central nervous system. Corticotropin-releasing factor is predominantly synthesized in the medial parvicellular neurons of the hypothalamic paraventricular nucleus (PVN), from which the majority of cells project to the median eminence. This CRF pathway comprises the hypothalamic portion of the endocrine stress axis. CRF is transported via the hypothalamo-pituitary circulation into the anterior pituitary where it acts upon corticotrope cells in the pituitary.

CRF-containing neurons have also been demonstrated in significant amounts in the following hypothalamic nuclei: the periventricular nucleus, the preoptic nucleus and the suprachiasmatic nucleus. Besides the hypothalamus, CRF-containing neurons occur in some additional regions of the central nervous system (Fig. 4.16). These areas belong preferentially to limbic structures and structures which are involved in autonomic functions and in the processing of sensory information.

CRF-containing interneurons are widely distributed in the neocortex and are believed to be important in several behavioral influences of CRF, including effects on cognitive processing. CRF perikarya in the central nucleus of the amygdala project terminals to the parabrachial nucleus of the brain stem as well as to the bed nucleus of the stria terminalis and the medial preoptic area.

The presence of CRF immunoreactivity in the raphe nuclei and locus coeruleus, the origin of the major serotonergic and noradrenergic pathways in brain, points to a role for CRF in modulating these monoaminergic systems, which have long been implicated in the pathophysiology of depression and anxiety disorders.

There is a considerable overlap in the occurrence of corticotropin-releasing factor and the distribution of the CRF receptors. To date, not all of the projections of CRF-containing neurons have been fully investigated. CRF-containing neurons of the PVN project to different nuclei of the brain stem. Further projections of CRF-containing neurons of the amygdala to the parabrachial nucleus and from the inferior olive to the cerebellum are known. Projections of CRF-

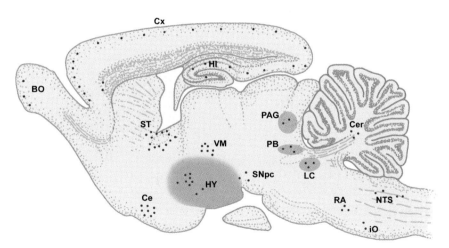

Fig. 4.16 Schematic representation of the distribution of CRF-expressing cells in the central nervous system. *Abbreviations*: BO=*bulbus olfactorius*; Ce=central nucleus of the amygdala; Cer=cerebellum; Cx=cortex; HI=hippocampus; HY=hypothalamus; LC=*locus coeruleus*; NTS=nucleus of the *tractus solitarius*; iO=olive; PAG=periaqueductal gray; PB=parabrachial nucleus; RA=raphe nuclei; SNpc=*substantia nigra (pars compacta)*; ST=*stria terminalis*; VM=ventro-medial nucleus of the thalamus.

containing neurons of the hypothalamus into the lateral septum have also been demonstrated.

Extrahypothalamic CRF does not contribute to the regulation of ACTH secretion. It seems more likely that this fraction is involved in the regulation of vegetative and behavioral functions.

Corticotropin-releasing factor is known to colocalize with other neuropeptides. Colocalizations of CRF with angiotensin II, CCK and enkephalin have been described, as well as colocalizations with neurotensin and VIP.

4.8.3
Biosynthesis and Degradation

The primary protein structure of CRH is highly conserved among vertebrates, with the peptide being identical in humans, rats and mice, and differing for sheep in only seven amino acids. The gene sequence of the human CRF gene and the CRF gene of rats share a homology of about 94%. The human CRF gene is located on chromosome 8.

The human CRF gene encodes a precursor (prepro-CRF) of 196 amino acids. The carboxy-terminal region of the precursor encodes the sequence of CRF. This sequence is flanked by a pair of basic amino acids at the 5′ end of the CRF sequence and by the amino acids Gly–Lys at the 3′ end. The prepro-CRF gives rise to pro-CRF. The precursor has been suggested to convey some biological activity itself; and it is able to induce the secretion of ACTH (Morrison et al. 1995).

The pro-CRF is processed to several subtypes of CRF, which are listed in a descending series, according to their biological activity (Fig. 4.17).

Since the PVN is innervated by serotonergic, cholinergic and catecholaminergic fibers, it seems likely that the synthesis or the secretion of CRF from the PVN is controlled by intricate neuronal mechanisms.

Little is known about the specific degradation of corticotropic-releasing factor. Peripherally, CRF binds rapidly to specific binding proteins which inactivate CRF function. These binding proteins regulate the concentration of the active fraction of peripheral CRF; however it is uncertain whether binding proteins also exist in the central nervous system and serve a similar function.

In addition to CRF, three CRF-related proteins have been identified in mammals, named urocortin, urocortin II (also known as stresscopin-related peptide) and urocortin III (stresscopin).

Urocortin II cDNA encodes a predicted 38-amino-acid peptide. Urocortin II binds selectively to the type 2 CRF receptor, with no appreciable activity on CRF receptor type 1. Transcripts encoding urocortin II are expressed in discrete re-

| CRF-(1-39), CRF-(1-41), CRF-(4-41) and CRF-(9-41) |

Fig. 4.17 The subtypes of CRF, listed according to their biological activity.

gions of the rodent central nervous system, including stress-related cell groups in the lateral septum, the hypothalamus (paraventricular and arcuate nuclei), the raphe nuclei, the locus coeruleus and the nucleus of the tractus solitarius.

Urocortin III binds with high affinity to the CRF type 2 receptor, but not to the type 1 receptor. Urocortin III-positive neurons are found predominately within the hypothalamus (median preoptic nucleus, rostral perifornical area, paravenricular nucleus), the medial amygdala and the bed nucleus of the stria terminalis.

4.8.4
Receptors and Signal Transduction

CRF interacts with the corticotropes of the anterior pituitary by binding to specific high-affinity membranous receptors. The CRF receptors are all members of the GPCR family, with the common heptaspan motif.

Two different CRF receptor types have been described: CRF1 and CRF2, which are both positively coupled to adenylate cyclase. The CRF2 receptor is currently known to exist in two different isoforms in both rat and humans ($CRH2_a$ and $CRH2_b$). The three receptors CRF1, $CRF2_a$ and $CRF2_b$, which encode 411-, 415- and 431-amino-acid proteins, transduce their signals via the stimulation of intracellular cAMP production. Thus, binding of CRF to its receptor leads to an activation of adenylate cyclase, which initiates the activation of PKA. This is accompanied by an influx of calcium into the cell, suggesting that PKA activation leads to phosphorylation of a calcium channel or associated peptides.

CRF receptors have been found in high densities in the cortex, the ventral tegmental area, the cerebellum, the inferior olive, the medial vestibular nucleus and the nucleus accumbens. Moderate densities are described in the hippocampus, the neostriatum, the stria terminalis, some thalamic nuclei, the paraventricular nucleus of the hypothalamus, the lateral hypothalamus, the inferior colliculus, the locus coeruleus and the nucleus of the tractus solitarius.

Besides these receptors, a 37-kDa CRF-binding protein (CRF-BP) binds CRF peptides with high affinity. CRF receptors and CRF-BP do not share a common amino acid sequence representing the ligand binding site. In view of the unusually slow offrate of CRF-BP, it is proposed that CRF-BP provides an efficient uptake of free extracellular CRF. Thus, the time of exposure of CRF to the CRF receptors can be limited.

An additional CRF-like peptide, urocortin, was recently identified through molecular cloning (Vaughan et al. 1995). Urocortin is a 40-amino-acid peptide with approximately 45% homology of its putative amino acid sequence to CRF. The peptide binds with equal affinity to both CRF receptor subtypes, but possesses much higher affinity for CRF2 receptors than CRF and is found in brain regions distinct from CRF. Although urocortin represents a potent agonist at the classic CRF1 receptor, the physiological role of urocortin and its involvement in the pathophysiology of psychiatric disorders remains disputable.

Urocortin has many of the effects of CRF but also is significantly more potent than CRF in decreasing feeding in both meal-deprived and free-feeding rats.

4.8.5
Biological Effects

Most of the endocrine, neurochemical, electrophysiological and behavioral effects which occur after intracerebral injection of CRF are comparable to those accompanying stress responses. These effects can be diminished or abolished by the administration of CRF antagonists. CRF is therefore considered to play a predominant role in the brain–pituitary–adrenal axis.

Besides a stimulatory effect of CRF on the secretion of ACTH, it also enhances the synthesis of proopiomelanocortin (the precursor of ACTH and several other opioidergic peptides). Furthermore, CRF is able to potentiate the effects of AVP and oxytocin upon the secretion of ACTH. CRF is also known to suppress the secretion of growth hormone (GH). This mechanism, however, seems to be indirect in nature since suppression of GH requires somatostatin.

Inhibitory effects of CRF on the secretion of the luteinizing hormone (LH) and gonadotropin-releasing hormone (GnRH) are further prominent functions of CRF.

The existence of direct synaptical connectivities between CRF-containing axons and dendrites of GnRH-containing neurons in the preoptic area of the rat led to the assumption that CRF influences directly the secretion or synthesis of GnRH.

A positive influence on the secretion of β-endorphin and dynorphin has also been elucidated, as is the case for tyrosine hydroxylase (the enzyme which is necessary for the biosynthesis of norepinephrine) within the locus coeruleus.

In contrast to the stimulatory effect on catecholamines, CRF seems to have no effect upon the metabolism of GABA or serotonin.

CRF was also shown to produce a long-lasting potentiation of synaptic efficacy in neurons of the rat dentate gyrus *in vivo*; and it has been shown that CRF mRNA level in the dentate gyrus were significantly increased after LTP *in vitro*. These data suggest that CRF is capable of modulating hippocampal-dependent learning processes. By behavioral studies in which CRF was directly administered into the dentate gyrus of Sprague–Dawley rats, a memory-enhancing effect of CRF in the hippocampus has been documented.

4.8.6
Neurological Disorders and Neurodegenerative Diseases

The most important function of the corticotropin-releasing factors is their involvement in stress. The release of ACTH from the corticotrophs is controlled principally by vasopressin and CRF. CRF is the most potent ACTH secretagogue in both rat and man. Its ability to stimulate ACTH secretion is potentiated several-fold by agonists such as angiotensin II, norepinephrine and epinephrine. Moreover, in the rat, CRF is the only hypothalamic neuropeptide known to increase POMC biosynthesis. After CRF-activation of corticotrophes, released

ACTH travels through the circulation and binds to its receptor on the surface of cells of the adrenal cortex. In response to receptor activation, cortical cells synthesize glucocorticoids.

Other targets are more related to peripheral effects of the stress response system. CRF causes an increase in the plasma concentrations of epinephrine and norepinephrine, which in turn enhances glycemia, increases the plasma concentration of glucagon and leads to higher oxygen consumption as well as an increase in blood pressure. These different effects on the autonomic nervous system seem to be consecutively induced by stimulation of the sympatic and parasympatic systems.

CRF plays a role in mediating not only the neuroendocrine, but also the autonomic and behavioral responses to stress. It has been demonstrated, for example, that administration of CRF produces physiological and behavioral changes which are almost identical to those observed in response to stress, including increased heart rate and mean arterial pressure, suppression of exploratory behavior in a new environment and increased conflict behavior. Moreover, centrally administered CRF has been shown to enhance behavioral responses to stressors, as evidenced by a reduction in exploratory behavior in a novel, presumably stressful environment and enhancement of stress-induced freezing behavior.

The behavioral effects of centrally administered CRF can be reversed by CRF-receptor antagonists; and a CRF-receptor antagonist can attenuate many of the behavioral consequences of stress, underscoring the role of endogenous CRF in mediating stress-induced behaviors.

A current hypothesis is that CRF plays a role in depression through hyper-secretion from hypothalamic as well as from extrahypothalamic neurons, resulting in hyperactivity of the brain–pituitary–adrenal axis and an elevation of its level in cerebrospinal fluid. This increase in activity of CRF neurons is believed to mediate some of the behavioral symptoms of depression involving sleep or appetite disturbances, reduced libido and psychomotor changes.

The administration of CRF in high doses leads to analgesic effects, which are thought to be coupled to a stimulation of the secretion of endogenous opiates (especially β-endorphin). Local application of CRF (e.g. by iontophoresis) stimulates neuronal activity of cortical and hippocampal neurons. Intraventricular application of high doses of CRF can also induce epileptiform activity.

Since central injection of CRF also induces anxiogenic reactions, which can be blocked by benzodiazepines, it seems likely that CRF is involved in behavior related to fear. Along this line, it has been shown that central administration of the CRF produces anxiogenic behavior. In addition, intracerebroventricular administration of urocortin-II induces anxiogenic-like behaviors in mice in a dose-dependent manner. This effect suggests a role for CRF and CRF-related proteins as mediators of some aspect of stress or anxiety.

Behavioral studies have shown that using an agonist (cortagine) of corticotropin-releasing factor receptor subtype 1 produces anxiogenic and anti-depressive effects in mice. Thus, both CRF (acting via CRF-1 receptors) and urocortin II (acting via CRF-2 receptors) may be involved in anxiogenic effects by acting via two different signaling pathways.

Since glucocorticoids convey a profound influence on the immune system, by suppressing the proliferation of lymphocytes, CRF is also an indirect immunomodulator. In addition, CRF itself seems to be able to exert a glucocorticoid-independent suppression of the immune system.

Further Reading

Antoni, F. A. (1986): Hypothalamic control of adrenocorticotropin secretion: advances since the discovery of 41-residue corticotropin-releasing factor. *Endocrinol. Rev.* **7**: 351–378.

Arborelius, L., Owens, M. J., Plotsky, P. M., Nemeroff, C. B. (1999): The role of corticotropin-releasing factor in depression and anxiety disorders. *J. Neuroendocrinol.* **160**: 1–12.

Dautzenberg, F. M., Hauger, R. L. (2002): The CRF peptide family and their receptors: yet more partners discovered. *Trends Pharmacol. Sci.* **23**: 71–77.

De Souza, E. B. (1995): Corticotropin-releasing factor receptors: physiology, pharmacology, biochemistry and role in central nervous system and immune disorders. *Psychoneuroendocrinology* **20**: 789–819.

Dunn, A. J., Berridge, C. W. (1990): Physiological and behavioral responses to corticotropin-releasing factor administration: is CRF a mediator of anxiety or stress responses? *Brain Res.* **15**: 71–100.

Dunn, A. J., Swiergiel, A. H., Palamarchouk, V. (2004): Brain circuits involved in corticotropin-releasing factor-norepinephrine interactions during stress. *Ann. N.Y. Acad. Sci.* **1018**: 25–34.

Eckart, K., Radulovic, J., Radulovic, M., Jahn, O., Blank, T., Stiedl, O., Spiess, J. (1999): Actions of CRF and its analogs. *Curr. Med. Chem.* **6**: 1035–1053.

Ehlers, C. L., Henriksen, S. J., Wang, M., Rivier, J., Vale, W., Bloom, F. E. (1983): Corticotropin-releasing factor produces increases in brain exitability and convulsive seizures in rats. *Brain Res.* **278**: 332–336.

Gravanis, A., Margioris, A. N. (2005): The corticotropin-releasing factor (CRF) family of neuropeptides in inflammation: potential therapeutic applications. *Curr. Med. Chem.* **12**:1503–1512.

Hsu, S. Y., Hsueh, A. J. (2001): Human stresscopin and stresscopin-related peptide are selective ligands for the type 2 corticotropin-releasing hormone receptor. *Nat. Med.* **7**: 605–611.

Irwin, M. (1993): Brain corticotropin-releasing hormone- and interleukin-1β-induced suppression of specific antibody production. *Endocrinology* **133**: 1352–1360.

Joseph, S. A., Pilcher, W. H., Knigge, K. M. (1985): Anatomy of the corticotropin-releasing factor and opiomelanocortin systems of the brain. *Fed. Proc.* **44**: 100–107.

Li, C., Vaughan, J., Sawchenko, P. E., Vale, W. W. (2002): Urocortin III-immunoreactive projections in rat brain: partial overlap with sites of type 2 corticotrophin-releasing factor receptor expression. *J. Neurosci.* **22**: 991–1001.

Matsumoto, Y., Abe, M., Watanabe, T., Adachi, Y., Yano, T., Takahashi, H., Sugo, T., Mori, M., Kitada, C., Kurokawa, T., Fujino, M. (2004): Intracerebroventricular administration of urotensin II promotes anxiogenic-like behaviors in rodents. *Neurosci. Lett.* **358**: 99–102.

Morrison, E., Tomasec, P., Linton, E. A., et al. (1995): Expression of biologically active procorticotropin-releasing hormone (ProCRH) in stably transfected CHO-K1 cells: characterization of nuclear ProCRH. *J. Neuroendocrinol.* **7**: 263–272.

Reul, J. M., Holsboer, F. (2002): Corticotropin-releasing factor receptors 1 and 2 in anxiety and depression. *Curr. Opin. Pharmacol.* **2**: 23–33.

Reyes, T. M., Lewis, K., Perrin, M. H., Kunitake, K. S., Vaughan, J., Arias, C. A., Hogenesch, J. B., Gulyas, J., Rivier, J., Vale, W. W., Sawchenko, P. E. (2001): Urocortin II: a member of the corticotropin-releasing factor (CRF) neuropeptide family that is selectively bound by type 2 CRF receptors. *Proc. Natl Acad. Sci. USA* **98**: 2843–2848.

Rivier, J., Rivier, C., Vale, W. (1984): Synthetic competitive antagonists of corticotropin-releasing factor: effect on ACTH secretion in the rat. *Science* **224**: 889–891.

Tezval, H., Jahn, O., Todorovic, C., Sasse, A., Eckart, K., Spiess, J. (2004): Cortagine, a specific agonist of corticotropin-releasing factor receptor subtype 1, is anxiogenic and antidepressive in the mouse model. *Proc. Natl Acad. Sci. USA* **101**: 9468–9473.

Vale, W., Spiess, J., Rivier, C., Rivier, J. (1981): Characterization of 41-residue bovine hypothalamic peptide that stimulates secretion of corticotropin and beta-endorphin. *Science* **213**: 1394–1397.

Vaughan, J., Donaldson, C., Bittencourt, J., Perrin, M.H., Lewis, K., Sutton, S., Chan, R., Turnbull, A.V., Lovejoy, D., Rivier, C., et al. (1995): Urocortin, a mammalian neuropeptide related to fish urotensin I and to corticotropin-releasing factor. *Nature* **378**: 287–292.

4.9 Dynorphin

4.9.1 General Aspects and History

Opioids are schematically divided into three different classes of neuroactive peptides, namely dynorphins, endorphins and enkephalins. The term dynorphin was coined by Goldstein et al. (1979), who discovered this group of neuropeptides. The acronym combines the prefix "dyn-", emphasizing the "extraordinary potency", with "-orphin", indicating the functional relationship to morphine. Dy-

Table 4.2 The amino acid sequences of metabolites of the precursor peptide prodynorphin. An obvious feature of the biologically active substances is that they all have the sequence of Leu-enkephalin (marked in bold letters) in common. Basic doublets, which represent potential cleavage sites for degradation, are indicated in italic letters.

Metabolite	Amino acid sequence
α-Neoendorphin	**Tyr-Gly-Gly-Phe-Leu**-*Arg-Lys*-Tyr-Pro-Lys
β-Neoendorphin	**Tyr-Gly-Gly-Phe-Leu**-*Arg-Lys*-Tyr-Pro
Dynorphin (1–32)[a]	**Tyr-Gly-Gly-Phe-Leu**-*Arg-Arg*-Ile-Arg-Pro-Lys-Leu-Lys-Trp-Asp-Asn-Gln-*Lys-Arg*-**Tyr-Gly-Gly-Phe-Leu**-*Arg-Arg*-Gln-Phe-Lys-Val-Val-Thr
Dynorphin A (1–17)	**Tyr-Gly-Gly-Phe-Leu**-*Arg-Arg*-Ile-Arg-Pro-Lys-Leu-Lys-Trp-Asp-Asn-Gln
Dynorphin A (1–13)	**Tyr-Gly-Gly-Phe-Leu**-*Arg-Arg*-Ile-Arg-Pro-Lys-Leu-Lys
Dynorphin A (1–8)	**Tyr-Gly-Gly-Phe-Leu**-*Arg-Arg*-Ile
Dynorphin B (1–29)	**Tyr-Gly-Gly-Phe-Leu**-*Arg-Arg*-Gln-Phe-Lys-Val-Val-Thr-Arg-Ser-Gln-Glu-Asp-Pro-Asn-Ala-Tyr-Ser-Gly-Glu-Leu-Phe-Asp-Ala
Dynorphin B (1–13)	**Tyr-Gly-Gly-Phe-Leu**-*Arg-Arg*-Gln-Phe-Lys-Val-Val-Thr
Leu-Enkephalin[b]	**Tyr-Gly-Gly-Phe-Leu**

a) Dynorphin (1–32) is a product of prodynorphin, which can be cleaved into dynorphin-A and dynorphin-B.
b) The amino acid sequence of Leu-enkephalin is found in all cleaved products of pro-dynorphin.

norphins possess a high affinity to specific receptors and exert physiological actions in the nanomolar range.

Goldstein's initial discovery led to the identification and purification of further peptides, which belong to the dynorphin family. These neuropeptides are dynorphin A-(1–17) and dynorphin A-(1–8). A third neuropeptide, dynorphin B-(1–13), has been isolated. Dynorphin B-(1–13) is composed of 13 amino acids and was purified from the pituitary. To date, seven members of the dynorphin family have been described, all of which are metabolized from a common precursor. This precursor, a 256-amino-acid peptide, is known as prodynorphin (or pro-enkephalin B; see Table 4.2).

4.9.2
Localization Within the Central Nervous System

Neuronal somata, which express prodynorphin, have been found in different areas of the central nervous system. In detail, prodynorphin-containing cells have been detected in the neocortex, the hypothalamus (paraventricular nucleus, supraoptic nucleus, arcuate nucleus, ventro-medial and dorso-medial nucleus of the hypothalamus) and in some nuclei of the amygdala, the caudate-putamen and the dentate gyrus.

Prodynorphin-containing neurons have also been detected in the periaqueductal gray matter and in several areas of the brain stem, as well as in the pituitary and the dorsal horn of the spinal cord (Fig. 4.18). Dynorphin immunoreactivity in the spinal cord is contained in both interneurons and projection neurons.

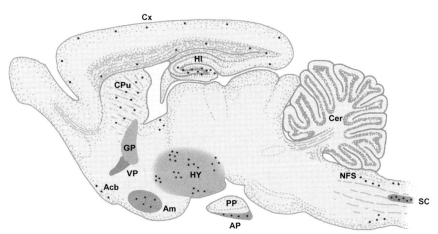

Fig. 4.18 Schematic overview of the distribution of dynorphin-containing neurons in the central nervous system. *Abbreviations:* Acb = *nucleus accumbens*; AP = anterior pituitary; CPu = caudate putamen; GP = *globus pallidus*; HI = hippocampus; HY = hypothalamus; NFS = nucleus of the *fasciculus solitarius*; PP = posterior pituitary; SC = spinal cord (dorsal horn); VP = ventral pallidum.

Areas with prominent dynorphinergic projections include the hippocampus, the globus pallidus, the substantia nigra, the ventral pallidum, the nucleus accumbens and the posterior pituitary. Outside the central nervous system, prodynorphin has been described in the autonomic nervous system and in several reproductive organs.

4.9.3
Biosynthesis and Degradation

The dynorphins are synthesized from a common precursor (pro-dynorphin or pro-enkephalin B; the term pro-enkephalin B was chosen because the precursor displays a striking homology with the amino acid sequence of pro-enkephalin). Prodynorphin harbors three segments containing copies of leucine–enkephalin (Leu-enkephalin; Fig. 4.19). Each of these copies is flanked by a pair of basic amino acids (Arg–Arg or Arg–Lys).

Pro-dynorphin can be found in high concentrations in the hypothalamus and the substantia nigra; and moderate concentrations have been found in the hippocampus and globus pallidus.

Interestingly, pro-dynorphin mRNA decreases significantly with age in the amygdala; while in contrast, pro-dynorphin mRNA increases significantly with age in the hippocampus, indicating an age-dependent modulation in dynorphin synthesis in brain region associated with learning and memory.

Hydrolysis of the Lys–Arg pair produces dynorphin A-(1–17), dynorphin B-(1–29) and α-neoendorphin. Cleavage at the Arg site leads to the formation of dynorphin A-(1–8) and dynorphin B-(1–13) (also known as rimorphin) and cleavage at the Lys residue allows the formation of β-neoendorphin from α-neoendorphin, as well as the formation of dynorphin A-(1–13) from dynorphin A-(1–17).

The levels of central dynorphins in descending concentrations are as follows: α-neoendorphin > β-neoendorphin > dynorphin B-(1–13) > dynorphin A-(1–17) > dynorphin A-(1–8).

Since there are consistently higher levels of neoendorphins than of dynorphin A-(1–17) and dynorphin B-(1–13) in the brain, it is speculated that the latter peptides are metabolized to shorter fragments, in particular to Leu-enkephalins.

The degradation steps of dynorphins processed from the precursor prodynorphin are not yet fully elucidated. Diverse aminopeptidases are involved in the degradation of dynorphin A-(1–13), dynorphin A-(1–8) and dynorphin A-(1–17). *In vitro* experiments have demonstrated that pro-dynorphin-derived peptides are de-

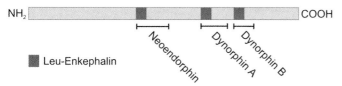

Fig. 4.19 Structure of the precursor prodynorphin.

graded by soluble and membrane-bound forms of the metalloendopeptidase (EC 3.4.24.15) and by enkephalinase (EC 3.4.24.11). The metalloendopeptidase EC 3.4.25.15, which constitutes a membranous peptidase in synaptosomal fractions, converts dynorphin A-(1–8) as well as α- and β-neoendorphin to Leu-enkephalin.

4.9.4
Receptors and Signal Transduction

At least three different families of opioid receptors have been discovered [mu (μ), delta (δ), and kappa (κ) receptors – these are discussed in detail in Section 4.11]. Dynorphin possess a high affinity for κ receptors and a low affinity to receptors of the μ and δ type.

The κ receptor is highly sensitive to the derivatives of pro-dynorphine. However, some effects of the short fragments of pro-dynorphin, like dynorphin A-(1–8) can be attributed to interactions with additional types of opioid receptor.

Internalization and down-regulation are important mechanisms in the modulation of receptor efficacy. Recently, Jordan and coworkers (2000) showed that bound dynorphin peptides promote the internalization of κ receptors.

The use of highly selective ligands for the κ receptor has enabled its distribution in the brain to be studied. The data obtained to date indicate that some regions of the brain exhibit colocalization of terminals of dynorphin-containing neurons and κ receptors, particularly in some nuclei of the hypothalamus, the amygdala, the striatum and the dorsal horn of the spinal cord.

In other brain areas, a differential distribution of ligands and receptors has been found, e. g. in the dentate gyrus abundant terminals of dynorphin-containing neurons are present with low numbers of kappa receptors, while in the pre-optic hypothalamic area this relationship is reversed.

At first glance, the lack of a corresponding relationship between receptor and dynorphin quantities seems paradoxical. A reasonable explanation derives from the high affinity of dynorphins to their receptors. Dynorphins can potentially diffuse over a long distance to reach their site of action (about 50–100 µm). This feature is not exclusive to dynorphins, because some other neuropeptides also interact with distant targets. The ability to act on distant targets is called "neurohumoral transmission" or volume transmission, to distinguish this mechanism from the "short-range" action of neuropeptides which exert their effect intersynaptically (so-called wiring transmission).

Interestingly, dynorphins are not only released from axon terminals but also from neuronal dendrites.

The binding of dynorphins to their receptors leads to an activation of G proteins. The downstream consequences of G protein activation are manifold:
- inhibition of adenylate cyclase (reducing the production of cAMP);
- inhibition of N-type calcium channels (decreasing the propagation of action potentials and reduction of transmitter release);
- activation of inward rectifying or delayed rectifying potassium channels (thus hyperpolarizing the cell and reducing neurotransmitter release).

4.9.5
Biological Effects

Morphines are alkaloids derived from plants and bind to opioidergic receptors. Dynorphins are peptidergic opiates which are synthesized by neurons and act as neuromodulators. Dynorphins seem not to bind to all receptors which are capable of binding morphines, but bind predominantly to κ receptors (see above). In addition to dynorphins, neurons also synthesize some additional endogenous opioids, like enkephalins and endorphins. All these endogenous opioidergic peptides are differently processed and possess particular features which allow them to bind to their specific opioidergic receptor complement.

In several studies, it has been demonstrated that dynorphins play a role in the control of a number of biological functions. Dynorphins are involved in the control of respiration, the control of the cardiovascular system and the control of gastrointestinal functions. Furthermore, dynorphins modulate motor functions and feeding behavior, and they are involved in regulating the secretion of hormones from the pituitary.

In addition, dynorphins may have a role in learning mechanisms. Dynorphin inhibits the induction of long-term potentiation (a model of synaptic plasticity) in the hippocampus and dynorphins could impair animal performance in behavioral tasks requiring spatial learning (a form of learning which is related to the hippocampus). Thus, rats trained in the Morris water task after microinjection with dynorphin B showed impaired spatial learning. Co-application of dynorphin B and κ receptor antagonists fully blocked the acquisition impairment caused by dynorphin B, indicating that dynorphin B acting through κ receptors impairs hippocampal-dependent spatial learning (Sandin et al. 1998).

4.9.6
Neurological Disorders and Neurodegenerative Diseases

The induction of a neuropathic state in rats by nerve ligation or constriction is associated with an up-regulation of dynorphin synthesis, reflected by enhanced expression of prepro-dynorphin protein and mRNA, increased dynorphin immunoreactivity and a greater percentage of spinal neurons receiving dynorphin-immunoreactive contacts. The presence of dynorphin receptors and dynorphins (and its increased synthesis in neuropathic states) in spinal pathways suggests a role for these peptides in nociception. This assumption is strengthened by the fact that agonists of κ receptors posses anti-nociceptive properties. Data on the effects of κ receptor agonists in the spinal cord are controversial. While agonists exert no effect on neurons of the dorsal horn, which are electrically stimulated, it has been shown that agonists inhibit depolarization induced by thermal or mechanical nociceptive stimuli. It has also been shown that the intraventricular application of κ receptor agonists or their direct application to the periaqueductal gray induces only subtle antinociceptive effects, whereas stimulation of supraspinal κ receptors is more effective in inhibiting thermal or mechanical nociceptive stimuli. Taken together, the data on κ receptors and their possible interaction with dynorphins indicate a considerable degree of functional diversification in nociceptive mechanisms.

It has been demonstrated that relatively low doses of dynorphin produce analgesia, whereas higher doses produce hyperalgesia that persists for greater then 60 days after a single intrathecal injection. This protracted effect appears to be independent of activation of opioid receptors. In addition it has been shown that, under pathological conditions resulting from injury to peripheral nerves, the up-regulation of spinal dynorphin is accompanied by the development of chronic pain states. Thus, the development of chronic pain states can be blocked by anti-dynorphin antiserum (Lai et al. 2001). Thus, dynorphin can have both nociceptive and antinociceptive properties. It is thought that low levels of dynorphin, acting via κ receptors, induce analgesia. Higher doses of dynorphin allows dynorphin to interact with multiple sites on the NMDA receptor complex and, thereby, to produce excitatory responses resulting in nociceptive and even toxic effects (Laughlin et al. 2001).

Dynorphins may also play a role in stress, since stress increases dynorphin immunoreactivity in different limbic brain regions and since dynorphin antagonism produces antidepressant-like effects in rats exposed to learned helplessness (an animal model of depression).

It was found that dynorphin modulates neuronal activity in *in vitro* brain slices of the hippocampus and that dynorphins are released in the hippocampus during complex partial seizures. Since dynorphin potentiates endogenous anti-ictal processes, it was suggested that dynorphin plays a role as an endogenous anticonvulsant. Thus, there seemed to exist also a role of dynorphins in epilepsy.

Further Reading

Dubner, R., Ruda, M.A. (1992): Activity-dependent neuronal plasticity following tissue injury and inflammation. *Trends Neurosci.* **15**: 96–103.

Elde, R., Arvidsson, U., Riedl, M., Vulchanova, L., Lee, J.H., Dado, R., Nakano, A., Chakrabarti, S., Zhang, X., Loh, H.H. (1995): Distribution of neuropeptide receptors. New views of peptidergic neurotransmission made possible by antibodies to opioid receptors. *Ann. N.Y. Acad. Sci.* **757**: 390–404.

Fallon, J.H., Leslie, F.M. (1986): Distribution of dynorphin and enkephalin peptides in the rat brain. *J. Comp. Neurol.* **249**: 293–336.

Goldstein, A., Tachibana, S., Lowney, L.I., Hunkapiller, M., Hood, L. (1979): Dynorphin-(1–13), an extraordinarily potent opioid peptide. *Proc. Natl Acad. Sci. USA* **76**: 6666–6670.

Jordan, B.A., Cvejic, S., Devi, L.A. (2000): Kappa opioid receptor endocytosis by dynorphin peptides. *DNA. Cell Biol.* **19**: 19–27.

Kotz, C.M., Weldon, D., Billington, C.J., Levine, A.S. (2004): Age-related changes in brain prodynorphin gene expression in the rat. *Neurobiol. Aging* **25**: 1343–1247.

Lai, J., Ossipov, M.H., Vanderah, T.W., Malan, T.P. Jr, Porreca, F. (2001): Neuropathic pain: the paradox of dynorphin. *Mol. Interv.* **1**: 160–167.

Laughlin,T.M., Larson, A.A., Wilcox, G.L. (2001): Mechanisms of induction of persistent nociception by dynorphin. *J. Pharmacol. Exp. Ther.* **299**: 6–11.

Millan, M.J. (1990): κ-Opioid receptors and analgesia. *Trends Biochem.* **11**: 70–76.

Morris, B.J., Johnston, H.M. (1995): A role for hippocampal opioids in long term functional plasticity. *Trends Neurosci.* **18**: 350–354.

Nahin, R. L., Hylden, J. L., Humphrey, E. (1992): Demonstration of dynorphin A 1–8 immunoreactive axons contacting spinal cord projection neurons in a rat model of peripheral inflammation and hyperalgesia. *Pain* **51**:135–143.

Narita, M., Tseng, L. F. (1998): Evidence for the existence of the beta-endorphin-sensitive "epsilon-opioid receptor" in the brain: the mechanisms of epsilon-mediated antinociception. *Jpn J. Pharmacol.* **76**: 233–253.

Sandin, J., Nylander, I., Georgieva, J., Schott, P. A., Ogren, S. O., Terenius, L. (1998): Hippocampal dynorphin B injections impair spatial learning in rats: a kappa-opioid receptor-mediated effect. *Neuroscience* **85**: 375–382.

Shikla, V. K., Lemaire, S. (1994): Non-opioid effects of dynorphins: possible role of the NMDA receptor. *Trends Pharmacol. Sci.* **15**: 420–424.

Shirayama, Y., Ishida, H., Iwata, M., Hazama, G. I., Kawahara, R., Duman, R. S. (2004): Stress increases dynorphin immunoreactivity in limbic brain regions and dynorphin antagonism produces antidepressant-like effects. *J. Neurochem.* **90**: 1258–1268.

Smith, A. P., Lee, N. M. (1988): Pharmacology of dynorphin. *Ann. Rev. Pharmacol. Toxicol.* **28**: 123–140.

Solbrig, M. V., Koob, G. F. (2004): Epilepsy, CNS viral injury and dynorphin. *Trends Pharmacol. Sci.* **25**:98–104.

Steiner, H., Gerfen, C. R. (1998): Role of dynorphin and enkephalin in the regulation of striatal output pathways and behavior. *Exp. Brain Res.* **123**: 60–76.

Qu, Z.-X., Isaac, L. (1993): Dynorphin A (1–13) potentiates dynorphin A (1–17) on loss of the tail-flick reflex after intrathecal injection in the rat. *Brain Res.* **610**: 340–343.

4.10
Eicosanoids and Arachidonic Acid

4.10.1
General and History Aspects

Eicosanoids (from the greek *eikosi*=twenty) comprise a family of biologically active molecules which derive from fatty acids with 20 carbon atoms that contain three to five double bonds. Eicosanoids are strikingly potent and are able to elicit profound physiological effects at very low concentrations.

Their major precursor is arachidonic acid, which is also known as eicosatetraenoic acid 20:4. Arachidonic acid is an essential fatty acid and is produced in literally all mammalian cells except erythrocytes.

Eicosanoids consist of different groups of biologically active unsaturated fatty acids: prostaglandins (PGs), prostacyclin, thromboxanes (TXs) and leukotrienes (LTs). Prostaglandins, prostacyclin and thromboxanes are also collectively labeled as prostanoids.

A particular member of eicosanoids is anandamide. Anandamide binds to specific receptors in brain tissue; and the effects of anandamide are comparable to those of the cannabinoids. Since anadamide is a potent neuromodulator, a separate section is dedicated to this eicosanoid (Section 4.2).

4.10.2
Biosynthesis and Degradation

Cell membranes in the central nervous system contain abundant phospholipids, one of which is provided by esterified arachidonic acid. Under normal conditions, the level of arachidonic acid within the cytoplasm is low, since acetyltransferases rapidly esterify arachidonic acid.

The rate-determining step in the generation of eicosanoids in the central nervous system is the release of arachidonic acid through the action of cytosolic phospholipase A2 (cPLA2). The activation of cPLA2 is under the control of phospholipase A2-activating protein (PLAP). For example, in the hippocampus, cPLA2 can be activated by receptor activation of the NMDA or 5-HT2 receptor-type.

Arachidonic acid release is locally restricted, as is its transformation into different metabolites. Since inactivation of the eicosanoids is fast, their effects are restricted to sites close to the formation site. Thus, eicosanoids represent prototypes of autocrine or paracrine factors.

Three different enzymatic pathways are known to metabolize arachidonic acid (Fig. 4.20) starting from cyclooxygenase (prostaglandin G/H synthase), which catalyzes the formation of prostaglandins [prostaglandin G2 (PGG2) and prostaglandin H2 (PGH2)] and thromboxanes.

Enzymatic isomerization of PGH2 gives rise to the prostaglandins PGD2, PGE2 and PGF2α. PGH2 can also be metabolized to the thromboxane A2 (TXA2) and prostacyclin (PGI2). Both metabolites possess very short lifetimes and they are metabolized rapidly to thromboxane B2 (TXB2) and 6-keto-prostaglandine F1α. Two different forms of cycloxigenases have been identified, a constitutive and an inducible form, which are encoded by two different genes (Cox1, alternatively PGHS-1; and Cox2, alternatively PGHS-2). COX-2 is the inducible isoform, rapidly expressed in several cell types in response to growth factors, cytokines and pro-inflammatory molecules. COX-2 expression in brain has been associated with pro-inflammatory activities, thought to be instrumental in neurodegenerative processes of several acute and chronic diseases. Growth factors and cytokines induce COX-2 synthesis, while the transcription of the COX-2 gene is inhibited by glucocorticoids.

Starting from lipoxygenases, which catalyzes the synthesis of leukotrienes, lipoxines and other specific hydroperoxy- and hydroxyeicosanoides, three different lipoxygenases are present in central nervous tissue: the 5-lipoxygenase, the 12-lipoxygenase and the 15-lipoxygenase. The 5-lipoxygenase is a cytosolic dioxygenase that catalyzes the formation of di-5(S)-hydroperoxy-eicosatetraenoic acid (5-HPETE) from arachidonic acid. This substrate is further metabolized to the leukotrienes LTA$_4$, LTB$_4$, LTC$_4$ and LTD$_4$. 5-Lipoxygenase has been identified within the cerebellum and the hippocampus. 12-Lipoxygenase has been found in neurons, but is also expressed in glia and in endothelial cells. It converts arachidonic acid to 12(S)-hydroperoxy-eicosatetraenoic acid (12-HPETE), which is further metabolized to 12(S)-hydroxy-eicosatetraenoic acid (12-HETE). 12-HETE is the

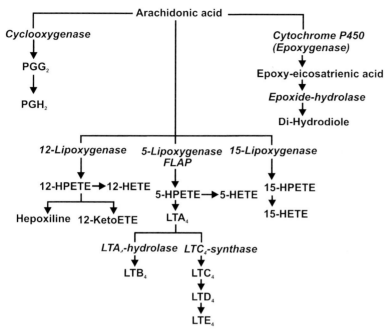

Fig. 4.20 Metabolism of arachidonic acid by the activity of cyclooxygenases, lipoxygenases and epoxygenases which gives rise to different derivatives.

most prominent arachidonic acid derivative identified in central nervous tissues. 15-Lipoxygenase metabolizes arachidonic acid to 15(S)-hydroxy-eicosatetraenoic acid (15-HETE).

Starting from epoxygenase (cytochrome P_{450}), which is involved in the generation of epoxyeicosanoide and dihydroxyeicosanoide, arachidonic acid is metabolized by the epoxygenase to epoxyeicosatrienoic acid. This in turn is further processed by epoxide hydrolases to dihydrodiols.

A further metabolic pathway gives rise to anandamide from its precursor N-arachidonyl-phosphatidyl-ethanolamine (NAPE), which is catalyzed by the enzyme phospholipase D (see Section 4.2). Furthermore, arachidonic acid can be peroxidized by free radicals to yield isoprostane. Isoprostane also possesses some biological activity and, unlike prostaglandins, is esterified to a phospholipid.

4.10.3
Receptors and Signal Transduction

Free arachidonic acid exerts a potent action on protein kinase C (PKC), as well as on Ca^{2+} and K^+ channels. It is thought that arachidonic acid itself serves as a retrograde messenger during long-term potentiation. Alternatively, arachidonic acid gives rise to biologically active eicosanoids.

Receptors for prostaglandine PGD2 and PGE2 (both are cyklooxygenase products) have been identified in the peripheral and in the central nervous system.

Prostaglandins exert their effects by activating rhodopsin-like seven transmembrane-spanning G protein-coupled receptors (GPCRs). The prostanoid receptor subfamily is comprised of nine members (DP, EP1-4, FP, IP, CRTH2 and TP). The EP1, EP2, EP3 and EP4 receptors are coupled to G proteins and they activate or inhibit adenylate cyclase. The entire receptor complement is expressed in the spinal cord, the medulla, the pons, the diencephalon, the hippocampus, the cerebral cortex, the cerebellum and the pituitary. Concentrations of EP3 receptors have been identified in the dorsal motor nucleus of the vagus, the spinal intermediolateral nucleus and the sacral parasympathetic nucleus, as well as in the median preoptic nucleus, the medial preoptic area, the dorsomedial hypothalamic nucleus and the premammillary nucleus.

12-HPETE (a product of the 12-lipoxygenase activity) opens K^+ channels in *Aplysia*. It is thought that 12-HPETE directly interacts with this receptor. In addition, 12-lipoxygenase products are able to inhibit the activity of calcium/calmodulin-dependent protein kinase II.

4.10.4
Biological Effects

Eicosanoids are involved in a wide range of biological processes, such as inflammatory responses, the duration and intensity of pain and fever, reproduction and blood pressure.

Eicosanoids are capable of modifying the activity of ligand-gated ion channels, activating protein kinase C and inhibiting Ca^{2+}/calmodulin-dependent protein kinase 2.

Arachidonic acid is able by itself to inhibit the activity of the Na^+/K^+ ATPase and to inhibit well neurotransmitters re-uptake systems.

The release of arachidonic acid can be stimulated by NMDA receptor activation, but not by kainate receptor activation. The non-selective NMDA antagonist MK-801 inhibits the release of arachidonic acid as well as the induction of LTP within the hippocampus. These data lend support to the idea that arachidonic acid may play a role as a retrograde messenger – as is thought in the case of nitric oxide (NO) – in mechanisms associated with learning and memory.

4.10.5
Neurological Disorders and Neurodegenerative Diseases

With respect to the cyclooxygenase products, it is of interest that some non-steroid anti-inflammatory drugs (such as acetylsalicylic acid) inhibit cyclooxygenase activity and thus reduce the formation of prostaglandins.

Prostaglandin receptors have been identified within diverse brain areas; and it has been supposed that prostaglandins are involved in regulation of circadian rhythm and body temperature. These ideas are corroborated by the fact that

PGE2 induces hyperthermia in response to administration of cytokines like interleukin1 (IL-1).

PGE2 in the ventromedial hypothalamus has also been shown to be involved in the regulation of pain, since PGE2 has an anti-nociceptive effect through its actions on specific EP1 receptors. Within the hypothalamus, PGD2 suppresses the release of luteinizing hormone. Oligodendrocytes synthesize and release prostaglandins of the E- and F-types. The production and release of oligodendroglial prostaglandins indicate that oligodendrocytes themselves can modulate inflammatory processes in the white matter of the central nervous system.

Neuroinflammation is thought to play a key role in some neurodegenerative disease, including amyotrophic lateral sclerosis (ALS). In this context, it is unclear whether ALS is propagated through inflammation, or whether the accompanying inflammation reflects an attempt to protect against further cellular injury. Inflammatory pathways involving COX enzymes and subsequent generation of prostaglandins are potential target sites for treatments to halt the progression of ALS. In the central nervous system, COX enzymes are localized to neurons, astrocytes and microglia and can be induced under various conditions. In addition, there appears to be a dual role for the prostaglandin products of COX enzymes in the nervous system. Thus, the inducible isoform of cyclooxygenase, COX-2, promotes motor neuron loss in rodent models of ALS, whereas PGE2 paradoxically protects motor neurons at physiological concentrations in this model.

The role of thromboxanes in the central nervous system is not well understood, however it has been shown that intraventricular application of thromboxane A2 promotes cardiovascular and hormonal responses.

The leukotrienes (LTB_4, LTC_4, LTD_4 and LTE4) are synthesized in small amounts in the brain. It is suspected that the leukotrienes play a role in pathological processes, as in inflammation, brain injury or vasogenic edema. In the pituitary, gonadotropin-releasing hormone (GRH) stimulates the formation of the leukotriene LTB_4 and the leukotriene LTC_4.

Further Reading

Althaus, H. H., Siepl, C. (1997): Oligodendrocytes isolated from adult pig brain synthesise and release prostaglandins. *Cell Tissue Res.* **287**: 135–141.

Bilak, M., Wu, L., Wang, Q., Haughey, N., Conant, K., St Hillaire, C., Andreasson, K. (2004): PGE2 receptors rescue motor neurons in a model of amyotrophic lateral sclerosis. *Ann. Neurol.* **56**: 240–248.

Carlson, R. O., Levitan, I. B. (1990): Regulation of intracellular free arachidonic acid in *Aplysia* nervous system. *J. Membrane Biol.* **116**: 249–260.

Clarke, S. D., Jump, D. B. (1993): Regulation of gene transcription by polyunsaturated fatty acids. *Prog. Cem.* **269**: 13057–13060.

Consilvio, C., Vincent, A. M., Feldman, E. L. (2004): Neuroinflammation, COX-2, and ALS – a dual role? *Exp. Neurol.* **187**: 1–10.

Cudd, T. A. (1998): Thromboxane A2 acts on the brain to mediate hemodynamic, adrenocorticotropin, and cortisol responses. *Am. J. Physiol.* **274**: R1353–R1360.

Ford-Hutchinson, A. W., Gresser, M., Young, R. N. (1994): 5-Lipoxygenase. *Annu. Rev. Biochem.* **63**: 383–417.

Hata, A. N., Breyer, R. M. (2004): Pharmacology and signaling of prostaglandin receptors: multiple roles in inflammation and immune modulation. *Pharmacol. Ther.* **103**: 147–166.

Hayaishi, O. (1991): Molecular mechanisms of sleep–wake regulation: roles of prostaglandins D2 and E2. *FASEB J.* **5**: 2575–2581.

Hunter, S. A., Burstein, S. H. (1997): Receptor mediation in cannabinoid stimulated arachidonic acid mobilization and anandamide synthesis. *Life Sci.* **60**: 1563–1573.

Koutek, B., Prestwich, G. D., Howlett, A. C., Chin, S. A., Salehani, D., Akhavan, N., Deutsch, D. G. (1994): Inhibitors of arachidonoyl ethanolamide hydrolysis. *J. Biol. Chem.* **269**: 22937–22940.

Leslie, J. B., Watkins, W. D. (1985): Eicosanoids in the central nervous system. *J. Neurosurg.* **63**: 659–668.

Minghetti, L. (2004): Cyclooxygenase-2 (COX-2) in inflammatory and degenerative brain diseases. *J. Neuropathol. Exp. Neurol.* **63**: 901–910.

Nakamura, K., Kaneko, T., Yamashita, Y., Hasegawa, H., Katoh, H., Ichikawa, A., Negishi, M. (1999): Immunocytochemical localization of prostaglandin EP3 receptor in the rat hypothalamus. *Neurosci. Lett.* **260**: 117–120.

Nakamura, K., Kaneko, T., Yamashita, Y., Hasegawa, H., Katoh, H., Negishi, M. (2000): Immunohistochemical localization of prostaglandin EP3 receptor in the rat nervous system. *J. Comp. Neurol.* **421**: 543–569.

Negishi, M., Sugimoto, Y., Ichikawa, A. (1993): Prostanoid receptors and their biological action. *Prog. Lipid Res.* **32**: 417–434.

Okuda, S., Saito, H., Katsuki, H. (1994): Arachidonic acid: toxic and trophic effects on cultured hippocampal neurons. *Neuroscience* **63**: 691–699.

Ordway, R. W., Singer, J. J., Walsh, J. V. (1991): Direct regulation of ion chanels by fatty acids. *Trends Neurosci.* **14**: 96–100.

Pellerin, L., Wolfe, L. S. (1991): Release of arachidonic acid by NMDA-receptor activation in the rat hippocampus. *Neurochem. Res.* **16**: 983–989.

Piomelli, D. (1994): Eicosanoids in synaptic transmission. *Crit. Rev. Neurobiol.* **8**: 65–83.

Piomelli, D., Greengard, P. (1990): Lipoxygenase metabolites of arachidonic acid in neuronal transmembrane signalling. *Trends Pharmacol. Sci.* **11**: 367–373.

Samuelsson, B., Dahlen, S. E., Lindren, J. A., Rouzer, C. A., Serhan, C. N. (1987): Leukotrienes and lipoxins: structures, biosynthesis and biological effects. *Science* **237**: 1171–1176.

Shimizu, T., Wolfe, L. S. (1990): Arachidonic acid cascade and signal transduction. *J. Neurochem.* **55**: 1–15.

Sigal, E. (1991): The molecular biology of mammalian arachidonic acid metabolism. *Am. J. Physiol.* **260**: L13–L28.

Tai, T. C., Lye, S. J., Adamson, S. L. (1998): Expression of prostaglandin E2 receptor subtypes in the developing sheep brainstem. *Brain Res. Mol. Brain Res.* **57**: 161–166.

Yamamoto, S. (1992): Mammalian liopxygenases: molecular structure and functions. *Biochim. Biophys. Acta.* **1128**: 117–131.

Yu, N., Martin, J.-L., Stella, N., Magistretti, P. J. (1993): Arachidonic acid stimulates glucose uptake in cerebral cortical asterocytes. *Proc. Natl Acad. Sci. USA* **90**: 4042–4046.

4.11
Endorphin

4.11.1
General Aspects and History

The high pharmacological potency and specificity of morphine suggested that it binds to specific receptors in the nervous system to induce its biological effects. In the early 1970s, several groups identified opioid receptors in brain and peripheral tissues, which prompted extensive search for an intrinsic ligand responsible for the opiate effects. In the middle of the 1970s, several groups identified a number of candidates. Among these was beta-endorphin, which was discovered by Bradbury and coworkers in 1975.

Beta-endorphin (a peptide composed of 31 amino acids, also known as endomorphin) belongs to the group of proteolytic substrates derived from the precursor proopiomelanocortin (POMC). A detailed description of the processing of proopiomelanocortin is provided separately (Section 4.28).

Beta-endorphin, like other opiodergic substances, activates μ, δ and κ-opioid receptors.

The neuropeptide beta-endorphin elicits effects comparable to those of morphine: analgesia, catatonia, hypotension and respiratory depression.

Furthermore, β-endorphin also induces certain neuroendocrine reactions. For instance, β-endorphin is capable of stimulating the secretion of several pituitary gland hormones, as is the case for prolactin. It is also known that β-endorphin inhibits the release of dopamine.

4.11.2
Localization Within the Central Nervous System

Beta-endorphin occurs in various areas of the central nervous system and is divergently transported by neurons to different target areas. The localization of proopiomelanocortin, the precursor of beta-endorphin, is described in Section 4.28.

4.11.3
Biosynthesis and Degradation

Proopiomelanocortin is the collective precursor of endorphins, but is also the precursor of some additional neuromodulators. One of the endproducts of the proteolytic processing of proopiomelanocortin is β-endorphin (also known as β-lipotropin 61–91; Fig. 4.21).

The neuromodulator β-endorphin is cleaved at its carboxy-terminal site (at a Lys–Lys cleavage site) which gives rise to β-endorphin-(1-27) and β-endorphin-(1–26).

> H-Tyr-Gly-Gly-Phe-Met-Thr-Ser-Glu-Lys-Ser-Gln-Thr-Pro-Leu-Val-Thr-Leu-Phe-Lys-Asn-Ala-Ile-Lys-Asn-Ala-Tyr-Lys-Lys-Gly-Glu-OH

Fig. 4.21 Amino acid sequence of β-endorphin.

Beta-endorphin can be further processed to form γ-endorphin-(1–17) and α-endorphin-(1–16).

Receptors and signal transduction

Three main classes of opioid receptors have been indentified (μ, δ and κ), each class consisting of different subtypes. These receptors were found to be localized both within the central nervous system and in peripheral tissues. The opioid receptor family has some features in common with the somatostatin receptor family. All of the cloned opioid receptor types belong to the G_i/G_o-coupled superfamily of receptors and exhibit the seven transmembrane domain motif. None of the receptors forms a ligand-gated ion channel. Endorphins are equipotent at μ and δ receptors, but they have a lower affinity for κ receptors.

In addition to these receptors, a further sensitive site for beta-endorphins has been postulated to account for the unique characteristics of an opioid receptor, defined as a putative receptor.

The μ receptors

The alkaloid morphine has played a crucial role in the investigation of endogenously occurring opioids. The μ receptors bind morphine with high affinity. These receptors are involved in pain relief, in the induction of respiratory depression and in the inhibition of motility of the gastrointestinal tract.

The MOR-1 gene, which encodes one form of the μ receptor, shows approximately 50–70% homology to the genes encoding the other opioid receptors. Two splice variants of the MOR-1 gene have been found, differing in eight amino acids of the carboxy-terminal region ($\mu 1/\mu 2$). Both splice variants exhibit differences in their rate of onset and recovery from agonist-induced internalization, but their pharmacology does not appear to differ in ligand-binding assays.

The μ receptors, like other opioid receptors, inhibit the activity of the adenylate cyclase. The distribution of μ receptors in the central nervous system has been determined by autoradiography. They are found in brain areas known to be involved in the control of nociception. Furthermore, μ receptors are found in the basal ganglia, thalamus, mesencephalon and metencephalon. High densities of μ receptors have been found in the neocortex, caudate-putamen, nucleus accumbens, hippocampus and amygdala, as well as in the nucleus of the tractus solitarius and in the colliculi.

Two subtypes, $\mu 1$ and $\mu 2$, can be distinguished in binding assays because the isoforms bind different opioidergic peptides with different affinities. Both forms, however, reveal nearly identical affinities to morphine.

Many of the morphine-mediated effects depend on the activation of the $\mu 1$ receptor. One prominent example is the central analgesic effect of morphine, in contrast to the effects of the $\mu2$ receptor, which mediates analgesic effects primarily in the spinal chord in addition to the morphine-dependent inhibition of respiration and motility of the gastrointestinal tract.

The δ receptors
Shortly after the discovery of the enkephalins, a receptor type was described which selectively binds opioid peptides. This receptor was named the δ receptor.

The only δ receptor gene which has been cloned to date is the DOR-1 gene. However, the receptor is commonly subdivided into $\delta1$ and $\delta2$ forms. This subdivision was originally proposed on the basis of pharmacological studies and it is believed that $\delta2$-receptors initiate analgesia in the spinal cord, whereas $\delta1$-receptors mediate analgesia in the brain. The δ-receptors are also members of the G protein-coupled receptor superfamily.

Within the brain, δ receptors are most common in olfactory regions, in the neocortex, the caudate-putamen, the amygdala and in the interpeduncular nucleus. In addition, their presence has been demonstrated in the CA3 area of the hippocampus, the bed nucleus of the stria terminalis, the preoptic area, the nucleus accumbens, the medial thalamus and in the spinal cord.

The functional properties of δ receptors in these brain regions are not well understood, but in all likelihood they are involved in cognition, olfaction and motor processing, as well as in some forms of analgesia and in the modulation of autonomic and endocrine functions.

The κ receptors
In 1993, the cDNA sequence of the κ receptor was identified by molecular cloning. The cDNA encodes for a protein which consists of 380 amino acids. The corresponding gene was labeled KOR-1. Like the other β-endorphine receptors, different subtypes of κ receptors seem to exist. These have been named $\kappa1$, $\kappa2$ and $\kappa3$; and the $\kappa1$ and $\kappa2$ groups can be further subdivided into different isoforms: $\kappa1A$, $\kappa1B$, $\kappa2A$ and $\kappa2B$. However, because of the absence of subtype-specific antagonists, definitive functional pharmacological support in favor of the $\kappa1$ and $\kappa2$ receptor group is lacking. The κ receptors have been found within the neocortex, the hippocampus, the amygdala, the hypothalamus, the preoptic area, the eminentia mediana, the locus coeruleus, the parabrachial nucleus, the paraventricular nucleus of the thalamus, the nucleus of the tractus solitarius and in the pituitary gland.

The existence of $\kappa1$ and $\kappa3$ receptors has been evidenced by pharmacological studies and binding assays. Receptors of the type $\kappa2$ have been identified in binding assays, but the specific pharmacology of the receptor is largely unknown.

The $\kappa1$ and $\kappa3$ receptors participate in analgesia, since specific antagonists are available; and their effects can be distinguished from the effects of other opioid receptors. The pharmacological characterization of the κ receptor family has been facilitated through the development of effective ligands.

Two types of κ receptor agonist have been developed:

- non-peptidergic agonists, like PD117302, U69593 or U50488H;
- derivates of dynorphin A, like DALKI or D-Pro^{10}-dynorphin (1–11).

The selective antagonist Nor-binaltorphimine and the selective agonist U50488H bind exclusively to κ1 receptors and do not interact with the other receptor subtypes of this family.

The question as to whether β-endorphin can bind to an additional receptor type is controversial. This putative binding site has been named ε-receptor, but it has not been characterized at the molecular level.

4.11.4
Biological Effects

Endorphins are involved in a variety of biological functions, since they interact with the peripheral hormonal system as well as with numerous central neuroregulatory functions via a variety of different neuromodulators and neurotransmitters. For instance, β-endorphin stimulates the secretion of several pituitary hormones, like prolactin.

The endorphins have many effects which are comparable to those of morphine. As mentioned above, their predominant effects center on analgesia, catatonia, hypotension and respiratory depression. In addition, the endorphins are involved in stress response and seem to play a role in alcoholism.

In 1980, Schulz and coworkers showed that, following long-term alcohol ingestion in rodents, the synthesis of brain endorphins and enkephalins is suppressed. Withdrawal of ethanol resulted in a complete recovery of endorphin levels in brain and pituitary within two weeks.

Results obtained in genetic mouse models with a high preference for alcohol support strengthen the view that alcohol intake depends on the activity of the endogenous opioid reward system and that alcohol consumption may serve to compensate for inherent deficits in this system. These experimental data accord well with those obtained from clinical studies in which opioid antagonists have been used to prevent relapse in alcoholics.

In the circulating blood, beta-endorphin shows anti-nocicepitive properties; and an increase in the level of β-endorphin correlates well with an increase in pain tolerance. Beta-endorphin is not only interactive with several neuroactive substances, but shows direct vasoactive effects on cerebral arterioles.

In the periphery, the activation of κ receptors by β-endorphin induces an inhibition of the release of acetylcholine in parasympathctic terminals.

Some evidence speaks in favor of a functional relationship between the immune system and the neuroendocrine system, in which β-endorphin plays a significant role. In this context, it seems noteworthy that an increase in cytokines influences the production of β-endorphins in the pituitary and in lymphocytes, consecutively, the increase in the level of β-endorphin apparently enhancing the production of antibodies.

4.11.5
Neurological Disorders and Neurodegenerative Diseases

Under pathophysiological conditions, i.e. autoimmune diseases, the enhanced production of cytokines can further the production of endorphins and so ultimately enhance the production of antibodies, including autoantibodies, with the consequence of aggravating the disease.

Further Reading

Bach, F. W. (1997): Beta-endorphin in the brain. A role in nociception. *Acta Anaesthesiol. Scand.* **41**: 133–140.

Bodnar, R. J., Klein, G. F. (2005): Endogeous opiates and behavior. 2004. *Peptides* **26**: 2629–2711.

Chen, Y., Mestek, A., Liu, J., Hurley, J. A., Yu, L. (1993): Molecular cloning and functional expression of a μ-opioid receptor from rat brain. *Mol. Pharmacol.* **44**: 8–12.

Dalayeun, J. F., Nores, J. M., Bergal, S. (1993): Physiology of beta-endorphins. A close-up view and a review of the literature. *Biomed. Pharmacother.* **47**: 311–320.

Evans, C. J., Keith, D. F., Morrison, H., Magendzo, K., Edwards, R. H. (1992): Cloning of a delta opioid receptor by functional expression. *Science* **258**: 1952–1955.

Goldfarb, A. H., Jamurtas, A. Z. (1997): Beta-endorphin response to exercise. An update. *Sports Med.* **24**: 8–16.

Herz, A. (1997): Endogenous opioid systems and alcohol addiction. *Psychopharm. Berl.* **129**: 99–111.

Min, B. H., Augustin, L. B., Felsheim, R. F., Fuchs, J. A., Loh, H. H. (1994): Genomic structure and analysis of promotor sequence of a mouse μ opioid receptor gene. *Proc. Natl Acad. Sci. USA* **91**: 9081–9085.

Morch, H., Pedersen, B. K. (1995): Beta-endorphin and the immune system – possible role in autoimmune diseases. *Autoimmunity* **21**: 161–171.

Narita, M., Tseng, L. F. (1998): Evidence for the existence of the β-endorphin-sensitive "ε-opioid receptor" in the brain: the mechanism of the -mediated antinociception. *Jpn J. Pharmacol.* **76**: 233–253.

Sellinger-Barnette, M., Weiss, B. (1982): Interaction of β-endorphin and other opioid peptides with calmodulin. *Mol. Pharmacol.* **21**: 86–91.

Smith, E. M., Harbour McMenamin, D., Blalock, J. E. (1985): Lymphocyte production of endorphins and endorphin-mediated immunoregulatory activity. *J. Immunol.* **135**: 779s–782s.

Rossier, J., Bloom, F. (1979): Central neuropharmacology of endorphins. *Adv. Biochem. Physiopharmacol.* **20**: 165–185.

Tang, F. (1991): Endocrine control of hypothalamic and pituitary met-enkephalin and beta-endorphin contents. *Neuroendocrinol.* **53**[Suppl 1]: 68–76.

Wang, J. B., Imai, Y., Eppler, C. M., Gregor, P., Spivak, C. E., Uhl, G. R. (1993): μ Opiate receptor: cDNA cloning and expression. *Proc. Natl Acad. Sci. USA* **90**: 10230–10234.

4.12
Enkephalin

4.12.1
General Aspects and History

Preparations of the opium poppy *Papaver somniferum* have been used for hundreds of years to relieve pain. In 1803, Sertürner isolated a crystalline sample of the main constituent, the alkaloid morphine, which was later shown to be almost entirely responsible for the analgesic effect of crude opium. The rigid structural and stereochemical pattern of this substance is essential for the analgesic action of morphine and its related opioids. The high-affinity binding of the opioids led to the idea that they produce their effects by interacting with specific endogenous receptors. The effects of morphine in reducing pain and inhibiting intestinal motility and secretion continued to be exploited clinically, but the presence of undesirable side-effects – depression of respiration, development of tolerance and dependence – stimulated the search for analogues with effects restricted to analgesia. For example, a semi-synthetic diacetylated analogue of morphine was introduced in the 19th century in the mistaken belief that this compound (heroin) possessed the specific analgesic effect without the deficits of the side-effects. Although heroin proved to be significantly more effective than morphine in terms of analgesia, it was soon discovered that this drug was disqualified by a major disadvantage, namely its highly addictive properties.

The idea that there is more than one type of opioid receptor arose initially to explain the dual actions of the synthetic opioid nalorphine, which antagonizes the analgesic effect of morphine in man but also acts as an analgesic drug by itself. Experimental evidence for the existence of multiple opioid receptors was obtained by the demonstration that the agonists morphine, ketazocine and N-allylnormetazocine show different pharmacological profiles. The identification of endogenous binding sites was facilitated by the development of specific opioid derivates.

Ultimately, in 1973, the presence of high-affinity binding sites in the central nervous system was definitely proven. The group of Hughes and colleagues (1975) managed to identify two pentapeptides, Leu-enkephalin (amino acid sequence: Tyr-Gly-Gly-Phe-Leu) and Met-enkephalin (amino acid sequence: Tyr-Gly-Gly-Phe-Met), which displayed opiate-agonistic properties. The primary structures of these peptidergic substances are very different from the structure of morphine. However, their tertiary conformation allows them to bind to the opioid receptors.

The discovery of the endogenous enkephalins led to the idea that neuropeptides themselves might represent a means of developing a new class of opioid agonist without the addictive properties of morphine. Although much knowledge has been gathered as to the functional characteristics of enkephalins over the years, no direct benefit in the form of a pharmacologically applicable drug has been achieved.

4.12.2
Localization Within the Central Nervous System

Enkephalin is heterogeneously distributed in the central nervous system. It has been shown to occur in the cortex, the hippocampus, the globus pallidus, the caudate-putamen, the amygdala, the hypothalamus, the periaqueductal gray, the interpeduncular nucleus, the parabrachial nucleus, the nucleus of the fasciculus solitarius, the area postrema and the substantia nigra. In addition, cells containing an enkephalin-like agent have been demonstrated in various other brain areas and in the spinal cord, as indicated in Fig. 4.22.

Fibers and terminals of the enkephalin-containing neurons have been identified in the same brain regions as the enkephalin-containing neuronal somata. This pattern indicates that enkephalinergic neurons belong to the class of short projecting nerve cells. However, two exceptions do exist. The first one is found in the striato-pallidal complex, where long projections from the caudate-puta-

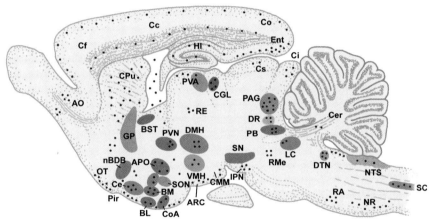

Fig. 4.22 Schematic representation of a sagittal section through the rat brain. The distribution of cells which express the precursor of enkephalin (proenkephalin) is outlined schematically. Abbreviations: AO=anterior olfactory nucleus; APO=preoptic area; ARC=arcuate nucleus; BL=basolateral nucleus of the amygdala; BM=basomedial nucleus of the amygdala; BST=bed nucleus of the stria terminalis; Cc=cingular cortex; Ce=central nucleus of the amygdala; Cf=frontal cortex; CGL=lateral corpus geniculatum; Ci=inferior colliculus; CMM=medial mammillary complex; Co=occipital cortex; CoA=cortical nucleus of the amygdala; Cs=superior colliculus; CPu=caudate-putamen; DMH=dorsomedial nucleus of the hypothalamus; DTN=dorsal tegmental nucleus; DR=dorsal raphe; Ent=entorhinal cortex; HI=hippocampus; IPN=interpeduncular nucleus; LC=locus coeruleus; nBDB=nucleus of the diagonal band of Broca; NR=nucleus reticularis; NTS=nucleus of the tractus solitarius; OT=olfactory tubercle; PAG=periaqueductal gray; PB=parabrachial nucleus; Pir=piriform cortex; PVA=periventricular nucleus of the thalamus; PVN=paraventricular nucleus of the hypothalamus; RA=raphe nuclei; RE=nucleus reuniens; RMe=medial raphe; SC=spinal cord (dorsal horn); SON=supraoptic nucleus; VMH=ventromedial nucleus of the hypothalamus.

men to the globus pallidus are consistently found. The second is in the hypothalamus, where enkephalin-containing neurons of the supraoptic nucleus and the paraventricular nucleus project to the pituitary gland.

4.12.3
Biosynthesis and Degradation

Enkephalins are processed by the proteolysis of a high molecular precursor, pro-enkephalin. Pro-enkephalin is also known as pro-enkephalin A, whereas pro-enkephalin B is not a direct precursor of enkephalins, but the precursor of dynorphins. The precursor protein was first discovered in 1980, in the adrenal gland by Lewis and coworkers, who unraveled its role in the biosynthesis of enkephalins.

The gene which encodes for the precursor consists of four exons separated by three introns. The four exons give rise to a 1.2-kb mRNA from which the product is generated by several posttranslational processes, including enzymatic cleavage. The final pro-enkephalin is a protein with a predicted molecular mass of 31 kDa (Fig. 4.23).

Proenkephalin is not only found in mammals, but also in a variety other species, e.g. annelids, mollusks and amphibians.

The precursor possesses several cysteine residues at its amino-terminal site: it contains four copies of Met-enkephalin, one copy of Leu-enkephalin, one copy of the heptapeptide Met-Enkephalin-Arg^6-Phe^7 and a final copy of the octapeptide Met-Enkephalin-Arg^6-Gly^7-Leu^8.

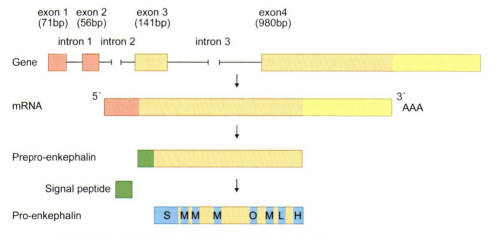

Fig. 4.23 Maturation of proenkephalin and biosynthesis of enkephalins. *Abbreviations*: bp=base pairs; H=heptapeptide (Met-ENK-Arg-Phe); L=Leu-enkephalin; M=Met-enkephalin (Met-ENK); O=octapeptide (Met-ENK-Arg-Gly-Leu); S=synenkephalin.

The individual sequences of these peptides are separated by dibasic peptides, which contain targets for proteolytic cleavage. The final cleavage products are generated by proteolysis in the endoplasmatic reticulum.

Several areas of the central nervous system exhibit intraneuronal localization of the neuroactive endproducts of pro-enkephalin (Met-enkephalin, Leu-enkephalin, heptapeptide, octapeptid and synenkephalin). In some brain areas, like the substantia nigra, Leu-enkephalin seems to be formed at an extracellular site. It is considered that the synthesis of the latter form depends on the hydrolysis of dynorphins.

Released enkephalins can be degraded by two main pathways. One of these pathways involves enzymatic cleavage by aminopeptidase M (EC 3.4.24.2), which hydrolyses enkephalins at the tyrosine–glycine site.

The second mechanism involves degradation through the activity of the enkephalinase (EC 3.4.24.11). This enzyme hydrolyzes enkephalins at the glycine–phenylalanine site.

4.12.4
Receptors and Signal Transduction

The enkephalins have a high affinity to δ receptors, a moderate affinity to μ receptors and a very low affinity to κ receptors. These receptors are discussed in more detail in Section 4.11 (Endorphin).

4.12.5
Biological Effects

Inhibition of the hydrolysis of enkephalins produces an analgesic effect, which is 30–40% weaker than the analgesic effect induced by morphine. On the supraspinal level, the analgesic effects of enkephalin seem to be coupled to the activation of μ receptors, whereas at spinal level they seem to depend on the activation of δ receptors.

Enkephalins are involved in pain regulation at supraspinal and spinal levels. For example, Met-enkephalin is released from the adrenal medulla in response to stress. This release modulates the anti-nociception that accompanies stress.

Met-enkephalin and Leu-enkephalin were initially proposed and are still considered as the endogenous ligands of the δ-opioid receptor; and evidence for δ-opioid receptor-mediated antinociception was provided by the introduction of pharmacological antagonists with relative selectivity for δ-opioid receptors. In addition, pretreatment of animals with antisense oligodeoxynucleotides to the δ-opioid receptor resulted in reduced antinociceptive response to δ- but not κ- or μ-selective receptor agonists.

Enkephalins, like morphine, appear capable of reducing norepinephrine turnover while increasing the speed of dopamine renewal. These effects are directly mediated through the activation of enkephalin-binding receptors located on the catecholaminergic neurons. Colocalization of dopaminergic receptors and enkephalin expression is common in striatal neurons. Stimulation of these neurons by dopamine enhances the synthesis and release of enkephalin.

4.12.6
Neurological Disorders and Neurodegenerative Diseases

Inhibition of the degradation of enkephalins in the striatum (which is strongly innervated by the dopaminergic system arising from the substantia nigra) can induce severe forms of hyperlocomotion. The latter is blocked by the administration of antagonists towards the dopamine receptor D2, indicative of an interaction between the dopaminergic and the enkephalinergic system. Disturbance of this interaction may be causally related to locomotor dysfunctions, as seen in Parkinsonism, schizophrenia and depression.

Further Reading

Abood, M. E., Noel, M. A., Farnsworth, J. S., Tao, Q. (1994): Molecular cloning and expression of a -opioid receptor from rat brain. *J. Neurosci. Res.* **37**: 714–719

Atweh, S., Kuhar, M. (1977): Autoradiographic localization of opiate receptors in rat brain. *Brain Res.* **129**: 1–12.

Belluzzi, J. D., Grant, N., Garsky, V., Sarantakis, D., Wise, C. D., Stein, L. (1976): Analgesia induced *in vivo* by central administration of enkephalin in rat. *Nature* **260**: 625–626.

Evans, C. J., Keith, D. E., Morrison, H., Magendzo, K., Edwards, R. H. (1992): Cloning of a delta receptor by functional expression. *Science* **258**: 1952–1955.

Finley, J. C. W., Maderdrut, J. L., Petrusz, P. (1981): The imunocytochemical localization of enkephalin in the central nervous system of the rat. *J. Comp. Neurol.* **198**: 541–565.

Hughes, J., Smith, T. W., Kosterlitz, H. W., Fothergill, L. A., Morgan, B. A., Morris, H. R. (1975): Identification of two related pentapeptides from brain with potent opiate agonist activity. *Nature* **258**: 577–579.

Hughes, J., Kosterlitz, H. W., Smith, T. W. (1977): The distribution of methionine-enkephaline and leicine-enkephaline in the brain and peripherial tissue. *Br. J. Pharmacol.* **61**: 639–647.

Miller, R. J., Cang, K.-J., Cooper, B., Cuatrecasas, P. (1978): Radioimmunoassay and characterization of enkephalins in rat tissue. *J. Biol. Chem.* **253**: 531–538.

Milligan, G. (2005): Opiod receptors and their interacting proteins. *Neuromol. Med.* **7**: 51–59.

Pasternak, G. W., Goodman, R., Snyder, S. H. (1975): An endogenous morphin-like factor in mammalian brain. *Life Sci.* **16**: 1765–1769.

Salzet, M. (2001): Neuroimmunology of opioids from invertebrates to humans. *Neuroendocrinol. Lett.* **22**: 467–474.

Steiner, H., Gerfen, C. R. (1998): Role of dynorphin and enkephalin in the regulation of striatal output pathways and behavior. *Exp. Brain. Res.* **123**: 60–76.

Vanderah, T., Takemori, A. E., Sultana, M., Portoghese, P. S., Mosberg, H. I., Hruby, V. J., Haaseth, R. C., Matsunaga, T. O., Porreca, F. (1994) Interaction of (D-Pen2,D-Pen5)enkephalin and (D-Ala2,Glu4)deltorphin with μ-opioid receptor subtypes *in vivo*. *Eur. J. Pharmacol.* **252**: 133–137.

Wang, J. B., Imai, Y., Eppler, M., Gregor, P., Spivak, C. E., Uhl, G. R. (1993): μ Opiate receptor: cDNA cloning and expression. *Proc. Natl Acad. Sci. USA* **90**: 10230–10234.

4.13
Fibroblast Growth Factors

4.13.1
General Aspects and History

In 1986, the first member of the fibroblast growth factor (FGF) family, FGF-2, was isolated and cloned. Since that time, the FGF family has increased in number and complexity. A total of 23 family members are known to date. Not all FGF subtypes are present in all species, including humans and mice. For example, FGF-15 has not been identified in the human genome; and FGF-19 could not be detected in mice, resulting in a total of only 22 FGF family members in each of these species.

Concerning their evolutionary development, FGFs are comparably old molecules that evolved in invertebrates. With regard to *Drosophila* and *Caenorhabditis elegans*, only one FGF-like sequence (branchless) has been found in *Drosophila* and two (egl-17 and let-756) in *C. elegans*. Across species, FGF proteins are highly conserved and share greater than 90% amino acid sequence homology. To date, four FGFs have been identified in zebrafish (FGF-3, -8, -17, and -18), six in the clawed toad *Xenopus* (FGF-3, Fgfi, Fgfii, FGF-8, -9, and -20), and seven in chicken (FGF-2, -4, -8, -12, -14, -18, and -19).

With regard to functions of FGFs in the CNS, important roles for FGFs have been discovered in neurogenesis, differentiation, axonal branching and neuron survival. Moreover, FGFs are important in repair processes following different types of brain and peripheral nerve lesions and degenerative disorders; and, in addition, FGFs are involved in cognitive processes, including learning and memory.

4.13.2
Localization Within the Central Nervous System

Different members of the FGF family are highly expressed during embryonic development. The expression decreases during postnatal stages. FGF-2 and FGF-15 are generally expressed throughout the developing CNS, whereas FGF-8 and FGF-17 are tightly localized to specific regions of the developing brain and are only expressed in the embryo during the early phases of proliferation and neurogenesis. Moreover, FGF-3, FGF-6 and FGF-7 are more highly expressed in late embryonic stages than in postnatal stages, suggesting that these members are involved in the late stages of brain development. In contrast, expression of FGF-1 and FGF-5 increases postnatally.

In the adult CNS, FGF-1 is found in neurons of the cerebral cortex, the hippocampus, the striatum, the nucleus basalis of Meynert, the magnocellular preoptic area, the diagonal bands of Broca, the medial septum, the thalamus, the hypothalamus, the substantia nigra, the ventral tegmental area, the reticular formation, the locus coeruleus, the lateral geniculate nucleus, the cerebellum, the pons, the oculomotor nucleus, in motoneurons and in sensory ganglia of the brainstem.

FGF-2 mRNA is distributed widely throughout the adult CNS and has been detected in the olfactory bulb, the cerebral cortex, the hippocampus, the thalamus, the substantia nigra, the striatum, the colliculi and the pons. Furthermore, FGF-2 has been found to be present in motor and sensory nuclei as well as in the neural and anterior lobes of the pituitary. FGF-2 has been found in glial cells and it is assumed that FGF-2 is predominantly synthesized by astrocytes.

FGF-5 mRNA reveals a broad distribution in the adult rodent CNS. The olfactory bulb, the hippocampus, the cerebral cortex and the spinal cord display higher levels of FGF-5 mRNA than the basal ganglia, the cerebellum, the midbrain and the hindbrain. In addition, FGF-5 mRNA occurs in several hypothalamic nuclei (e.g. the paraventricular nucleus, supraoptic nucleus, arcuate nucleus, median eminence and ventromedial hypothalamic nucleus).

Expression of FGF-8 has been shown by immunohistochemistry in the following areas of the adult human and rodent brain: the cortex, the hippocampus, the thalamus, the substantia nigra, cranial nerve nuclei and Purkinje's cells of the cerebellum.

Within the adult brain, FGF-9 mRNA is moderately or weakly expressed in widespread regions, including the olfactory bulb, the caudate putamen, the cerebral cortex, the hippocampus, the thalamus, the hypothalamus, the midbrain, the brainstem and the cerebellum and in motor neurons of the spinal chord. FGF-9 mRNA is strongly expressed in different nuclei, including the red nucleus and oculomotor nucleus in the midbrain, the vestibular nucleus and facial nucleus in the brainstem and in the medial cerebellar nucleus, interposed cerebellar nucleus and lateral cerebellar nucleus of the cerebellum. FGF-9 mRNA was found to be expressed preferentially in neurons but, like FGF-2, FGF-9 has also been detected in glial cells. For example, FGF-9 has been detected in GFAP-positive astrocytes in the white matter tracts of the spinal cord and in the brainstem of adult rats. In addition, FGF-9 has been be detected in CNPase-positive oligodendrocytes in rat cerebellum and corpus callosum.

FGF-10 mRNA is although expressed in the brain. However, in contrast to the lung, only low levels of FGF-10 mRNA expression have been found. FGF-10 mRNA showed spatially restricted low expression in some regions of the brain, including the hippocampus, thalamus, midbrain and brainstem. A somewhat higher expression level can be found in the motor nuclei, including the oculomotor nucleus, dorsal motor nucleus of vagus, motor trigeminal nucleus, facial nucleus and hypoglossal nucleus. The cellular localization of FGF-10 mRNA shows that the mRNA in the brain is preferentially expressed in neurons but not in glial cells.

FGF-12 and FGF-13 are expressed in the embryonic nervous system, but they seem also to occur in the adult brain. By Northern hybridization analysis, a prominent expression of FGF-13 in the adult human brain, particularly in the cerebellum and cortex, has been demonstrated. However, as in case for many other members of the FGF-family, data from detailed analyses of the distribution pattern of mRNAs or proteins are not available.

To mention a few further examples: FGF-18 is transiently expressed at early postnatal stages in various regions of the rat brain, including the cerebral cortex and hippocampus, with a preferential expression in neurons. With regard to FGF-20, it has been claimed that FGF-20 is expressed in the normal brain, particularly in the cerebellum and in the substantia nigra pars compacta. Concerning the last member of the FGF-family, FGF-23 mRNA has been examined by *in situ* hybridization and found preferentially in the ventrolateral thalamic nucleus.

4.13.3
Biosynthesis and Degradation

With the exception of FGF-16, the chromosomal localization for all human FGF genes is known. Several human FGF genes are clustered on distinct chromosomal regions. For example, genes encoding FGF-3, FGF-4 and FGF-19 are located on chromosome 11q13, being separated by only 40 kb and 10 kb, respectively. Furthermore, the FGF-6 and FGF-23 genes are located in a small, 55-kb fragment of chromosome 12p13. The FGF-17 gene and FGF-20 gene also reside together on chromosome 8p21–8p22.

For the mouse, localization of 16 FGF genes has been determined, revealing some similarities to their arrangement on human chromosomes. For example, the genes encoding FGF-3, FGF-4 and FGF-19 are linked on an 80-kb stretch of chromosome 7F; and FGF-6 and FGF-23 are on chromosome 6F3-G1.

All known FGF genes consist of three coding exons with exon 1 containing the start codon. However, some FGF genes, such as FGF-2 and FGF-3, contain additional 5′ untranscribed regions initiating from upstream CUG codons. In certain FGF subfamilies, exon 1 is subdivided into two or four alternatively spliced sub-exons (1A-1D for FGF-8), with a single initiation codon residing in exon 1A. Other family members (e.g. FGF-11 to FGF-14) have alternatively spliced amino-terminal regions resulting from the usage of alternate 5′ exons.

The different FGFs display a 13–71% amino acid identity. The defining features of the family are a high affinity for heparin and heparin-like glycosaminoglycans (HLGAGs) and a central core that is highly homologous between the family members.

The majority of FGFs (FGF-3 to FGF-8, FGF-10, FGF-15, FGF-17 to FGF-19 and FGF-21 to FGF-23) possess amino-terminal signal peptides and, therefore, can be assumed to be readily secreted from cells. However, FGF-1, FGF-2, FGF-9, FGF-16 and FGF-20 lack conventional signal peptides but, nevertheless, are secreted into the extracellular space.

An important feature of FGF biology involves the interaction between FGF and heparin or heparan sulfate proteoglycan. These interactions stabilize FGFs to thermal denaturation and proteolysis and may severely limit their diffusion and release into interstitial spaces. FGFs must saturate nearby heparan sulfate-binding sites before becoming effective for remote tissues, or alternatively must become mobilized by heparin/heparan sulfate-degrading enzymes. The interaction between FGFs and heparan sulfate results in the formation of dimers and higher-order oligomers. Although the biologically active form of FGF is poorly defined, it has been established that heparin is required for FGFs to effectively activate the FGF receptors.

4.13.4
Receptors and Signal Transduction

To date, four different subtypes of FGF receptors (FGFR) have been identified. The affinity of the FGF receptors for their ligands is highly diverse, with different affinities for each member of the FGF-family (Table 4.3).

The diversity of FGF receptors beyond the four receptor subtypes is achieved by the generation of alternative splice variants of the FGF receptor genes. The protein region with the highest impact on FGF receptor-binding specificity is a portion of the IgIII domain for which three different splice variants have been identified so far, termed IgIIIa, IgIIIb and IgIIIc. While for FGFR1 and for FGFR2 all three splice variants could be detected, FGFR3 occurs only as the IGIIIb and IGIIIc variants, and FGFR4 exclusively as the IgIIIc variant.

FGFRs contain an extracellular ligand-binding domain, a single transmembrane domain and an intracellular tyrosine kinase domain. The ligand-binding specificity of FGFRs depends on the third extracellular Ig-like domain. Ligand binding to FGF receptors leads to the formation of a receptor complex consisting of two FGF molecules bound to a receptor, linked by a heparan sulfate proteoglycan molecule, e.g. heparin.

Table 4.3 Receptor binding of FGF family members to the four FGF receptor subtypes (modified from Ornitz et al. 1996). *Code*: += receptor binding, −=no receptor binding, ?=not investigated.

	FGFR-1	FGFR-2	FGFR-3	FGFR-4
FGF-1	+	+	+	+
FGF-2	+	+	+	+
FGF-3	+	+	−	−
FGF-4	+	+	+	+
FGF-5	+	+	−	−
FGF-6	+	+	−	+
FGF-7	−	+	−	−
FGF-8	+	+	+	+
FGF-9	−	+	+	+
FGF-10	−	+	−	−
FGF-11 to FGF-16	?	?	?	?
FGF-17	−	+	+	+
FGF-18	−	+	+	+
FGF-19 to FGF-23	?	?	?	?

Activation of FGFRs triggers several intracellular signaling cascades. These cascades include the phosphorylation of src and PLCγ, leading finally to activation of PKC, as well as activation of Crk and Shc.

SNT/FRS2 serves as an alternative link of FGFRs to the activation of PKC and, in addition, activates the Ras signaling cascade. Ras directly interacts with and activates Raf, which in turn phosphorylates and activates MEK, which then phosphorylates and activates the MAP kinases, including ERK1 and ERK2.

In the adult CNS, the different subtypes of FGF receptors have been described: FGFR-1, FGFR-2 and FGFR-3 are found to be highly expressed in the diencephalon and telencephalon and moderately expressed in the mesencephalon and metencephalon, while expression of these FGFRs is relatively low in the myelencephalon. FGFR-1 is expressed in widespread but specific neuronal populations in the adult CNS. It is not confined to neurons and has been detected in astrocytes of white matter tracts. In contrast to FGFR-1, which is predominantly expressed on neurons, FGFR-2 occurs primarily in glial cells, as is the case for FGFR-3.

The fourth member of the FGF receptors (FGFR-4) is not detectable in adult CNS.

4.13.5
Biological Effects

FGFs are vital intercellular signaling molecules which regulate numerous processes during embryogenesis and organogenesis. FGF signaling is required for cell proliferation/survival at the time of mouse embryonic implantation. FGFs regulate the development of the brain, teeth, limbs, lungs, kidneys and many other organs at the later stages of embryogenesis. FGFs exert diverse effects on the development and maintenance of neurons, including effects on fate determination, migration and differentiation, as well as on cell survival. Starting at the earliest stages of brain development, both FGF-1 and FGF-2 have been shown to be expressed in distinct expression patterns, with their expression persisting even in the adult CNS.

Proliferation and differentiation

The effects of the FGF family members on neuron differentiation greatly depend on the developmental time-point at which a factor is applied. For example, it has been shown *in vitro* that, at an early developmental time-point, FGF-2 is able to expand the period of dopamine precursor division in conjunction with a delay in differentiation. The proliferation-stimulating effects of FGF-2 are apparently not restricted to dopaminergic neurons, but have been described for GABAergic neurons as well.

FGF-2 is an important regulator of prenatal, postnatal and adult neurogenesis which induces proliferation of neural progenitor cells in the hippocampus and in the subventricular zone. Subcutaneous injection of FGF-2 at P1 increases [^{3}H]thymidine incorporation by 70% in hippocampal and subventricular zones and elicits a two-fold increase in mitotic nuclei in the dentate gyrus and the dorsolateral subventricular zone, suggesting that FGF-2 penetrates the blood–brain barrier to regulate adult neurogenesis. Moreover, cultured hippocampal cells from adult rats are able to proliferate and to generate neurons in defined medium containing FGF-2. These cells have been shown to express various neuronal and glial markers, like O4, NSE, MAP2, NF150, GAD and calretinin. Two months after transplantation to the adult rat hippocampus, descendents of these cells can be found in the dentate gyrus, where they reveal transdifferentiation into neurons exclusively in the granule cell layer.

Morphogenesis
Besides their effects on proliferation and differentiation of neural precursor cells, FGFs have been found to affect neuronal morphogenesis. Many growth factors, like the neurotrophins BDNF and NT-4, have been shown to influence axon branching. However, one of the most effective regulators of this process is FGF-2. Within the hippocampus, FGF-2 promotes selectively bifurcation and growth of axonal branches without affecting the elongation rate of primary axons, resulting in increased complexity of axonal trees. This accelerated axonal branch formation in the presence of FGF-2 can be restored to the basal rate following removal of FGF-2, indicating that the action of FGF-2 is reversible and continuous presence of the factor is required for prolonged effectiveness.

Within the hippocampus, multiple factors enhance branching of axons, but not dendrites of the same neurons. The most effective factor in axonal branching has been found to be FGF-2, whereby other factors have been shown to be less effective.

Learning and memory
A number of studies suggest that FGFs play a role in certain brain functions, even in the adult brain, especially in processes attributed to learning and memory.

FGF-1 has been shown to enhance of the magnitude of short-term potentiation and facilitates the generation of long-term potentiation (LTP). Since LTP is thought to be linked to memory formation and learning, it is speculated that FGF-1 might play a role in mechanisms underlying learning and memory. Evidence supporting this view came from a study using intracysternal binjections of FGF-1. LTP can be induced using subthreshold stimulations in combination with administration of FGF-1; however, LTP can not be induced using the subthreshold stimulations alone.

In addition to FGF-1, FGF-2 seems to be involved in neuronal signaling. In the dentate gyrus, subthreshold stimulation (20 pulses at 60 Hz) normally fails to induce LTP; however, after administration of FGF-2, LTP can be induced using the same protocol.

Neuroprotection and lesion repair

Considerable evidence suggests that FGFs are potent trophic factors for many different neuronal populations *in vitro* and following brain lesions. For example, it has been shown that FGF-2 decreases glutamate-induced neuronal cell death in the hypocampus by regulating glutamate receptor subunits, leading to suppression of the 71-kDa NMDA receptor protein (NMDARP-71) but not of the AMPA/kainate receptor GluR1. Furthermore, FGF-2 potentiates quisqualate-induced inositol phosphate formation in hippocampal cultures from day 1 up to 10 days. This effect can be blocked by addition of the AMPA/kainate receptor antagonist 6,7-dinitro-quinoline-2,3-dione (DNQX), suggesting an involvement of an AMPA/kainate receptor subtype distinct from GluR1. FGF-2 can also promote the survival of septal cholinergic and non-cholinergic neurons following fimbria-fornix transection. However, effects of FGF-2 on cholinergic neuron survival seem to be indirect, involving stimulation and expansion of glial cells as a source for "secondary" survival factors. Another neuron population for which FGF-2 can act as a trophic factor is represented by the mesencephalic dopaminergic neurons. Survival of dopaminergic neurons in cultures from embryonic midbrain exposed either to the neurotoxic substance 1-methyl-4-phenyl-1,2,3,6-tetrahydropyridine (MPTP) or to its active form, the methyl pyridinium ion (MPP$^+$), was significantly augmented by treatment with FGF-2.

A prominent feature of FGF is the mediation of neuroprotective effects. As indicated above for cholinergic neurons of the hippocampus and for dopaminergic nigrostriatal neurons, FGF-2 is also able to prevent neuronal death after fiber transection or chemical injury. In accordance with these findings, endogenous FGF has been found to be greatly upregulated after lesioning, e.g. following cortical lesions. Interestingly, it has been shown that FGF-2-mediated repair processes can also improve behavioral scores of mice after lesioning, indicating a prominent role of FGF-2 in brain repair.

Ischemic insult results in the destruction of distinct brain regions, depending on the type of vascular occlusion applied. Ischemia has been shown to cause rapid neuronal cell death, which can be overcome by neuroprotective growth factors, including FGF-1, FGF-2 and FGF-7. Thus, systemic administration of FGF-2 can ameliorate acute focal ischemic injury in the cerebral cortex without increasing blood flow following occlusion of the middle cerebral artery (MCAO). Moreover, FGF-2 knockout mice have an enlarged infarct size after MCAO.

The mechanisms underlying the neuroprotective capacity of the FGFs are still enigmatic. It has been speculated that the neuroprotective effects of FGF-2 results, in part, from a prevention of attenuation of oxidative damage, possibly by suppressing oxidative impairment of synaptic transporter functions.

4.13.6
Neurological Disorders and Neurodegenerative Diseases

Whether FGF-2 has anticonvulsant or proconvulsant properties is still a matter of debate. FGF-2 does not induce major anticonvulsive effects when administered prior to or after kainic acid induced seizures. However, FGF-2 has been found to induce seizures on its own after unilateral injection into the dentate gyrus.

In contrast, exogenous application of FGF-1 decreases convulsions in the kainate model, indicating that FGF-1 may have anticonvulsant properties. In addition, it prevents cell loss in the hippocampus, indicating that FGF-1 has a substantial neuroprotective effect in the kainate model.

Further Reading

Abraham, J.A., Mergia, A., Whang, J.L., Tumolo, A., Friedman, J., Hjerrild, K.A., Gospodarowicz, D., Fiddes, J.C. (1986): Nucleotide sequence of a bovine clone encoding the angiogenic protein, basic fibroblast growth factor. *Science* **233**: 545–548.

Asai, T., Wanaka, A., Kato, H., Masana, Y., Seo, M., Tohyama, M. (1993): Differential expression of two members of FGF receptor gene family, FGFR-1 and FGFR-2 mRNA, in the adult rat central nervous system. *Brain Res. Mol. Brain Res.* **17**: 174–178.

Bean, A.J., Elde, R., Cao, Y.H., Oellig, C., Tamminga, C., Goldstein, M., Pettersson, R.F., Hokfelt, T. (1991): Expression of acidic and basic fibroblast growth factors in the substantia nigra of rat, monkey, and human. *Proc. Natl Acad. Sci. USA* **88**: 10237–10241.

Ernfors, P., Lonnerberg, P., Ayer-LeLievre, C., Persson, H. (1990): Developmental and regional expression of basic fibroblast growth factor mRNA in the rat central nervous system. *J. Neurosci. Res.* **27**: 10–15.

Gomez-Pinilla, F., Cotman, C.W. (1993): Distribution of fibroblast growth factor 5 mRNA in the rat brain: an *in situ* hybridization study. *Brain Res.* **606**: 79–86.

Grothe, C., Janet, T. (1995): Expression of FGF-2 and FGF receptor type 1 in the adult rat brainstem: effect of colchicine. *J. Comp. Neurol.* **353**: 18–24.

Ishihara, A., Saito, H., Nishiyama, N. (1992): Basic fibroblast growth factor ameliorates learning deficits in basal forebrain-lesioned mice. *Jpn J. Pharmacol.* **59**: 7–13.

Johnson, D.E., Lee, P.L., Lu, J., Williams, L.T. (1990): Diverse forms of a receptor for acidic and basic fibroblast growth factors. *Mol. Cell Biol.* **10**: 4728–4736.

Johnson, D.E., Williams, L.T. (1993): Structural and functional diversity in the FGF receptor multigene family. *Adv. Cancer Res.* **60**: 1–41.

Kiprianova, I., Schindowski, K., von Bohlen und Halbach, O., Krause, S., Dono, R., Schwaninger, M., Unsicker, K. (2004): Enlarged infract volume and loss of BDNF mRNA induction following brain ischemia in mice lacking FGF-2. *Exp. Neurology* **189**: 252–260

Lee, P.L., Johnson, D.E., Cousens, L.S., Fried, V.A., Williams, L.T. (1989): Purification and complementary DNA cloning of a receptor for basic fibroblast growth factor. *Science* **245**: 57–60.

Miyakawa, K., Hatsuzawa, K., Kurokawa, T., Asada, M., Kuroiwa, T., Imamura, T. (1999): A hydrophobic region locating at the center of fibroblast growth factor-9 is crucial for its secretion. *J. Biol. Chem.* **274**: 29352–29357.

Nakamura, S., Todo, T., Motoi, Y., Haga, S., Aizawa, T., Ueki, A., Ikeda, K. (1999): Glial expression of fibroblast growth factor-9 in rat central nervous system. *Glia* **28**: 53–65.

Nishimura, T., Nakatake, Y., Konishi, M., Itoh, N. (2000): Identification of a novel FGF, FGF-21, preferentially expressed in the liver. *Biochim. Biophys. Acta* **1492**: 203–206.

Ornitz, D.M., Xu, J., Colvin, J.S., McEwen, D.G., MacArthur, C.A., Coulier, F., Gao, G., Goldfarb, M. (1996): Receptor specificity of the fibroblast growth factor family. *J. Biol. Chem.* **271**: 15292–15297.

Ornitz, D.M., Itoh, N. (2001): Fibroblast growth factors. *Genome Biol.* **2**: 3005.1–3005.12.

Powers, C.J., McLeskey, S.W., Wellstein, A. (2000): Fibroblast growth factors, their receptors and signaling. *Endocr. Rel. Cancer* **7**: 165–197.

Raffioni, S., Thomas, D., Foehr, E.D., Thompson, L.M., Bradshaw, R.A. (1999): Comparison of the intracellular signaling responses by three chimeric fibroblast growth factor receptors in PC12 cells. *Proc. Natl Acad. Sci. USA* **96**: 7178–7183.

Reuss, B., von Bohlen und Halbach, O. (2003): Fibroblast growth factors and their receptors in the central nervous system. *Cell Tissue Res.* **313**: 139–157.

Riva, M. A., Gale, K., Mocchetti, I. (1992): Basic fibroblast growth factor mRNA increases in specific brain regions following convulsive seizures. *Brain Res. Mol. Brain Res.* **15**: 311–318.

Vainikka, S., Partanen, J., Bellosta, P., Coulier, F., Birnbaum, D., Basilico, C., Jaye, M., Alitalo, K. (1992): Fibroblast growth factor receptor-4 shows novel features in genomic structure, ligand binding and signal transduction. *EMBO J.* **11**: 4273–4280.

Wilcox, B. J., Unnerstall, J. R. (1991): Expression of acidic fibroblast growth factor mRNA in the developing and adult rat brain. *Neuron* **6**: 397–409.

4.14
Galanin

4.14.1
General Aspects and History

In 1983, Mutt and coworkers discovered the neuropeptide galanin. Galanin and galanin message-associated peptide (GMAP) are the two established biologically active fragments of the galanin precursor molecule.

In most mammals, the galanin sequence consists of 29 amino acids and is amidated at the carboxyl terminus. In humans, galanin consists of 30 amino acids and is not amidated at the carboxyl terminus.

Galanin is a phylogenetically old peptide which is also found in invertebrates. The amino-terminal sequence of galanin (1–13) is highly homologous among all species, while the carboxyl terminus varies considerably, indicating the existence of species-specific isoforms of galanin.

4.14.2
Localization Within the Central Nervous System

Galanin-containing cells have been found in the gut as well as in the central nervous system. A high density of galanin-containing neurons is present in some brain areas. For example, galanin-containing neurons are found in significant numbers in the septum and the hypothalamus and have also been detected in the amygdala, the prefrontal cortex and in some thalamic nuclei. Within the brain stem, galanin-containing neurons are found in the nucleus of the tractus solitarius, the locus coeruleus and in some raphe nuclei (Fig. 4.24).

4.14.3
Biosynthesis and Degradation

A single gene consisting of six exons encodes the galanin precursor preprogalanin, which has a length of 123–124 amino acids (Fig. 4.25).

Different species-specific forms of galanin have been described; and differential cleavage gives rise to different short forms: galanin 1–19 when cleaved from

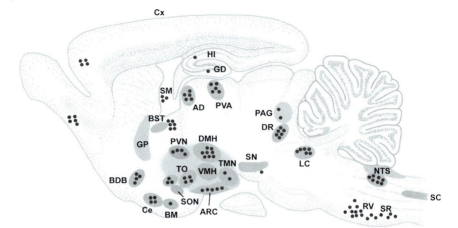

Fig. 4.24 Schematic drawing of the distribution of galanin-containing neurons in the brain of rats. *Abbreviations*: AD = antero-dorsal nucleus of the thalamus; ARC = arcuate nucleus; BDB = diagonal band of Broca; BM = basomedial nucleus of the amygdala; Ce = central nucleus of the amygdala; Cx = cortex; DMH = dorso-medial nucleus of the hypothalamus; GD = gyrus dentatus; GP = *globus pallidus*; HI = hippocampus; LC = *locus coeruleus*; SON = supraoptic nucleus of the hypothalamus; NTS = nucleus of the *tractus solitarius*; PAG = periaqueductal gray; PVA = paraventricular nucleus of the thalamus; PVN = paraventricular nucleus of the hypothalamus; DR = dorsal raphe; RV = ventral raphe; SM = medial septum; SC = spinal cord (dorsal horn); SN = *substantia nigra*; SR = *substantia reticularis*; TMN = tuberomammillary nucleus; TO = *tractus opticus*; VMH = ventromedial nucleus of the hypothalamus.

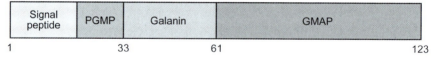

Fig. 4.25 Preprogalanin. Galanin is synthesized from the precursor preprogalanin. In addition, the galanin message-associated peptide (GMAP) and the pregalanin message-associated peptide PGMP are processed from this precursor.

the carboxy-terminal region and galanin 5–29 following cleavage at a more amino-terminal portion.

The enzymatic steps necessary for the biosynthesis and degradation of galanin are largely unknown. However, by using galanin as a substrate, a membrane-bound galanin-inactivating metallo-peptidase from bovine spinal cord could be purified (Jureus and coworkers 1998). This enzyme constitutes a novel 70-kDa, Zn^{2+}-dependent metallo-peptidase.

Galanin seems to exhibit sexual dimorphism, since females express galanin at higher concentrations, suggesting that steroids modulate the synthesis of galanin. This idea is supported by the finding that administration of estrogen specifically elevates the galanin concentrations in some cells of the anterior pituitary and in neurons of the median eminence.

4.14.4
Receptors and Signal Transduction

Specific galanin-binding receptors seem to belong to the superfamily of G protein-coupled receptors: thus, they can activate or inhibit adenylate cyclase or phospholipase C.

The galanin receptors in the central nervous system display high affinities for the amino-terminal sequence of galanin. The fragments Gal (1–19) or Gal (1–16) are also capable of reproducing some of the effects induced by the complete galanin molecule. Among these effects, the stimulation of the secretion of growth hormones, inhibition of acetylcholine release in the hippocampus and the stimulation of food intake are the most prominent features. These findings have permitted the development of several galanin-specific antagonists, which include chimeric peptides with an amino-terminal sequence of galanin (1–13) and variable carboxy-terminal fragments. For example, the chimeric compound M15 [Galanin (1–13)-substance P (5–11)] is an antagonist which inhibits galanin-induced central or peripheral effects.

Moreover, an endogenous galanin-like peptide (GALP, a 60-amino-acid peptide) which shares sequence homology with galanin and competes with galanin at GAL2 receptor sites, was isolated from the hypothalamus. The biosynthesis of this neuroactive substance is largely unknown, but it seems to be regulated by leptins.

A total of three G protein-coupled receptor (GPCR) subtypes for galanin have been identified and cloned from rat, mouse and human.

A stretch of about 83 amino acids is conserved in all three rat galanin receptor subtypes, yielding a shared amino acid homology of 23%. The galanin receptor proteins show a striking degree of similarity (30–38%) with the somatostatin sst 4 and sst 5 receptor subtypes and with the ORL1 (nociceptin) receptor.

Galanin-like immunoreactivity, binding assays and autoradiography for the receptor subtypes GAL1, GAL2 and GAL3, as well as *in situ* hybridization, reveal that the receptors are widely distributed in the central and peripheral nervous systems of mammals.

In 2004, a novel G protein-coupled receptor with homology to galanin receptors was cloned and characterized. This new receptor was named the galanin receptor-like receptor (GalRL). The genes of GalRL are localized to chromosome 5q32 in humans and to chromosome 18B3 in mice. GalRL is expressed in the CNS, with a distinct localization in the habenular complex. The ligand of that receptor is unknown; however, since a weak activation of GalRL by galanin has been observed, it is suggested that the endogenous ligand shares structural features with galanin.

GAL1 receptors

The first galanin receptor GAL1 was isolated from human tissues. The human GAL1 receptor contains 349 amino acids, with structural features of a G protein-coupled receptor. Amino acid similarities (conserved amino acid exchanges are enclosed) are found with the human GAL2 receptor (42%) and with the human GAL3 receptor (38%). The rat GAL1 receptor has a putative length of 346 amino acids and a homology of about 92% with the human GAL1 receptor. The human and rat GAL1 share the same consensus sites for N-linked glycosylation and for intracellular phosphorylation, with the exception of two additional phosphorylation sites in the human GAL1 carboxy-terminal domain. The gene which encodes for the human GAL-1 receptor is located on human chromosome 18q23.

The activation of GAL1 receptors reduces the concentration of cAMP, opens G protein-coupled inwardly rectifying K^+ channels and stimulates MAP kinase activity in a manner that is sensitive to pertussis toxin. This is consistent with a coupling to G_o-type G proteins.

In the central nervous system, the distribution of GAL1 mRNA, as evidenced by *in situ* hybridization, is in accord with galanin expression sites. GAL-1 mRNA has been found in the hippocampus, amygdala, thalamus, hypothalamus, brain stem (for example in the locus coeruleus and in the parabrachial nucleus) and spinal cord (dorsal horn).

GAL2 receptors

The cloned human GAL2 receptor consists of 387 amino acids, having 15 amino acids more than the rat GAL2 receptor. The gene which encodes the human GAL2 receptor is located on chromosome 17.

Activation of GAL2 receptors elicits multiple intracellular reactions involving phospholipase C, intracellular Ca^{2+} mobilization and Ca^{2+}-dependent Cl^- channel activation.

The GAL2 receptor belongs to the superfamily of G protein-coupled receptors, but it is not clear whether the receptor is coupled to G proteins of the G_q type or the G_o type.

Unlike the mRNA of the GAL1 receptor, the mRNA encoding the rat GAL2 receptor is widely distributed throughout the body, including the brain (with the highest levels found in the hypothalamus, hippocampus, amygdala and cortex).

GAL3 receptors

A third cloned galanin receptor subtype is the GAL3 receptor. The cDNA of the rat GAL3 receptor encodes a protein of 370 residues, with an amino acid similarity to rat GAL1 receptor of 36% and to rat GAL2 receptor of 55%. The human GAL3 has a putative length of 368 of about 90% to the amino acid sequence of the rat GAL3 receptor. The gene which encodes the human GAL3 receptor is located on chromosome22. Activation of the GAL3 receptors stimulates G proteins of the $G_{i/o}$ type.

The GAL3 receptor is found in different regions of the central nervous system, including the olfactory bulb, the cerebral cortex, the caudate-putamen, the hypothalamus, the medulla oblongata, the cerebellum, the spinal cord and the pituitary.

4.14.5
Biological Effects

The physiological and behavioral effects of galanin include release of pituitary hormones, gastrointestinal and cardiovascular effects. Galanin is up-regulated following neuronal injury and exhibits inhibitory effects on neuronal firing and neurotransmitter release as well as on pain transmission and on learning and memory.

The administration of galanin *in vivo* modifies endocrine functions, with effects on the secretion of pituitary modulators. Immuno-neutralization of endogenous central nervous galanin revealed that galanin tonically stimulates the secretion of growth hormone and prolactin, exerts a tonic inhibition of ACTH and thyreotropic hormone (TH) and stimulates the secretion of luteinising hormone (LH).

In the hypothalamus, galanin is involved in the release of serotonin and the secretion of growth hormone-releasing hormone (GHRH). It also inhibits the release of the neurotransmitters norepinephrine, dopamine and histamine.

Galanin is co-expressed in a subset of gonadotropin-releasing hormone (GnRH) and growth hormone-releasing hormone (GHRH) neurons in the brain and has an important role in the neuroendocrine regulation of gonadotropin and growth hormone secretion. Galanin may act as a regulator of gonadotropin-releasing hormone (GnRH) neurons, since it may co-release with GnRH and acts presynaptically on GnRH neurons, potentiating the GnRH-induced secretion of LH. Thus, galanin may coordinate GnRH release in a pulsative fashion. In the absence of galanin, GnRH is released in a continuous manner, which is ineffective in evoking the LH surge.

Galanin-mediated effects on the release of acetylcholine can be demonstrated in the cerebral cortex and in the ventral hippocampus where they include a decrease in basal acetylcholine release. Furthermore, galanin infusion into the ventral hippocampus has been shown to impair the acquisition of spatial learning. This observation, coupled with the demonstration that galanin inhibits the release of acetylcholine, has led to speculation that a pharmacologically applied galanin receptor antagonist might enhance cognitive behavior. Along this line, galanin transgenic mice (overexpressing endogenous galanin) show deficits in hippocampus-dependent learning such as the Morris water maze task.

Galanin stimulates food intake in rats. This positive effect on appetite was achieved after injection of galanin into the ventricles, the paraventricular nucleus of the hypothalamus or the amygdala of rats.

Galanin also has nociceptive properties. In response to nociceptive stimuli, intrathecally applied galanin facilitates nociceptive reflexes at a low dose, but inhibits them at a high dose. The facilitatory action is blocked by galanin antagonists.

4.14.6
Neurological Disorders and Neurodegenerative Diseases

Overexpression of galanin has been found to modulate seizures, anxiety-related behavior, depression-related and sexual-related behavior, as well as feeding.

A study by Mazarati and coworkers (2000) indicates that hippocampal galanin acts as an endogenous anticonvulsant, suggesting that genetically induced changes in galanin expression modulate hippocampal excitability and predispose to epileptic seizures. Moreover, it has been shown that galanin acts as an endogenous neuroprotective factor in the hippocampus, implying that galanin agonists might have therapeutic uses in some forms of brain injury.

Furthermore, galanin-expressing neurons are substrates of structural changes in Alzheimer's disease. In this condition, galanin-containing neuronal projections exhibit hyperinnervation of acetylcholine-containing neurons of the basal forebrain. A selective increase in the concentration of galanin in the basal nucleus of Meynert has also been demonstrated. This increase is due to an enhanced generation of galaninergic terminals in this brain area. It has been found that galanin and GAL receptors are overexpressed in limbic brain regions associated with cognition in Alzheimer's disease. However, the functional consequences of this overexpression are unclear. Since it has been shown that galanin can act as a neuroprotective factor in the hippocampus, it might be possible that increased galanin production might act as a neuroprotective peptide in Alzheimer's disease (Counts et al. 2003).

Further Reading

Bedecs, K., Berthold, M., Bartfai, T. (1995): Galanin – 10 years with a neuroendocrine peptide. *Int. J. Biochem. Cell Biol.* **27**: 337–349.

Branchek, T.A., Smith, K.E., Walker, M.W. (1998): Molecular biology and pharmacology of galanin receptors. *Ann. N.Y. Acad. Sci.* **863**: 94–107.

Branchek, T.A., Smith, K.E., Gerald, C., Walker, M.W. (2000): Galanin receptor subtypes. *Trends Pharmacol. Sci.* **21**: 109–117.

Brewer, A., Echevarria, D.J., Langel, U., Robinson, J.K. (2005): Assessment of new functional roles for galanin in the CNS. *Neuropeptides* **39**: 323–326.

Counts, S.E., Perez, S.E., Ginsberg, S.D., De Lacalle, S., Mufson, E.J. (2003): Galanin in Alzheimer disease. *Mol. Intervent.* **3**: 137–156.

Crawley, J.N., Wenk, G.L. (1989): Co-existance of galanin and acetylcholin: is galanin involved in memory processes and dementia? *Trends Neurosci.* **2**: 278–282.

Elliott-Hunt C.R., Marsh, B., Bacon, A., Pope, R., Vanderplank, P., Wynick, D. (2004): Galanin acts as a neuroprotective factor to the hippocampus. *Proc. Natl Acad. Sci. USA* **10**: 5105–5110.

Gundlach, A.L. (2002): Galanin/GALP and galanin receptors: role in central control of feeding, body weight/obesity and reproduction? *Eur. J. Pharmacol.* **440**: 255–268.

Habert-Ortoli, E., Amiranoff, B., Loquet, I., Laburthe, M., Mayeux, J.F. (1994): Molecular cloning of a functional human galanin receptor. *Proc. Natl Acad. Sci. USA* **91**: 9780–9783.

Hohmann, J.G., Clifton, D.K., Steiner, R.A. (1998): Galanin: analysis of its coexpression in gonadotropin-releasing hormone and growth hormone-releasing hormone neurons. *Ann. N.Y. Acad. Sci.* **863**: 221–235.

Ignatov, A., Hermans-Borgmeyer, I., Schaller, H.C. (2004): Cloning and characterization of a novel G-protein-coupled receptor with homology to galanin receptors. *Neuropharmacology* **46**: 1114–1120.

Jureus, A., Lindgren, M., Langel, U., Bartfai, T. (1998): Purification of a galanin degrading 70 kDa metallo-peptidase from bovine spinal cord. *Neuropeptides* **32**: 453–460.

Jureus, A., Cunningham, M. J., McClain, M. E., Clifton, D. K., Steiner, R. A. (2000): Galanin-like peptide (GALP) is a target for regulation by leptin in the hypothalamus of the rat. *Endocrinology* **141**: 2703–2706.

Karelson, E., Langel, U. (1998): Galaninergic signalling and adenylate cyclase. *Neuropeptides* **32**: 197–210.

Larm, J. A., Gundlach, A. L. (2000): Galanin-like peptide (GALP) mRNA expression is restricted to arcuate nucleus of hypothalamus in adult male rat brain. *Neuroendocrinology* **72**: 67–71.

Lee, M. C., Schiffman, S. S., Pappas, T. N. (1994): Role of neuropeptides in the regulation of feeding behavior: a review of cholecystokinin, bombesin, neuropeptide Y, and galanin. *Neurosci. Biobehav. Rev.* **18**: 313–323.

Mazarati, A. M., Hohmann, J. G., Bacon, A., Liu, H., Sankar, R., Steiner, R. A., Wynick, D., Wasterlain, C. G. (2000): Modulation of hippocampal excitability and seizures by galanin. *J. Neurosci.* **20**: 6276–6281.

Merchenthaler, I., Lopez, F. J., Negro-Vilar, A. (1993): Anatomy and physiology of central galanin-containing neurons. *Prog. Neurobiol.* **40**: 769–771.

Orgen, S. O., Kehr, J., Schott, P. A. (1996): Effects of ventral hippocampal galanin on spatial learning and on *in vivo* acetylcholine release in the rat. *Neuroscience* **75**: 1127–1140.

Ogren, S. O., Schott, P. A., Kehr, J., Misane, I., Razani, H. (1999): Galanin and learning. *Brain Res.* **848**: 174–182.

Rossmanith, W. G., Clifton, D. K., Steiner, R. A. (1996): Galanin gene expression in GnRH neurons of the rat: a model for autocrine regulation. *Horm. Metab. Res.* **28**: 257–266.

Rustay, N. R., Wrenn, C. C., Kinney, J. W., Holmes, A., Bailey, K. R., Sullivan, T. L., Harris, A. P., Long, K. C., Saavedra, M. C., Starosta, G., Innerfield, C. E., Yang, R. J., Dreiling, J. L., Crawley, J. N. (2005): Galanin impairs performance on learning and memory tasks: findings from galanin transgenic and GAL-R1 knockout mice. *Neuropeptides* **39**:239–243.

Veening, J. G., Coolen, L. M., (1998): Neural activation following sexual behavior in the male and female rat brain. *Behav. Brain Res.* **92**: 181–193.

Vrontakis, M. E., Torsello, A., Friesen, H. G. (1991): Galanin. *J. Endorinol. Invest.* **14**: 785–794.

4.15
Ghrelin

4.15.1
General Aspects and History

Ghrelin was initially identified in the stomach. The first recognized effect of ghrelin, however, was its impact upon growth hormone release directly from the pituitary as well as through a hypothalamic action. The strong GH-releasing activity of ghrelin is mediated by the activation of the so-called GH secretagogue (GHS) receptor type 1a (GHS-R 1a). Before the discovery of ghrelin, this orphan receptor had been shown to be specific for a family of peptidyl, nonpeptidyl and synthetic GHS. These GHS receptors are concentrated in the hypothalamus–pituitary axis, but can also be found in some other central and peripheral tissues. Aside from mammals, ghrelin has been detected in a variety of species, including birds, amphibians and fishes.

4.15.2
Localization Within the Central Nervous System

Ghrelin-expressing cells are found within the CNS. Some ghrelin-positive cells have been detected in the nucleus arcuatus of the hypothalamus and in a distinct group of neurons located close to the third ventricle.

However, the location of these few ghrelin-positive neurons identified by immunohistochemistry was not confirmed by real-time polymerase chain reaction (RT-PCR). By RT-PCR, no detectable levels of ghrelin were found in the cerebral cortex or hypothalamus of rhesus monkey (Katakami et al. 2004). Therefore, it cannot be ruled out that ghrelin found in the hypothalamus may possibly derive from the periphery. This assumption is based on the fact that acylated ghrelin crosses the blood–brain barrier in both directions using a saturable transport system that requires the presence of the unique octanoyl residue of the ghrelin molecule. In contrast, desacyl-ghrelin crosses the blood–brain barrier by nonsaturable passive mechanisms and – once within the central nervous system – it is retained within the brain (Banks et al. 2002).

4.15.3
Biosynthesis and Degradation

Ghrelin is a motilin-related, growth hormone-releasing and orexigenic peptide that was originally isolated from the stomach by Kojima and colleagues (1999). The ghrelin peptide is a 28-amino-acid protein with a fatty-chain modification on the amino-terminal third amino acid, which seems to be important for some but not all of its biological functions (Fig. 4.26). Ghrelin (molecular weight: 3314) displays a high degree of homology in various mammals. Ghrelin is derived from a 117-amino-acid precursor peptide, named preproghrelin.

The strong GH-releasing activity of ghrelin is mediated by the activation of the growth hormone secretagogue (GHS) receptor type 1a (GHS-R1a). Interestingly, ghrelin mainly circulates as des-octanoyl ghrelin (i.e. without an esterification of Ser3), a form of the protein that is unable to stimulate GHS-R1a. This non-acylated ghrelin, which is present in human serum in far greater quantities than acylated ghrelin, seems to be devoid of any endocrine action, with the exception of some non-endocrine actions, including cardiovascular and antiproliferative effects that are probably due to an interaction with different GHS-R subtypes or receptor families (van der Lely et al. 2004).

Shortly after the identification of ghrelin, a further protein was isolated and named des-Gln14-ghrelin. This protein is a 27-amino-acid protein with a sequence similar to that of ghrelin. The only difference is that it lacks Gln in position 14. Similar to ghrelin, this molecule also requires the n-octanoylation of Ser3 for its biological activity. These two peptides, ghrelin and des-Gln14-ghrelin, are derived from a single gene which produces two alternative, distinct mRNAs. des-Gln14-ghrelin, like ghrelin, is an endogenous ligand for GHS-R 1a and it seems to possess the same biological activities as ghrelin. However, des-

Fig. 4.26 Structure of the human 28-amino-acid ghrelin, in which Ser3 is modified by a fatty acid, primarily *n*-octanoic acid. This modification is essential for the activity of ghrelin.

Gln14-ghrelin seems to be present only in low amounts in the stomach. In 2003, several other ghrelin-derived molecules were identified which can be classified into four different groups by the type of acylation observed at the serine-3 position: non-acylated, octanoylated (C8:0), decanoylated (C10:0) and possibly decenoylated (C10:1). All peptides found were either 27 or 28 amino acids in length. The 27-amino-acid isoforms lack the carboxy-terminal Arg28 and are derived from the same ghrelin precursor through alternative processing.

Interestingly, the motilin-related peptides have been found to display the same amino acid sequence as ghrelin. Based on structural and effect-related similarities, motilin and ghrelin are considered to represent a novel peptide superfamily that may have evolved from a common ancestral gene.

4.15.4
Receptors and Signal Transduction

Some, but not all biological effects of ghrelin, are mediated by activation of the growth hormone secretagogue (GHS) receptor (GHS-R). GHS-R is expressed by a single gene found on the human chromosome 3 (q26.2). Two types of GHS-R cDNAs exist, which are presumably the result of an alternate processing of a pre-mRNA. The two types have been designated as receptor 1a and 1b. In this context it is important to note that, unlike GHS-R 1a, GHS-R 1b seems not to be activated by ghrelin and its functional role is unknown.

The binding of ghrelin to GHS-R 1a activates the phospholipase C signaling pathway, thereby leading to an increased inositol phosphate turnover and protein kinase C activation, followed by the release of Ca^{2+} from intracellular stores. GHS-R activation also leads to an inhibition of K^+ channels, allowing the entry of calcium through voltage-gated L-type channels.

Since several effects of ghrelin can not be explained by an interaction of ghrelin with GHS-R 1a, it is suggestive that a group of receptors exists which are capable of binding ghrelin. In addition, the motilin receptor is also a member of the GHS-R family. However, unlike ghrelin, acylation of motilin is not needed for activation of its receptor. A further important feature is that prepromotilin, which is also produced by the enteroendocrine cells of the stomach, is almost identical with human preproghrelin, except for the serine-26 residue that is not octanoylated in the prepromotilin-related peptides. In contrast to ghrelin and its receptor, motilin and motilin receptors have been characterized in dogs and humans, but rodents seem not to possess motilin receptors.

The expression of GHS-R 1a within the central nervous system is mainly restricted to the hypothalamus (arcuate nucleus) and anterior pituitary gland. GHS-R 1a mRNA was also demonstrated in various extrahypothalamic areas, such as the hippocampus, including areas CA2-CA3 and the dentate gyrus. Moreover, GHS-R 1a mRNA was detected in the substantia nigra pars compacta, the ventral tegmental area, the dorsal and medial raphe nuclei, as well as in the pons and the medulla oblongata. Aside from this, GHS receptors are also expressed by neurons of the gut and stomach and in peripheral organs, including the adrenal, thyroid, spleen, pancreas, myocard and ovary.

4.15.5
Biological Effects

It has been reported that ghrelin stimulates prolactin, arginine–vasopressin and, to a lesser extent, corticotropin-releasing hormone release. Ghrelin possesses a strong and dose-related GH-releasing activity whereby the most potent GH-releasing activity was found in humans as compared to several other mammalian species. There is also evidence that ghrelin can act as a functional somatostatin antagonist in the pituitary and within the hypothalamus.

Aside from the known effect of ghrelin upon the release of growth hormone, ghrelin was found to stimulate appetite via the secretion of the hypothalamic hypocretinergic (oxigeneric) hormones, such as neuropeptide Y (NPY) and orexin and the inhibition of pro-opiomelanocortin/alpha-melanocyte-stimulating hormone. However, the source of ghrelin in this context is largely enigmatic, since it is still not clear whether circulating ghrelin reaches the hypothalamus or whether it is produced by specific neurons in the hypothalamus. Nevertheless, ghrelin is an important neuromodulator affecting food intake. Ghrelin administration in rodents causes weight gain. The effects of ghrelin in this context are surprisingly fast, since changes in body weight induced by ghrelin administration become significant in rodents after no more than 48 h. This effect is dose-dependent and central administration is more effective than peripheral administration, suggesting a central mechanism of action.

Besides regulating food intake, it has been shown that ghrelin also affects anxiety-like behavior in mice. Intracerebroventricular and intraperitoneal administration of ghrelin induces anxiogenic activities, wherby administration of a CRH receptor antagonist inhibits the ghrelin-induced anxiogenic effects.

4.15.6
Neurological Disorders and Neurodegenerative Diseases

It has been shown that ghrelin increases ACTH-release. The stimulatory effect of ghrelin in this context is augmented and higher than that of human CRH in patients with pituitary ACTH-dependent Cushing's disease, a fact which may reflect a direct action of ghrelin on pituitary ACTH-secreting tumor cells.

Further Reading

Arvat, E., Giordano, R., Ramunni, J., Arnaldi, G., Colao, A., Deghenghi, R., Lombardi, G., Mantero, F., Camanni, F., Ghigo, E. (1998): Adrenocorticotropin and cortisol hyperresponsiveness to hexarelin in patients with Cushing's disease bearing a pituitary microadenoma, but not in those with macroadenoma. *J. Clin. Endocrinol. Metab.* **83**: 4207–4211.

Asakawa, A., Inui, A., Kaga, T., Yuzuriha, H., Nagata, T., Fujimiya, M., Katsuura, G., Makino, S., Fujino, M.A., Kasuga, M. (2001): A role of ghrelin in neuroendocrine and behavioral responses to stress in mice. *Neuroendocrinology* **74**: 143–147.

Banks, W.A., Tschop, M., Robinson, S.M., Heiman, M.L. (2002): Extent and direction of ghrelin transport across the blood–brain barrier is determined by its unique primary structure. *J. Pharmacol. Exp. Ther.* **302**: 822–827.

Hosoda, H., Kojima, M., Matsuo, H., Kangawa, K. (2000): Purification and characterization of rat des-Gln14-Ghrelin, a second endogenous ligand for the growth hormone secretagogue receptor. *J. Biol. Chem.* **275**: 21995–212000.

Hosoda, H., Kojima, M., Mizushima, T., Shimizu, S., Kangawa, K. (2003): Structural divergence of human ghrelin. Identification of multiple ghrelin-derived molecules produced by post-translational processing. *J. Biol. Chem.* **278**: 64–70.

Howard, A.D., Feighner, S.D., Cully, D.F., Arena, J.P., Liberator, P.A., Rosenblum, C.I., Hamelin, M., Hreniuk, D.L., Palyha, O.C., Anderson, J., Paress, P.S., Diaz, C., Chou, M., Liu, K.K., McKee, K.K., Pong, S.S., Chaung, L.Y., Elbrecht, A., Dashkevicz, M., Heavens, R., Rigby, M., Sirinathsinghji, D.J., Dean, D.C., Melillo, D.G., Van der Ploeg, L.H. (1996): A receptor in pituitary and hypothalamus that functions in growth hormone release. *Science* **273**: 974–977

Katakami, H., Shimizu, K., Kimura, N., Ashida, S., Terasawa, E. (2004): Cloning and characterization of ghrelin and GHRH in the rhesus monkey, *Macaca mulatta*. In: *Proc. Endocrine Soc. Annu. Meet.* **86**

Kojima, M., Hosoda, H., Date, Y., Nakazato, M., Matsuo, H., Kangawa, K. (1999): Ghrelin is a growth hormone-releasing acylated peptide from stomach. *Nature* **402**: 656–660.

Kojima, M., Kangawa, K. (2005): Ghrelin: structure and function. *Physiol. Rev.* **85**: 495–522.

Korbonits, M., Grossman, A.B. (2004): Ghrelin: update on a novel hormonal system. *Eur. J. Endocrin.* **151**: S67–S70.

Spiegelman, B.M., Flier, J.S. (2001): Obesity and the regulation of energy balance. *Cell* **104**: 531–543

Tannenbaum, G.S., Bowers, C.Y. (2001): Interactions of growth hormone secretagogues and growth hormone-releasing hormone/somatostatin. *Endocrine* **14**: 21–27.

Tschop, M., Smiley, D.L., Heiman, M.L. (2000): Ghrelin induces adiposity in rodents. *Nature* **407**: 908–913

van der Lely, A.J., Tschop, M., Heiman, M.L., Ghigo, E. (2004): Biological, physiological, pathophysiological, and pharmacological aspects of ghrelin. *Endocrinol. Rev.* **25**: 426–457

4.16
Gonadotropin-releasing hormone

4.16.1
General Aspects and History

Gonadotropin-releasing hormone (GnRH) is also known as gonadoliberin or luteinizing hormone-releasing hormone (LHRH). This substance was isolated and characterized in 1971 by two groups (the group of Schally and the group of Amos). This neuropeptide is composed of ten amino acids (sequence: pGlu-His-Trp-Ser-Tyr-Gly-Leu-Arg-Pro-Gly-NH$_2$).

In mammals, only one form of GnRH is found, but different forms of GnRH have been found in the brain of non-mammalian vertebrates (nine forms), which vary in their composition by one or more amino acids.

GnRH is not only produced in neuronal tissues of the central nervous system, but also in some peripheral tissues, including the gonads, the placenta, the pancreas and some other tissues.

The best known function of GnRH is the stimulator effect on the release of luteinizing hormone (LH) and follicle-stimulating hormone (FSH) from the pituitary gland in a pulsative fashion.

The maturation of the GnRH precursor leads to the synthesis not only of GnRH, but also of a 56-residue GnRH-associated peptide (GAP). This peptide is also a neuroactive substance, since it has been shown, for example, to inhibit prolactin secretion. The inhibition of prolactin secretion appears to be the primary action of GAP and, therefore, this peptide is also termed the prolactin secretion-inhibiting factor.

4.16.2
Localization Within the Central Nervous System

The highest immunoreactivity for GnRH is found in the basomedial hypothalamus and in the arcuate nucleus of the hypothalamus.

In addition, GnRH-expressing cells have been found in other areas, such as the diagonal band of Broca, the preoptic area, the periventricular nucleus of the hypothalamus, the anterior hypothalamic area, the olfactory bulbus, the stria terminalis, the cortical nucleus of the amygdala, the medial nucleus of the amygdala and the hippocampus.

Most of the hypothalamic GnRH neurons project to the eminentia mediana. In addition, GnRH neurons from the preoptic area are also known to innervate the organum vasculosum of the lamina terminalis (OVLT). Furthermore, GnRH-carrying terminals have been found in the stria terminalis, the amygdala and in the ventral tegmental area.

4.16.3
Biosynthesis and Degradation

The GnRH gene is composed of four exons. The first exon contains the 5′ untranslated region and the second codes for the signal peptide, the GnRH peptide, an enzymatic amidation site and the precursor-processing site. In addition, this exon codes for the amino-terminal amino acids of the 56-amino-acid GnRH-associated peptide (GAP). The third exon codes for the next 32 amino acids of the GAP peptide. The fourth exon codes the last 13 amino acid of GAP, the termination codon and the 3′ untranslated region.

The precursor of GnRH has a molecular mass of 10 kDa. The human GnRH precursor gene encodes a 92-residue polypeptide chain consisting of a signal peptide, the gonadotropin-releasing hormone (GnRH) and the 56-residue GnRH-associated peptide (GAP; Fig. 4.27). A dibasic recognition sequence for a subtilisin-type protease separates GnRH and GAP in the precursor to facilitate posttranslational cleavage.

GnRH and GAP are the main products of the maturation of the precursor. Beside these two substances, different intermediate forms were generated, like (Gln) or (pGlu)GnRH-(1–13) and (1–12), (pGlu)GnRH-(1–11), GAP(1–32), GAP(34–56), GAP(1–36) and GAP(38–56). The GAP is present in GnRH-containing neurons of the rat brain and it coexists with GnRH in secretory granules of nerve terminals in the median eminence.

GnRH is secreted from the neurons in a pulsative manner. This pulsative secretion can then induce a pulsative release of LH. The pulsative release of

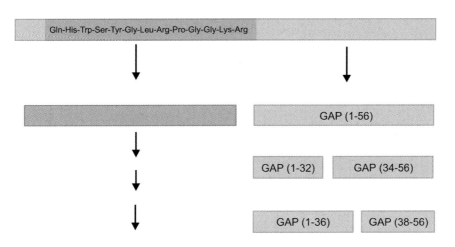

Fig. 4.27 The GnRH-precursor is not only the precursor of GnRH, but also for of a couple of other peptides, which belong to the gonadotropin-associated peptide family (see text).

GnRH can be influenced by steroids, neuropeptides and neurotransmitters. Hormones play a crucial role in GnRH expression and secretion. Glucocorticoid repression of GnRH transcription has been localized to the GnRH promoter.

There is physiological evidence that GnRH itself has a role in tonic inhibition or modulation of GnRH function by short feedback mechanisms. Ultrastructural studies, for example, have revealed GnRH synapses on GnRH neurons and their processes.

The blockage of the carboxy-terminal and the amino-terminal region of GnRH by pGlu and $CONH_2$ prevents degradation by exopeptidases. However, GnRH can be degraded by endopeptidases. Among the peptidases which are capable of degrading GnRH are pyroglutamate aminopeptidase and the post-proline-cleavage enzyme (PPCE).

4.16.4
Receptors and Signal Transduction

A single receptor type, with a molecular mass of 60 kDa, has been identified in the anterior pituitary, the amygdala and the hippocampus, as well as in lymphocytes.

In some other tissues, in addition to the first receptor, a second type of GnRH receptor has been identified, which has a molecular mass of 136 kDa. It is assumed that this receptor exists as a multiunit complex.

In gonadotroph cells of the pituitary, binding of GnRH to its receptor initiates the opening of the calcium channel and thus initiates Ca^{2+} influx. This calcium than interacts with intracellular calmodulin.

The absence of calcium or the prevention of calcium entry into the cell prevents the binding of GnRH to its receptor from inducing any biological effects. Therefore, calcium is necessary as a second messenger in GnRH signaling.

The interactions of GnRH with its receptor are accompanied by the formation of phosphoinositol and diacylglycerol (by the activity of phospholipase C). Phosphoinositol leads to the mobilization of intracellular calcium, whereas diacylglycerol acts as an activator of proteinkinase C.

GnRH can regulate its own receptors, which are initially downregulated by continuously administration of GnRH, followed by recovery and up-regulation. This autoregulation seems to depend on calcium.

4.16.5
Biological Effects

Gonadotropin-releasing hormone exerts most of its biological functions by acting within the anterior pituitary. GnRH is expressed in specific hypothalamic neurons and secreted into the hypophyseal portal venous circulation in a pulsative manner. The secreted hormone travels to the anterior pituitary, where it binds to GnRH receptors on the pituitary gonadotrope. Occupation of these receptors activates signal transduction cascades, finally causing the release of the gonadotropins, luteinizing hormone (LH) and follicle-stimulating hormone (FSH). LH and FSH play a role in gonadal maturation, onset of puberty and ovulation. Changes in GnRH pulse frequency and amplitude regulate gonadotropin release; and GnRH is regulated at the level of transcription, translation and secretion to produce these physiological effects.

The release of LH and FSH is nearly completely dependent upon the pulsative release of GnRH. A continuous infusion of GnRH, for example, suppresses the secretion of LH or FSH.

The innate pulsative secretory pattern of the GnRH pulse generator network in the hypothalamus is under the control of two hypothalamic neuropeptides, NPY and galanin. Furthermore, GnRH can modulate its own release by an ultrashort loop feedback mechanism (autoregulation), but the pulsatory secretion seems also to depend on further internal factors (e.g. emotions) as well as external factors (e.g. light).

Since GnRH controls the release of LH from the pituitary, it has been hypothesized that LH can also control GnRH production by a short feedback loop.

The additional neuroactive product of the precursor pre-GnRH, namely GAP, seems to influence the secretion of gonadotropins by a mechanism, which is independent from the activation of GnRH receptors.

The presence of GnRH in neurons outside the hypothalamic–hypophysial system indicates that GnRH plays a role as a neuromodulator. However, only a little experimental evidence on this matter exists. In behavioral studies, it has been demonstrated that GnRH induces mating behavior in different species. Since gonadotropins alone are unable to induce this behavior, it is likely that this GnRH-induced behavior is a direct effect.

4.16.6
Neurological Disorders and Neurodegenerative Diseases

Gonadotropins alone have no effect on mating behavior, but GnRH can induce this behavior, even in hypophysectomized animals. GnRH seems to be able to coordinate behavior involved in reproduction in both females and males. Thus, GnRH agonists and antagonists can be used in a variety of clinical conditions involving reproduction. A significant involvement in neurological disorders or neurodegenerative diseases, however, has not yet been discovered.

Further Reading
Chavali, G.B., Nagpal, S., Majumdar, S.S., Singh, O., Salunke, D.M. (1997): Helix-loop-helix motif in GnRH-associated peptide is critical for negative regulation of prolactin secretion. *J. Mol. Biol.* **272**: 731–740.
Daikoku-Ishido, H., Okamura, Y., Yanaihara, N., Daikoku, S. (1990): Development of the hypothalamic luteinizing hormone-releasing hormone-containing neuron system in the rat: *in vivo* and in transplantation studies. *Dev. Biol.* **140**: 374–387.
Hattori, M., Ishii, S. (1984): Stimulation of FSH and LH release by two chicken LHRHs and mammalian LHRH *in vitro* and *in vivo*. *J. Steroid Biochem.* **20**: 1548.
Hazum, E., Conn, P.M. (1988): Molecular mechanism of gonadotropin-releasing hormone (GnRH) action. I. The GnRH receptor. *Endocrinol. Rev.* **9**: 379–386.

Huckle, W. R., Conn, P. M. (1988): Molecular mechanism of gonadotropin-releasing hormone (GnRH) action. II. The effector system. *Endocrinol. Rev.* **9**: 387–395.

Karten, M. J., Rivier, J. E. (1986): Gonadotropin-releasing hormone analog design. Structure–function studies towards the development of agonists and antagonists: rationale and perspective. *Endocrinol. Rev.* **7**: 44–66.

Knobil, E. (1989): The electrophysiology of the GnRH pulse generator. *J. Steroid Biochem. Mol. Biol.* **33**: 669–671.

Millar, R. P., King, J. A. (1988): Evolution of a gonadotropin-releasing hormone: multiple usage of a peptide. *News Physiol. Sci.* **3**: 49–53.

Nelson, S. B., Eraly, S. A., Mellon, P. L. (1998): The GnRH promoter: target of transcription factors, hormones, and signaling pathways. *Mol. Cell Endocrinol.* **140**: 151–155.

Schwanzel-Fukuda, M., Silverman, A. J. (1980): The nervus terminalis of the guinea pig: a new luteinizing hormone releasing-hormone (LHRH) neuronal system. *J. Comp. Neurol.* **191**: 213–225.

Schwanzel-Fukuda, M., Pfaff, D. W. (1989): Origin of luteinizing hormone-releasing hormone neurons. *Nature* **338**: 161–164.

Sherwood, N. M., Lovejoy, D. A., Coe, I. R. (1993): Origin of mammalian gonadotropin-releasing hormones. *Endocrinol. Rev.* **14**: 241–254.

Witkin, J. W. (1999): Synchronized neuronal networks: the GnRH system. *Microsc. Res. Tech.* **44**: 11–18.

4.17
Growth Hormone-releasing Hormone

4.17.1
General Aspects and History

The existence of a hypothalamic factor which stimulates the secretion of growth hormone was proposed in the 1950s. However, it was not until 1982 that the growth hormone-releasing hormone (GHRH) was isolated [in two isoforms, GHRH (1–40) and GHRH (81–44)-NH$_2$)] by the group of Guillemin and Rivier.

GHRH, also known as somatoliberin, is a hypothalamic peptide which specifically stimulates the secretion of growth hormone (GH, also known as somatotropin). Analysis of the sequence of GHRH revealed that it belongs to a family of structurally and functionally closely related proteohormones. Further members of this family are glucagon, secretin, corticotropin-releasing factor (CRF), pituitary adenylate cyclase-activating peptide (PACAP) and intestinal vasoactive polypeptide (VIP).

4.17.2
Localization Within the Central Nervous System

By using specific antibodies towards GHRH, the distribution of GHRH-containing cells in the central nervous system has been elaborated. Most GHRH-containing neurons are found in the arcuate nucleus of the hypothalamus and in the pituitary. The periventricular nucleus of the hypothalamus and some neurons in the neighborhood of the ventromedial hypothalamic nucleus also pos-

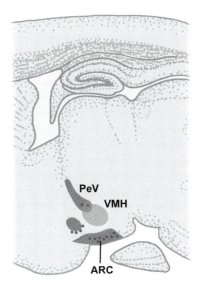

Fig. 4.28 Schematic representation of GHRH-containing neurons in the rat central nervous system. *Abbreviations*: ARC = arcuate nucleus; PeV = periventricular nucleus of the hypothalamus; VMH = ventromedial nucleus of the hypothalamus.

sess GHRH-containing neurons (see Fig. 4.28). Neurons containing GHRH project preferentially into the median eminence. In addition, extrahypothalamic fibers containing GHRH extend through the cerebral cortex, the limbic system and the hindbrain.

GHRH colocalizes with one or more neuromodulators or neurotransmitters, including acetylcholine, dopamine, enkephalin, GABA, galanin, α-MSH, NPY, somatostatin and substance P.

4.17.3
Biosynthesis and Degradation

GHRH belongs to the secretin–glucagon family, which also includes VIP (see above). VIP and GHRH share nine residues (positions 3, 4, 6, 7, 10, 16, 21, 23, 26) which cluster in the amino-terminal regions of the peptides.

The human gene which encodes for GHRH is located on chromosome 20 and was cloned by Mayo and coworkers in 1985. The 9-kb gene is composed of five exons and four introns. The third exon codes for the biologically active form of GHRH.

The precursor prepro-GHRH consists of 107 or 108 amino acids. The 108-amino-acid isoform of prepro-GHRH shows an additional serine in the carboxy-terminal region. Alternative splicing produces the two isoforms.

The prepro-GHGHs are processed by enzymatic cleavage of the signal sequence and by proteolysis of the amino- and carboxy-terminal regions.

Both forms are present in the hypothalamus in nearly equal amounts and they both have similar potencies in regulating GH secretion.

Besides the human form of GHRH (GHRH$_{44NH2}$ is composed of 44 amino acids) and the rat GHRH (GHRH$_{40OH}$ is composed of 40 amino acids), a 29- or

30-amino-acid peptide is cleaved from the precursors. However, its physiological function remains unknown.

GHRH is under the control of somatostatin and growth hormone by feedback loops. It is inactivated by peptidases, which upon proeteolytic cleavage produce the biologically inactive form $GHRH_{3\text{-}44NH2}$.

4.17.4
Receptors and Signal Transduction

The gene which encodes the human GHRH receptor, is located on chromosome 6 and is composed of 13 exons, with an alternative exon located between exons 10 and 11. The GHRH receptor is a member of the GPCR family. The receptor shares some structural similarities with VIP receptors and with the receptors for calcitonin and secretin.

The stimulation of GHRH receptors enhances the accumulation of intracellular cAMP, a process that involves a G_s protein-linked receptor/adenylate cyclase pathway. Subsequently, the cAMP increase activates protein kinase A.

The signal transduction pathway also impinges on the phosphatidylinositol system and the Ca^{2+}/calmodulin system.

Furthermore, GHRH stimulates MAP kinase activity. This activation appears to be independent of the G_s/cAMP/PKA pathway and is more likely to be mediated by the $\beta\gamma$-subunits of the G protein which is linked with a Ras-dependent pathway.

GHRH receptors are expressed in the adenohypophysis. Estrogen and glucocorticoids can modulate the expression of GHRH receptors in this area. Bolus administration of any one of these substances causes a decrease in the amount of mRNA coding for GHRH receptors, while continuous administration (rather than pulsative administration, which occurs under *in vivo* conditions) reduces the response of GHRH receptors to the binding of GHRH. This effect is either mediated by an uncoupling of the G protein from the receptor or by down-regulation of the receptor. Such desensibilization of GHRH receptors can be prevented by the administration of somatostatin (although somatostatin inhibits GHRH).

4.17.5
Biological Effects

The best known effect of GHRH is stimulation of the synthesis and secretion of growth hormone (GH). Pituitary somatotropes are stimulated to proliferate and secrete growth hormone by GHRH via its hypothalamic secretion. This is a direct effect, since it is mediated through GHRH receptors in the pituitary. The secretion of growth hormone from the pituitary is under the control loop of GHRH (stimulation) and somatostatin (inhibition). GHRH not only stimulates GH synthesis and release but also proliferation and differentiation of the pituitary somatotrophs. Retarded growth due to a lack of GH is often caused by a hypothalamic deficiency of GHRH. Clinical studies have shown an acceleration of growth in children when GHRH was given in a pulsatory manner, but after one year of therapy, the growth slows down (Vance and Thorner 1988).

The action of both GHRH and somatostatin causes a pulsative release of growth hormone. The stimulation of GH synthesis by GHRH is transcriptionally controlled. This regulation of transcription involves elements in the promoter region of the GH gene which are sensitive to cAMP. The action of GHRH is necessary for normal development. In mammals, middle and late adulthood is characterized by a decrease in spontaneous and stimulated GH secretion and it is thought that the age-related decline of GH secretion is governed by a decrease in the responsiveness of somatotroph cells to GHRH.

The destruction of GHRH-containing neurons results in a delay of normal growth in rats. Likewise, the administration of anti-GHRH antibodies can reduce the normal growth of young rats.

In contrast, pulsative administration of GHRH enhances growth. GHRH not only stimulates growth of the whole body, it also stimulates the proliferation of GH-synthesizing pituitary cells (somatotrope cells). This effect is mediated by cAMP. In transgenic mice overexpressing GHRH, a good correlation has been found between hyperplasia of the pituitary and increase in the plasma levels of GH. Activation of c-fos by GHRH has been suggested to be responsible for this effect.

GHRH also participates in the regulation of sleep and feeding behavior. Central or peripheral administration of GHRH induces sleep. This effect seems to be directly linked to GHRH, since coapplication of GHRH antagonists or immuno-neutralization of GHRH prevents sleep induction. The intracerebroventricular administration of GHRH also stimulates food intake in rats. This effect can be blocked by the administration of anti-GHRH antibodies.

4.17.6
Neurological Disorders and Neurodegenerative Diseases

GHRH regulates growth hormone release from the pituitary and, in addition to this neuroendocrine actions, much evidence implies an additional role for GHRH in carcinogenesis in non-pituitary tissues. Moreover, hypothalamic tumors (as e.g. hamartomas, choristomas, gliomas and gangliocitomas) may produce excessive GHRH. This increased production may lead to subsequent GH hypersecretion, resulting in acromegaly. Immunoreactivity for GHRH is present in several tumors, including carcinoid tumors, pancreatic cell tumors, small-cell lung cancers, adrenal adenomas and pheochromocitomas which have been reported to secrete GHRH.

A potential therapeutic treatment to suppress the effects of GHRH-overexpression is the use of GHRH antagonists. Moreover, several GHRH antagonists have antiproliferative effects in many tumor models. However, GHRH antagonists that should act within the hypothalamus must cross the blood–brain barrier. Recently, one GHRH antagonist was found to cross the blood–brain barrier, indicating that GHRH antagonists may provide a potential treatment for malignant glioblastomas.

However, to date, no major contribution of GHRH in neurological disorders or neurodegenerative diseases has been documented.

Further Reading

Bertherat, J., Bluet-Pajot, M.T., Epelbaum, J. (1995): Neuroendocrine regulation of growth hormone. *Eur. J. Endocrinol.* **132**: 12–24.

Doga, M., Bonadonna, S., Burattin, A., Giustina, A. (2001): Ectopic secretion of growth hormone-releasing hormone (GHRH) in neuroendocrine tumors: relevant clinical aspects. *Ann. Oncol.* **12**: S89–S94.

Frohman, L.A., Downs, T.R., Chomczynski, P. (1992): Regulation of growth hormone secretion. *Front. Neuroendocrinol.* **13**: 344–405.

Jaeger, L.B., Banks, W.A., Varga, J.L., Schally, A.V. (2005): Antagonists of growth hormone-releasing hormone cross the blood–brain barrier: a potential applicability to treatment of brain tumors. *Proc. Natl Acad. Sci. USA* **102**:12495–12500.

Kiaris, H, Koutsilieris, M., Kalofoutis, A., Schally, A.V. (2003): Growth hormone-releasing hormone and extra-pituitary tumorigenesis: therapeutic and diagnostic applications of growth hormone-releasing hormone antagonists. *Expert Opin. Investig. Drugs* **12**: 1385–1394.

Krueger, J.M., Obal, F. Jr (1993): Growth hormone-releasing hormone and interleukin-1 in sleep regulation. *FASEB J.* **7**: 645–652.

Locatelli, V., Torsello, A. (1997): Growth hormone secretagogues: focus on the growth hormone-releasing peptides. *Pharmacol. Res.* **36**: 415–423.

Müller, E.E., Rolla, M., Ghigo, E., Belliti, D., Arvat, E., Andreoni, A., Torsello, A., Locatelli, V., Camanni, F. (1995): Involvement of brain catecholamines and acetylcholine in growth hormone hypersecretory states. Pathophysiological diagnostic and therapeutic implications. *Drugs* **50**: 805–837.

Pombo, C.M., Zalvide, J., Gaylinn, B.D., Dieguez, C. (2000): Growth hormone-releasing hormone stimulates mitogen-activated protein kinase. *Endocrinology* **141**: 2113–2119.

Rosskamp, R. (1988): Wachstumshormon-Releasinghormon. Übersicht [Growth hormone-releasing hormone. Review]. *Klin. Padiatr.* **200**: 81–88.

Vance, M.L. (1990): Growth hormone-releasing hormone. *Clin. Chem.* **36**: 415–420.

Vance, M.L., Thorner, M.O. (1988): Some clinical considerations of growth hormone and growth hormone releasing hormone. *Front. Neuroendocrinol.* **10**: 279–294.

Veldhuis, J.D., Iranmanesh, A., Weltman, A. (1997): Elements in the pathophysiology of diminished growth hormone (GH) secretion in aging humans. *J. Pediatr. Endocrinol. Metab.* **7**: 41–48.

4.18
Hypocretin (Orexin)

4.18.1
General Aspects and History

In 1998, two independent research groups simultaneously discovered two new neuropeptides, named orexin A (or hypocretin 1) and orexin B (or hypocretin 2). One group observed an appetite-stimulating effect of these peptides and coined the term orexin (Sakuria et al. 1998), derived from the Greek word "*or-*

exis" (appetite). The Sutcliff group, using the technique of substractive RNA hybridization, identified a hypothamalus-specific mRNA (which they coined preprohypocretin) that represented the precursor of two peptides which they termed hypocretins (Hcrt) from their *hypo*thalamic localization and their structural similarities to the gut hormone se*cretin* (De Lecea et al. 1998). In the following, we prefer the term hypocretin since accruing evidence made it doubtful that these peptides are cirtically important in appetite regulation. However, there is little doubt that Hcrts are vitally important in behavioral regulations in which the hypothalamus plays a key role.

Hcrts are produced by neurons localized in the hypothalamus. They show a wide-spread projection into many areas of the brain, including systems that are known to regulate sleep and wakefulness. Hcrts seem to play an essential role in driving the arousal system; and the lack of hcrts-producing neurons seems to be causually linked to the pathophysiology of narcolepsy–cataplexy.

4.18.2
Localization Within the Central Nervous System

In situ hybridization and immunohistochemistry showed hcrt mRNA and protein exclusively in a restricted area of the hypothalamus. These hcrt-expressing neurons were mainly found in the perifornical nucleus and the dorsomedial hypothalamic nucleus, and in the dorsal and lateral hypothalamic areas. Moreover, a few cells were observed in the posterior hypothalamic area and at the junction of the thalamus and hypothalamus. Hcrt neurons are variable in size and shape (spherical, fusiform, multipolar). Within the rat brain, the total number of cells has been estimated by different groups at about 1000 to 3500, indicating that the number of neurons is low. Despite this local confinement, Hcrt neurons exhibit a strong innervation of the hypothalamus and widespread projections throughput the brain and spinal cord. Thus, Hcrt-immunoreactive fibers are found throughout the entire hypothalamus. Their projecting tracts can be subdivided into four different pathways: dorsal and ventral ascending pathways as well as dorsal and ventral descending pathways (Peyron et al. 1998).

The dorsal ascending pathway consists of fibers emanating from neurons which run through the zona incerta to the paraventricular nucleus of the thalamus, the central medial nucleus of the thalamus and the lateral habenula. Hcrt-positive fibers are also found in the dorsal anterior nucleus of the olfactory bulb, the medial and lateral septal nuclei, the bed nucleus of the stria terminalis and the substantia innominata. In addition, fibers innervating the cortex and the medial amygdaloid nuclei belong to the system of the dorsal ascending pathway.

The ventral ascending pathway is composed of fibers which project to the ventral pallidum, the vertical and horizontal limb of the diagonal band of Broca, the accumbens nucleus and the olfactory bulb.

The dorsal descending pathway innervates the colliculi and the pontine central gray, particularly the locus coeruleus, the dorsal raphe nucleus and the later-

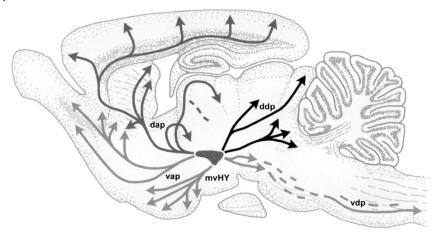

Fig. 4.29 Schematic summary drawing of pathways taken by hypocretin (orexin) processes that widely innervate rat brain. *Abbreviations*: mvHY=medio-ventral hypothalamus; dap=dorsal ascending pathway; vap=ventral ascending pathway; ddp=dorsal descending pathway; vdp=ventral descending pathway.

odorsal tegmental nucleus. Hcrt-positive fibers which run through the dorsal tegmental area to the pedunculopontine nucleus, the parabrachial nucleus and the subcoeruleus area also belong to the dorsal descending pathway.

The ventral descending pathway comprises fibers which ramify to the interpeduncular nucleus, the ventral tegmental area and the substantia nigra pars compacta. Fibers in the reticular formation of the pons, the ventral medulla oblongata and the raphe nuclei also belong to the ventral descending pathway (Fig. 4.29).

Hcrt peptides and their receptors are also found outside the central nervous system, e.g. the gastrointestinal tract, the pancreas and the adrenal gland.

4.18.3
Biosynthesis and Degradation

The mammalian preprohypocretin is composed of 130–131 amino acids. The human preprohypocretin gene is located on chromosome 17. It is composed of two exons, the latter of which includes the entire coding region of the final peptides. The cleavage of one molecule of the precursor and further modification leads to the production of Hctr-1 (orexin A) and Hctr-2 (orexin B).

The Hctr-1 peptides in humans, rats and mice are identical, whereas Hctr-2 in rats and mice differ by two amino acids from the human peptide. Mammalian Hctr-1 is a 33-amino-acid protein that contains two intrachain disulfide bridges, whereas Hctr-2 is a linear peptide composed of 28 amino acids.

Hctr-1 seems to exhibit much higher stability than Hctr-2 under physiological conditions, which may explain why Hctr-1 is more readily detectable in cerebrospinal fluid than Hctr-2. Moreover, Hctr-1 also displays higher lipid solubility than the Hctr-2 form. This fact may explain why Hctr-1, in contrast to Hctr-2, can cross the blood–brain barrier (Kastin and Akerstrom 1999).

Interestingly, Hctr-1 and preprohypocretin mRNA concentrations show circadian variation in rats in a cycle of 12 h light/12 h dark. Preprohypocretin mRNA and Hctr-1 concentrations in the hypothalamus reach a maximum around the light onset and a minimum at the light offset.

4.18.4
Receptors and Signal Transduction

Two receptors have been cloned that are capable of binding Hctrs. These receptors (Hctr receptor-1 and receptor-2, synonyms of OX1 and OX2 according to the orexin terminology) belong to the GPCR family. Both receptors reveal a 64% sequence identity. The gene encoding Hctr receptor-1 (a protein of 425 amino acids) is located on chromosome 1. The gene encoding Hctr receptor-2 (a 444-amino-acid protein) is located on human chromosome 6.

Similarly to the distribution of Hctrs, the distribution of their receptors has been investigated with molecular biological and immunocytochemical approaches. It has been shown that Hctr receptors are found at the projection sites of Hctr neurons, but the distribution of mRNAs of both receptors are strikingly different. Receptor-1 predominates in the following brain regions: the cingulate cortex, the anterior olfactory nucleus, prefrontal and infralimbic cortex, bed nucleus of stria terminalis, anterior hypothalamus and locus coeruleus. Receptor-2 is the predominant type in the medial septal nucleus, nucleus of the diagonal band of Broca, hippocampal CA3 field, arcuate nucleus of the hypothalamus, lateral hypothalamus and the tuberomamillary nucleus.

4.18.5
Biological Effects

The hypothalamic Hctr system directly and strongly innervates and excites the monoaminergic system and cholinergic neurons. It also plays a major role in modulating the release of glutamate and other amino acid transmitters. A close look to the different projecting sites provides some clues for the general functional implications of the Hctr system.
- There exists a strong innervation of the monoaminergic system including serotonergic and noradrenergic regions, which reveal descending pathways to centers for movement and muscle tone control.
- Ascending projections to forebrain regions involved in sensory intergration are eminent.
- The histaminergic hypothalamic system (which is involved in forebrain alerting) and the dopaminergic system are also strongly innervated by the Hctr projections.

- In addition, Hctr neurons project to the major cholinergic cells in the brainstem and basal forebrain. These regions are known to play a central role in the cortical EEG activation that characterizes waking.

Taken together, the most evident function of the Hctr system is to act as a kind of a servo-system that coordinates the activity of the above-mentioned arousal system with motor activity.

Animal studies indicate that hypocretins are also involved in the regulation of feeding behavior and in neuroendocrine and autonomic functions.

Indeed, it has been shown that intracerebroventricular injection of hypocretins (orexins) induces feeding in rats. Intraperitoneal injection of an OX1 receptor antagonist (SB-334867) inhibits baseline feeding, normal weight gain and the feeding response to orexin injection in the hypothalamus. However, Hcrt knockout animals are not emaciated. Although the food intake is somewhat reduced, their weight is normal. A current concept is that the reported effects of the Hctr system on food intake is related to manipulations of the motor activity of the animals.

4.18.6
Neurological Disorders and Neurodegenerative Diseases

Human narcolepsy is a chronic sleep disorder affecting about one in 2000 individuals. The disease is characterized by excessive daytime sleepiness, cataplexy and other abnormal manifestations of REM sleep, such as sleep paralysis and hypnagogic hallucinations.

Hctr deficiencies, as shown by low or absent concentrations in CSF, were found in 90% of patients with sporadic narcolepsy–cataplexy and less commonly in familial narcolepsy. For example, Hctr-1 is undetectable in the CSF of most human narcoleptic patients and the number of Hctr neurons in autoptic narcoleptic brains has been found to be reduced.

Hctr knockout mice and Hctr-receptor knockout mice as well as mice with a toxic destruction of Hctr cells by hypothalamic injection of Hctr-2/saporin conjugates display a phenotype reminiscent of human narcolepsy with cataplexy (e.g. fragmented behavioral states and episodes of sudden behavioral arrest).

In dogs, narcolepsy is mainly a familial disorder. In 1999, familial canine narcolepsy of Doberman pinschers and Labrador retrievers was shown to be caused by mutations in the hypocretin-2 receptor gene. In humans, mutations in Hctr- or Hctr-receptors genes are rarely associated with narcolepsy. Thus, there is only one case report (Peyron and coworkers 2000) of a mutation in the case of early onset narcolepsy. Some evidence speaks in favor of an autoimmune background of human narcolepsy, since 85–95% of human narcoleptics share a particular HLA antigen genotype.

Further Reading

Alam, M. N., Kumar, S., Bashir, T., Suntsova, N., Methippara, M. M., Szymusiak, R., McGinty, D. (2005): GABA-mediated control of hypocretin – but not melanin concentrating hormone-immunoreactive neurones during sleep in rats. *J. Physiol.* **563**: 569–582.

Baumann, C. R., Bassetti, C. L. (2005): Hypocretins (orexins) and sleep–wake disorders. *Lancet Neurol.* **4**: 673–682.

De Lecea, L., Kilduff, T. S., Peyron, C., Gao, X., Foye, P. E., Danielson P. E., Fukuhara, C., Battenberg, E. L., Gautvik, V. T., Bartlett, F. S., Frankel, W. N., van den Pol ,A. N., Bloom, F. E., Gautvik, K. M., Sutcliffe, J. G. (1998): The hypocretins: hypothalamus-specific peptides with neuroexcitatory activity. *Proc. Natl Acad. Sci. USA* **95**: 322–327.

Espana, R. A., Baldo, B. A., Kelley, A. E., Berridge, C. W. (2001): Wake-promoting and sleep-suppressing actions of hypocretin (orexin): basal forebrain sites of action. *Neuroscience* **106**: 699–715.

Fujiki, N., Yoshida, Y., Ripley, B., Honda, K., Mignot, E., Nishino, S. (2001): Changes in CSF hypocretin-1 (orexin A) levels in rats across 24 hours and in response to food deprivation. *Neuroreport* **12**: 993–997.

Gencik, M., Dahmen, N., Wieczorek, S., Kasten, M., Bierbrauer, J., Anghelescu, I., Szegedi, A., Menezes Saecker, A. M., Epplen J. T. (2001): A prepro-orexin gene polymorphism is associated with narcolepsy. *Neurology* **56**: 115–117.

Gerashchenko, D., Kohls, M. D., Greco, M., Waleh, N. S., Salin-Pascual, R., Kilduff, T. S., Lappi, D. A., Shiromani, P. J. (2001): Hypocretin-2-saporin lesions of the lateral hypothalamus produce narcoleptic-like sleep behavior in the rat. *J. Neurosci.* **21**: 7273–7283.

Kastin, A. J., Akerstrom, V. (1999): Orexin A but not orexin B rapidly enters brain from blood by simple diffusion. *J. Pharmacol. Exp. Ther.* **289**: 219–223.

Lin, L., Faraco, J., Li, R., Kadotani, H., Rogers, W., Lin, X., Qiu, X., de Jong, P. J., Nishino, S., Mignot, E. (1999): The sleep disorder canine narcolepsy is caused by a mutation in the hypocretin (orexin) receptor 2 gene. *Cell* **98**: 365–376.

Nishino, S, Kanbayashi, T. (2005): Symptomatic narcolepsy, cataplexy and hypersomnia, and their implications in the hypothalamic hypocretin/orexin system. *Sleep Med. Rev.* **9**: 269–310.

Nishino, S., Ripley, B., Overeem, S., Lammers, G. J., Mignot, E. (2000): Hypocretin (orexin) deficiency in human narcolepsy. *Lancet* **355**: 39–40.

Peyron, C., Faraco, J., Rogers, W., Ripley, B., Overeem, S., Charnay, Y., Nevsimalova, S., Aldrich, M., Reynolds, D., Albin, R., Li, R., Hungs, M., Pedrazzoli, M., Padigaru, M., Kucherlapati, M., Fan, J., Maki, R., Lammers, G. J., Bouras, C., Kucherlapati, R., Nishino, S., Mignot, E. (2000): A mutation in a case of early onset narcolepsy and a generalized absence of hypocretin peptides in human narcoleptic brains. *Nat. Med.* **6**: 991–997.

Peyron, C., Tighe, D. K., van den Pol, A. N., de Lecea, L., Heller, H. C., Sutcliffe, J. G., Kilduff, T. S. (1998): Neurons containing hypocretin (orexin) project to multiple neuronal systems. *J. Neurosci.* **18**: 9996–10015.

Preti A. (2002): Orexins (hypocretins): their role in appetite and arousal. *Curr. Opin. Invest. Drugs* **3**: 1199–1206.

Rodgers, R. J., Halford, J. C., Nunes de Souza, R. L., Canto de Souza, A. L., Piper, D. C., Arch, J. R., Upton, N., Porter, R. A., Johns, A., Blundell, J. E. (2001): SB-334867, a selective orexin-1 receptor antagonist, enhances behavioural satiety and blocks the hyperphagic effect of orexin-A in rats. *Eur. J. Neurosci.* **13**: 1444–1452.

Sakurai, T., Amemiya, A., Ishii, M., Matsuzaki, I., Chemelli, R. M., Tanaka, H., Williams, S. C., Richardson, J. A., Kozlowski, G. P., Wilson, S., Arch, J. R., Buckingham, R. E., Haynes, A. C., Carr, S. A., Annan, R. S., McNulty, D. E., Liu, W. S., Terrett, J. A., Elshourbagy, N. A., Bergsma, D. J., Yanagisawa, M. (1998): Orexins and orexin receptors: a family of hypothalamic neuropeptides and G protein-coupled receptors that regulate feeding behavior. *Cell* **92**: 573–585.

Siegel, J. M. (2004): Hypocretin (orexin): role in normal behavior and neuropathology. *Annu. Rev. Psychol.* **55**: 125–148.

Taheri, S., Sunter, D., Dakin, C., Moyes, S., Seal, L., Gardiner, J., Rossi, M., Ghatei, M., Bloom, S. (2000): Diurnal variation in orexin A immunoreactivity and prepro-orexin mRNA in the rat central nervous system. *Neurosci. Lett.* **279**: 109–112.

Thannickal, T. C., Moore, R. Y., Nienhuis, R., Ramanathan, L., Gulyani, S., Aldrich, M., Cornford, M., Siegel, J. M. (2000): Reduced number of hypocretin neurons in human narcolepsy. *Neuron* **27**: 469–474.

Van den Pol, A. N. (2000): Narcolepsy: a neurodegenerative disease of the hypocretin system? *Neuron* **27**: 415–418.

Van den Pol, A. N., Gao, X.-B., Obrietan, K., Kilduff, T. S., Belousov, A. B. (1999): Presynaptic and postsynaptic actions and modulations of neurons by a new hypothamalic peptide, hypocretin/orexin. *J. Neurosci.* **18**: 7962–7971.

4.19
Interleukin

4.19.1
General Aspects

The interleukins belong to the family of pro-inflammatory peptides, which become highly active under inflammatory conditions and are essential molecules involved in the regulation of immune responses. The general group is known as the cytokine family.

The cytokine family comprises the interleukins (e.g. IL-1α, IL-1β, IL2–IL8), some growth factors, the tumor necrosis factors TNF-α and TNF-β and the interferones (IFN). The interleukins IL-1, IL-2 and IL-6, as well as the tumor necrosis factors, play a significant role in the induction of fever. The most potent of these is IL-1. Peripheral or intraventricular administration of IL-1 results in an increase in body temperature, but it also induces additional effects, such as changes in behavior and modifications in circadian rhythms.

Ample evidence indicates that IL-1 acts directly in the central nervous system and is able to modulate interactions between the brain and the immune system.

In this context, it is of interest that interleukin-1 (as well as TNF-α) cross the blood–brain barrier via specific and selective transporters.

An additional interleukin, interleukin-6 (IL-6), has recently received considerable attention because of its putatative involvement in developmental mechanisms in the central nervous system and in some neurodegenerative diseases, i.e. it shows direct effects upon neuronal survival, protection and neuronal differentiation.

4.19.2
Localization Within the Central Nervous System

Since interleukins are produced by glial cells, they influence nearly all parts of the central nervous system which express interleukin-specific receptors. In addition to the well known synthesis of interleukins by glial cells, some neuronal populations are also known to produce and release interleukins. For example, it has been shown that some neurons of the dentate gyrus are able to synthesize IL-1 (Fig. 4.30). Furthermore, IL-2-positive neurons have been demonstrated in some hypothalamic areas. Although various cell types in the central nervous system can produce IL-6, especially in the hypothalamus, astrocytes seem to be the major source of this cytokine. Under normal conditions, the amount of synthesized and released interleukins is low.

Additional interleukins have been detected in the brain. For example, interleukin-16 (IL-16) expression by microglial cells has been demonstrated in the rat brain. In contrast to the human brain, IL-16 is not expressed constitutively in the rat brain.

In addition, interleukin-18 (also called interferon-gamma-inducing factor, IGIF) has been cloned from adult rat brain (Culhane et al. 1998). By RT-PCR, Culhane's group revealed evidence that IL-18 is expressed constitutively in the cerebellum, the hippocampus, the hypothalamus, the cortex and the striatum.

4.19.3
Biosynthesis and Degradation

An increase in IL-1 mRNA can be induced by stimulation of the immune system via peripheral application of bacterial lipopolysaccharides (LPS).

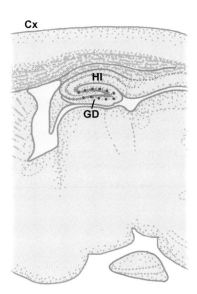

Fig. 4.30 Schematic representation of the distribution of IL-1-synthesizing neurons in the central nervous system. *Abbreviations*: Cx=cortex; GD=*gyrus dentatus*; HI=hippocampus.

LPS administration in the periphery also stimulates IL-1 production by hippocampal neurons. The main source of IL-1 derives from peripheral macrophages, which apparently enhance the production of IL-1 in the central nervous system. The mechanisms and pathways involved in the central response to a peripheral stimulus are not known.

Two distinct, but distantly related DNAs encode interleukin-1. The primary translation products of the genes have a length of 271 and 269 amino acids and have been named IL-alpha and IL-beta. The final expression products are IL-1-alpha, with a putative length of 159 amino acids (17 kDa), and IL-1-beta, which consists of 153 amino acids (17 kDa).

Interleukin 6 (IL-6; also referred to as interferon beta 2) is a multiple glycosylated, 22–27–kDa glycoprotein. The IL-6 gene consists of five exons and four introns. The transcript is translated as a 212-amino-acid molecule, composed of a signal sequence of 28 amino acids followed by further 184 amino acids.

Within the central nervous system, the predominant source of IL-6 seems to be activated astrocytes.

Interleukin 18 is a potent inflammatory cytokine which is synthesized as an inactive precursor (pro-IL-18) and which is further cleaved into its functionally active form by caspase-1.

4.19.4
Receptors and Signal Transduction

IL-1 binds to a set of specific receptors in the central nervous system. The IL-1 receptors are expressed in the choroid plexus and in ependymocytes; and IL-1 receptors have been identified in some parenchymal regions, including the frontal cortex, dentate gyrus, pituitary and, to a minor extent, in the hypothalamus.

The IL-1 receptors bind specifically IL-1α and IL-1β and are inhibited by the physiologically endogenous antagonist IL-1ra. The interleukin IL-1ra constitutes a peptide which is secreted by lymphocytes during an inflammatory process, thereby antagonizing the actions of IL-1α and IL-1β.

The biological effects of IL-1 are mediated by two distinct receptors: the p80 type I IL-1 and p68 type II IL-1 receptor proteins (IL-1RI and IL-1RII, respectively). Both isoforms are expressed in the central nervous system, particularly in neuronal somata of the granular layer of the dentate gyrus and pyramidal cells of the CA1–CA4 fields of the hippocampus, in epithelial cells of the chorioid plexus, in the ependyma and in somata of Purkinje cells of the cerebellum. The IL-1RII isoform, but not IL-1RI, is expressed in specific neuronal somata and proximal cell processes of the paraventricular gray matter of the hypothalamus.

IL-6 belongs to a subfamily of factors which includes ciliary neurotrophic factor (CNTF), IL-11, and cardiotrophin-1 (CT-1). Characteristic features of this family are pluripotency and the redundancy of biological responses elicited by its members. This redundancy derives from the convergence on a common signal-transducing receptor, gp130. All IL-6 type cytokines act via receptor complexes that contain gp130. As already mentioned, IL-6, IL-11, CNTF and possibly cardiotrophin-1 require a ligand-specific receptor. Interestingly, these receptors function in both membrane-bound and soluble forms.

Unlike most soluble receptors that act antagonistically to their ligands, the soluble receptors of the IL-6 family exhibit agonistic functions. For example, fibroblasts, osteoblasts, some neurons and astrocytes express gp130, but respond weakly to IL-6 alone. However, addition of the soluble IL-6 receptor greatly enhances IL-6 mediated responses.

The soluble IL-6 receptor is produced by two mechanisms, which involve either shedding or alternative splicing. Shedding results from limited proteolysis of the membrane-bound form (80 kDa) and produces the 45–55 kDa soluble form.

Analyses of the mRNA expression have shown that some cells transcribe two distinct mRNAs for the IL-6 receptors, one encoding the membrane-bound IL-6 receptor form, and a second, smaller transcript, which generates a soluble form of the IL-6 receptor.

Sequence analysis has revealed that the smaller, alternatively spliced IL-6 receptor mRNA lacks a complete region. This deletion causes a shift of the reading frame, which results in a protein with ten unique amino acids at the carboxyl terminus.

To date, several non-neuronal cell types (among which are leukocytes) are known to be capable of producing the soluble IL-6 receptor either by shedding, by alternative mRNA splicing, or by both.

Some reports indicate that the soluble IL-6 receptors can modulate IL-6-inducible responses in both astrocytes and neurons. However, a common source of the soluble IL-6 receptor in the central nervous system has not yet been identified.

IL-6 acts on target cells through a receptor complex composed of IL-6, gp130, and either membrane-bound IL-6 receptor or soluble IL-6 receptor. Receptor activation is believed to require the formation of a hexameric complex composed of two of the proteins mentioned above. IL-6 receptor formation leads to phosphorylation and activation of gp130-associated tyrosine kinases and JAK kinases (Janus kinases). JAK activation further leads to tyrosine phosphorylation of STATs ("signal transducer and activator of transcription"). The Jak-STAT pathway is a very rapid cytosol-to-nuclear signaling pathway that underscores how fast extracellular proinflammatory signals can be transmitted to the nucleus.

Concerning IL-18, it is known that it acts via a receptor complex that closely resembles that of IL-1 receptor complexes. The IL-18Rbeta seems to exist in two different splice variants and both have been found to be expressed in rat cortex, striatum, hypothalamus and hippocampus. IL-18Rbeta is not only expressed by neurons, but also by microglia and astrocytes.

4.19.5
Biological Effects

Peripheral immune responses generated by IL-1 are known to drive a number of acute systemic inflammatory processes. Within the central nervous system, microglia-derived IL-1 is believed to orchestrate molecular and cellular cascades, which are the equivalent of injury-induced inflammation and repair processes in the periphery.

IL-1 has been shown to elicit a number of trophic and toxic actions on cells of the central nervous system, including both autocrine and paracrine effects, which may be important in neurodegeneration. Relevant autocrine effects include promotion of microglial proliferation and promotion of increased microglial expression of IL-1 and IL-6.

Relevant paracrine effects of IL-1 include:
- induction of neuronal expression and processing of the β-amyloid precursor protein (β-APP), an injury response molecule of central importance in the pathophysiology of Alzheimer's disease;
- effects on astrocytes, including promotion of astrocytic activation and up-regulation of the glial-derived proteins apolipoprotein E (ApoE).

Under *in vitro* conditions, low levels of IL-1 enhance neuronal survival, while higher levels are neurotoxic.

Cytokines, including the interleukins, seem to act as neuromodulators in the central nervous system. Interactions of interleukins with the renin–angiotensin system are known as well as interactions with other neuromodulators. The intraventricular application of IL-1, for example, stimulates the expression of corticotropin-releasing hormone (CRH) in neurons of the paraventricular nucleus of the hypothalamus, with the consequence of an increase in adrenocorticotropic hormone (ACTH) and glucocorticoids. In addition, IL-1 can inhibit the release of luteinizing hormone (LH).

Intraventricularly applied IL-1 also induces immuno-suppression, indicative of a role for this cytokine in the modulation of the centrally regulated immune responses. A further indication that IL-1 is a major player in the neuro-immunological axis is the increase of IL-1 production in peripheral macrophages as well as centrally in the hippocampus, which follows after stimulation of the immune system by bacterial endotoxins, for example lipopolysaccharides (LPS).

Since IL-1 can not cross the blood–brain barrier, it seems unlikely that peripherally produced IL-1 acts directly on the central nervous system. The mechanisms by which peripheral stimuli activate the central nervous interleukin system are unknown.

It has been shown that IL-1 stimulates the proliferation of astrocytes and, for this reason, it was suggested that IL-1 is also involved in astrogliosis. In addition, IL-1 stimulates the proliferation and differentiation of cerebral endothelial cells *in vitro*. Further evidence indicates that, under the influence of IL-1, endothelial cells display altered chemotactic effects on neutrophils, macrophages and lymphocytes. These effects lend support to the idea that IL-1 may also contribute to the infiltration of lymphocytes into the central nervous system under sustained inflammatory conditions, specifically in certain demyelinization diseases, such as multiple sclerosis.

IL-1 alone, or together with TNF-α, can induce potent cytotoxic effects, making it likely that interleukins contribute to processes related to cell death.

The stimulation of glial cells by Il-1 or TNF-α is accompanied by augmentation of interferon (IFNγ) production. Moreover, it has been shown that interferon γ induces the expression of MHC antigens on glial cells and neurons.

Enhancement of interferon production has been suggested to be responsible for the acquisition of antigen-presenting phenotypes in endogenous cells of brain tissues. Under normal conditions, the capacity of antigen presentation may be relevant for the survival of cells and tissues and, thus, this process may provide the basis for autoimmune diseases under pathological conditions.

In the case of IL-6, numerous studies suggest that IL-6 is responsible for multiple effects in the central nervous system, including neuronal survival, protection and differentiation. For instance, IL-6 has been shown to induce neuronal differentiation of PC12 cells, to increase survival of primary neurons in culture and to protect primary neurons against glutamate induced excitotoxicity.

In glial cells, IL-6 promotes astrocytic proliferation, suggesting an involvement in astrogliosis.

Recently, Depino and coworkers have shown that IL-1alpha influences behavior and memory consolidation. Thus, hippocampal-dependent learning tasks induce IL-1alpha mRNA induction in this brain area. IL-6 seemed also to play a role in hippocampal-dependent memory. However, it does not facilitate LTP. Blockade of endogenous IL-6 induces a remarkable prolongation of LTP after stimulation. Furthermore, blockade of endogenous IL-6, 90 min after hippocampus-dependent spatial alternation learning resulted in a significant improvement of long-term memory (Balschun et al. 2004).

4.19.6
Neurological Disorders and Neurodegenerative Diseases

A body of evidence suggests a role for IL-1 as a key mediator of cell death in acute neurodegenerative conditions, such as stroke and head injury. Along this line, mice lacking both IL-1alpha and IL-1beta show dramatically reduced ischemic cell death.

Moreover, IL-1 has also been implicated in a number of chronic diseases, including Parkinson's and Alzheimer's disease, as well as epilepsy. Constitutive expression of IL-1 is very low in normal brain, but is up-regulated rapidly in response to local or peripheral insults. The mechanisms regulating the expression IL-1 are not well defined, but appear to involve multiple effects on neuronal, glial and endothelial cell function. Most of the neurodegenerative effects of IL-1 are thought to happen through IL-1beta.

Like IL-1, IL-6 levels are constitutively low in the normal brain, but increase under pathological conditions, such as injury, infection, stroke and inflammation. The significance of this increase is largely unknown.

IL-18, which is also expressed in the brain, is involved in mediating neuroinflammation and neurodegeneration in the CNS under pathological conditions, such as bacterial and viral infection, and in autoimmune demyelinating disease (Felderhoff-Mueser et al. 2005).

Enhanced interleukin production in the central nervous system has been observed following injuries or local infections and in neurodegenerative diseases, such as Alzheimer's disease and in trisomy 21. Enhanced IL-1 synthesis is thought to occur in early stages of Alzheimer's disease, provoking the induction of β-amyloid precursor synthesis and gliosis.

Some data indicate that interleukins are involved in epilepsy. Intracerebral application of IL-1beta has been found to enhance epileptic activity in animal models, while its naturally occurring receptor antagonist (IL-1Ra) mediates anticonvulsant actions. Furthermore, transgenic mice overexpressing IL-1Ra are less susceptible to seizures, indicating that endogenous IL-1 exhibits proconvulsant activity. Interestingly, polymorphism in the IL-1beta gene promoter has been found in patients suffering from temporal lobe epilepsy with hippocampal sclerosis.

Further Reading

Allan, S.M., Pinteaux, E. (2003): The interleukin-1 system: an attractive and viable therapeutic target in neurodegenerative disease. *Curr. Drug Targets CNS Neurol. Disord.* **2**: 293–302.

Andre, R., Wheeler, R.D., Collins, P.D., Luheshi, G.N., Pickering-Brown, S., Kimber, I., Rothwell, N.J., Pinteaux, E. (2003): Identification of a truncated IL-18R beta mRNA: a putative regulator of IL-18 expressed in rat brain. *J. Neuroimmunol.* **145**: 40–45.

Antonipillai, I., Wang, Y., Horton, R. (1990): Tumor necrosis factor and interleukin-1 may regulate renin secretion. *Endocrinology* **126**: 273–278.

Balschun, D., Wetzel, W., Del Rey, A., Pitossi, F., Schneider, H., Zuschratter, W., Besedovsky, H.O. (2004): Interleukin-6: a cytokine to forget. *FASEB J.* **18**: 1788–1790.

Ban, E., Milon, G., Prudhomme, N., Fillion, G., Haour, F. (1991): Receptors for interleukin-1 (α and β) in mouse brain: Mapping and neuronal localization in the hippocampus. *Neuroscience* **43**: 21–30.

Culhane, A.C., Hall, M.D., Rothwell, N.J., Luheshi, G.N. (1998): Cloning of rat brain interleukin-18 cDNA. *Mol. Psychiatry* **3**: 362–366.

Depino, A.M., Alonso, M., Ferrari, C., del Rey, A., Anthony, D., Besedovsky, H., Medina, J.H., Pitossi, F. (2004): Learning modulation by endogenous hippocampal IL-1: blockade of endogenous IL-1 facilitates memory formation. *Hippocampus* **14**: 526–535.

Felderhoff-Mueser, U., Schmidt, O.I., Oberholzer, A., Buhrer, C., Stahel, P.F.(2005): IL-18: a key player in neuroinflammation and neurodegeneration? *Trends Neurosci.* **28**: 487–493.

French, R.A., VanHoy, R.W., Chizzonite, R., Zachary, J.F., Dantzer, R., Parnet, P., Bluthe, R.M., Kelley, K.W. (1999): Expression and localization of p80 and p68 interleukin-1 receptor proteins in the brain of adult mice. *J. Neuroimmunol.* **93**: 194–202.

Guo, L.H., Mittelbronn, M., Brabeck, C., Mueller, C.A., Schluesener, H.J. (2004): Expression of interleukin-16 by microglial cells in inflammatory, autoimmune, and degenerative lesions of the rat brain. *J. Neuroimmunol.* **146**: 39–45.

March, C.J., Mosley, B., Larsen, A., Cerretti, D.P., Braedt, G., Price, V., Gillis, S., Henney, C.S., Kronheim, S.R., Grabstein, K., et al. (1985): Cloning, sequence and expression of two distinct human interleukin-1 complementary DNAs. *Nature* **315**: 641–647.

Patel, H.C., Boutin, H., Allan, S.M. (2003): Interleukin-1 in the brain: mechanisms of action in acute neurodegeneration. *Ann. N.Y. Acad. Sci.* **992**: 39–47.

Rothwell, N.J., Hopkins, S.J. (1995): Cytokines in the nervous system II: actions and mechanisms of action. *Trends Neurosci.* **18**: 130–136.

Schobitz, B., DeKloet, E. R., Holsboer, F. (1994): Gene expression and function of interleukin 1, interleukin 6 and tumor necrosis factor in the brain. *Prog. Neurobiol.* **44**: 397–432.

Van Wagoner, N. J., Benveniste, E. N. (1999): Interleukin-6 expression and regulation in astrocytes. *J. Neuroimmunol.* **100**: 124–139.

Vezzani, A., Moneta, D., Richichi, C., Perego, C., De Simoni, M. G. (2004): Functional role of proinflammatory and anti-inflammatory cytokines in seizures. *Adv. Exp. Med. Biol.* **548**: 123–133.

Viviani, B., Bartesaghi, S., Corsini, E., Galli, C. L., Marinovich, M. (2004): Cytokines role in neurodegenerative events. *Toxicol. Lett.* **149**: 85–89.

Watanabe, T., Saiki, Y., Sakata, Y. (1997): The effect of central angiotensin II receptor blockage on interleukin-1β- and prostaglandin E-induced fevers in rats: possible involvement of brain angiotensin II receptors in fever induction. *J. Pharmacol. Exp. Ther.* **282**: 873–881.

Yu, B., Shinnick Gallagher, P. (1994): Interleukin-1 beta inhibits synaptic transmission and induces membrane hyperpolarization in amygdala neurons. *J. Pharmacol. Exp. Ther.* **27**: 590–600.

4.20
Melanin-concentrating Hormone

4.20.1
General Aspects and History

Hogden and Slone in 1931 proposed the existence of two hormones believed to exert opposite effects upon the pigmentation of lower vertebrates. In the 1950s, a-melanocyte stimulating hormone (a-MSH) was identified and, in 1983, a functional endogenous antagonist of a-MSH was discovered. This substance, characterized by Kawauchi and coworkers (1983) in the salmon pituitary, was the melanin-concentrating hormone (MCH).

In teleost fish, MCH serves as a color-regulating hormone by antagonizing the melanin-dispersing action of a-MSH on skin melanophores. However, in mammalian brains, MCH and a-MSH (see Section 4.21) also have important functions as neuropeptides.

Mammalian MCH appears to have evolved as a major regulatory hormone, with a role in several behavior-related effects. Unlike teleosts, mammals do not express MCH in melanophores of normal skin.

MCH is a cyclic peptide composed of 17 amino acids in fish, and 19 amino acids in mammals (Fig. 4.31). Cloning of the cDNA and the related gene, which encodes for the precursor of MCH, has led to the discovery of other peptides derived from pro-MCH.

MCH of rats	NH$_2$-Asp-Phe-*Asp*-Met-Leu-*Arg*-*Cys*-*Met*-Leu-*Gly*-*Arg*-*Val*-*Tyr*-*Arg*-*Pro*-*Cys*-*Trp*-Gln-*Val*-COOH
MCH of teleosts	NH$_2$-*Asp*-*Thr*-Met-*Arg*-*Cys*-*Met*-Val-*Gly*-*Arg*-*Val*-*Tyr*-*Arg*-*Pro*-*Cys*-*Trp*-Glu-*Val*-COOH

Fig. 4.31 Amino acid sequence of the rat and teleost MCH.
Identical amino acids are marked in italics.

4.20.2
Localization Within the Central Nervous System

MCH-positive neurons are located mainly in the zona incerta and in the lateral hypothalamus, from which they project to most brain regions, including intrahypothalamic sites.

In addition, two small population of MCH-expressing neurons have been identified in the olfactory tubercle and in the reticular formation (Fig. 4.32).

In contrast to the restricted location of MCH perikarya, the MCH terminals are widely distributed throughout the central nervous system and include the cortex, the limbic system and the spinal chord.

4.20.3
Biosynthesis and Degradation

The MCH gene has been cloned in teleosts and mammals. In the former, two genes, which lack an intron, encode for MCH. In rodents, a single gene has been identified, composed of three exons and two introns.

This gene seems to be the ortholog of the human MCH gene. In addition to this MHC gene, a second gene has been identified in humans, revealing considerable homology with the original MCH gene and which is believed to encode a variant of MCH.

The predicted sequence of the precursor pro-MCH contains a hydrophobic sequence at the amino-terminal region. The MCH sequence is located in the car-

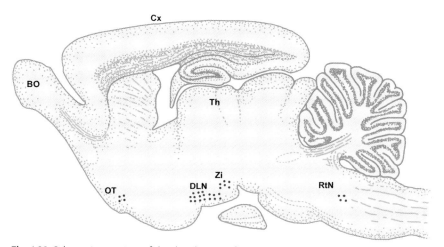

Fig. 4.32 Schematic overview of the distribution of MCH-expressing neurons in the central nervous system of the rat. *Abbreviations:* BO=*bulbus olfactorius*; Cx=cortex; DLN=dorsolateral nucleus of the hypothalamus; OT=olfactory tubercle; RtN=reticular nucleus; Th=thalamus; Zi=*Zona incerta*.

boxy-terminal region and is flanked by a doublet of basic amino acids, representing the cleavage sites for MCH-processing.

Within the sequence of the mammalian pro-MCH, two additional pairs of basic amino acids have been identified. These pairs enclose two further active peptides, namely neuropeptide glutamate-isoleucine (NEI) and neuropeptide glycine-glutamate (NG).

The common precursor pro-MCH seems to be differentially cleaved by proconvertases (PC), with proconvertase PC2 necessary for the generation of NEI, and other proconvertases for the biosynthesis of the additional neuroactive peptides which derive from pro-MCH.

The mammalian MCH is highly sensitive to degradation by endopeptidase 24.11, aminopeptidase M and carboxypeptidase A. In the presence of these peptidases, additional proteolytic enzymes seem to play a role in the degradation of MCH.

4.20.4
Receptors and Signal Transduction

The neuropeptide MCH acts via a unique receptor, which has been recently identified. The receptor is an orphan G protein-coupled receptor (SLC-1) which binds MCH with nanomolar affinity. The receptor has the typical seven transmembrane domains of G protein-coupled receptors. The MCH receptor of rats and humans shares significant homology and, in addition, reveals homology with the somatostatin receptor.

The distribution of SLC-1 mRNA and its corresponding protein have been examined by *in situ* hybridization and immunocytochemistry.

SLC-1 mRNA and SLC-1 protein are widely expressed throughout the brain. Strong immunoreactivity is observed in areas that overlap extensively with regions mapping positive for mRNA. SLC-1 signals are observed in the cerebral cortex, the olfactory tubercle, the hippocampal formation, the caudate-putamen, the nucleus accumbens, the thalamus and the hypothalamus (ventromedial and dorsomedial nuclei of the hypothalamus) as well as in various nuclei of the mesencephalon and rhombencephalon.

Interestingly, the MCH receptor has also been observed in areas associated with the dopaminergic system, such as the substantia nigra and the ventral tegmental area.

The binding of MCH to its receptor induces calcium mobilization and can inhibit the forskolin-induced accumulation of cAMP.

4.20.5
Biological Effects

Central melanin-concentrating hormone is involved in various functions, including feeding, reproduction, stress and behavior.

MCH works antagonistically to MSII in fish. The physiological properties of MCH in mammals are somewhat unclear, but recent investigations have demonstrated that MCH and α-MSH show antagonistic effects in the brain, with consequences for feeding behavior, aggression, anxiety, arousal and reproductive function through the release of luteinizing hormone (LH).

MCH is also implicated in the control of general arousal and goal-orientated behaviors in mammals and appears to be an essential messenger in the regulation of food intake.

The neuropeptides MCH, NEI and NG are expressed at significantly high levels in the early stages during development, and activation of the MCH gene has been assessed during fetal and postnatal development.

The level of MCH mRNA reveals a circadian rhythm. This variation parallels the rhythm of endogenous corticoid levels, suggesting that corticoids participate in the regulation of the MCH system.

Furthermore, it has been demonstrated that an elevation in the level of MCH mRNA correlates with locomotor activity and the characteristic feeding behavior of rats during the night. Moreover, a role for MCH-producing neurons in the central regulation of paradoxical sleep has been proposed, since MCH-containing neurons are strongly active during paradoxical sleep. Thus, intraventricular (icv) administration of MCH induces a dose-dependent increase in paradoxical sleep and slow-wave sleep quantities (Verret et al. 2003).

The behavioral and neurochemical effects of NEI, and its interaction with α-MSH or MCH, have been investigated in the ventromedial nucleus and the medial preoptic area in female rats (bilateral administration of 100 ng, in 0.5 µl per side). Administration of NEI to the ventromedial nucleus, but not to the preoptic area, stimulates exploratory behavior, increases anxiety and reduces dopamine and DOPAC release. The behavioral effects are antagonized by α-MSH. NEI administration to the medial preoptic area stimulates female sexual receptivity. Administration of NEI to the ventromedial nucleus has no effect on sexual activity, but partially antagonizes the stimulatory effect of MCH.

Whole-cell recording from neurons from the lateral hypothalamic area revealed that MCH decreases the amplitude of voltage-dependent calcium currents. This inhibition was found to desensitize rapidly and dose-dependently. Further experiments showed that MCH exerts an inhibitory effect on calcium currents via PTX-sensitive G protein pathways in LH neurons, with the main effect being related to N-type calcium currents (Gao and van den Pol 2002).

MCH peptide has been mapped to the hippocampus, a structure well known for its participation in learning and memory processes. By electrophysiological recordings from that region, it has been shown that MCH increases hippocampal synaptic transmission. Thus, perfusing rat hippocampal brain slices with MCH results in a long-lasting potentiation on the hippocampal-evoked response. Along that line, mice with a deletion of the MCH receptor display deficits in passive avoidance behavior.

These mice also show a further prominent alteration: the mice display an upregulation of mesolimbic dopamine receptors and norepinephrine transporter (NET), indicating that the MCH receptor may negatively modulate monoamine function.

4.20.6
Neurological Disorders and Neurodegenerative Diseases

Thirst, induced by high salt concentrations or by reduced availability of water, induces physiological stress which is coupled to changes in the osmolality of the organism. Stress induced by thirst also induces a diminished production of MCH mRNA. At the cellular level, changes in response to osmotic stimuli have been measured in those hypothalamic areas which posses MCH containing neurons, and to a lesser extent in the zona incerta.

Electroshocks given at irregular time-intervals can induce stress in animals, which persists almost a week. During this time, a reduction of mRNA encoding for MCH can be found. After the postictus interval, MCH mRNA recovers to normal levels. Adrenalectomized animals also display a decreased MCH mRNA level. This effect can be counteracted by application of synthetic glucocorticoids (dexamethasone). A plausible explanation for these data is that the expression of MCH mRNA is reduced during stress and then subsequently upregulated by circulating glucocorticoids.

Further Reading

Adamantidis, A., Thomas, E., Foidart, A., Tyhon, A., Coumans, B., Minet, A., Tirelli, E., Seutin, V., Grisar, T., Lakaye, B. (2005): Disrupting the melanin-concentrating hormone receptor 1 in mice leads to cognitive deficits and alterations of NMDA receptor function. *Eur. J. Neurosci.* **21**: 2837–2844.

Baker, B. B. (1994): Melanin-concentrating hormone updated. functional considerations. *Trends Endocrinol. Metab.* **5**: 120–126.

Bittencourt, J. C., Elias, C. F. (1998): Melanin-concentrating hormone and neuropeptide EI projections from the lateral hypothalamic area and zona incerta to the medial septal nucleus and spinal cord: a study using multiple neuronal tracers. *Brain Res.* **805**: 1–19.

Bittencourt, J. C., Frigo, L., Rissman, R. A., Casatti, C. A., Nahon, J. L., Bauer, J. A. (1998): The distribution of melanin-concentrating hormone in the monkey brain (*Cebus apella*). *Brain Res.* **804**: 140–143.

Eberle, A. N. (1988): Melanin-containing hormone. In: *The Melanotropins*. Karger, Basel, pp 321–332.

Gao, X. B., van den Pol, A. N. (2002): Melanin-concentrating hormone depresses L-, N-, and P/Q-type voltage-dependent calcium channels in rat lateral hypothalamic neurons. *J. Physiol.* **542**: 273–286.

Gonzalez, M. I., Baker, B. I., Hole, D. R., Wilson, C. A. (1998): Behavioral effects of neuropeptide E-I (NEI) in the female rat: interactions with alpha-MSH, MCH and dopamine. *Peptides* **19**: 1007–1016.

Hervieu, G. J., Cluderay, J. E., Harrison, D., Meakin, J., Maycox, P., Nasir, S., Leslie, R. A. (2000): The distribution of the mRNA and protein products of the melanin-concentrating hormone (MCH) receptor gene, slc-1, in the central nervous system of the rat. *Eur. J. Neurosci.* **12**: 1194–1216.

Kawauchi, H., Kawazoe, I., Tsubokawa, M., Kishida, M., Baker, B. I. (1983): Characterization of melanin-concentrating hormone in chum salmon pituitaries. *Nature* **305**: 321–323.

Miller, C. L., Hruby, V. J., Matsunga, T. O., Bickford, P. C. (1993): Alpha-MSH and MCH are functional antagonists in a CNS auditory gating paradigm. *Peptides* **14**: 431–440.

Nahon, J. L. (1994): The melanin-concentrating hormone: from the peptide to the gene. *Crit. Rev. Neurobiol.* **8**: 221–262.

Shimada, M., Tritos, N.A., Lowell, B.B., Flier, J.S., Maratos-Flier, E. (1998): Mice lacking melanin-concentrating hormone are hypophagic and lean. *Nature* **396**: 670–674.

Smith, D.G., Tzavara, E.T., Shaw, J., Luecke, S., Wade, M., Davis, R., Salhoff, C., Nomikos, G.G., Gehlert, D.R. (2005): Mesolimbic dopamine super-sensitivity in melanin-concentrating hormone-1 receptor-deficient mice. *J. Neurosci.* **25**: 914–922.

Strand, F.L., Rose, K.J., Zuccarelli, L.A., Kume, J., Alves, S.E., Antonawich, F.J., Garrett, L.Y. (1991): Neuropeptide hormones as neurotropic factors. *Physiol. Rev.* **71**: 1017–1046.

Varas, M., Perez, M., Ramirez, O., de Barioglio, S.R. (2002): Melanin concentrating hormone increase hippocampal synaptic transmission in the rat. *Peptides* **23**: 151–155.

Verret, L., Goutagny, R., For, T.P., Cagnon, L., Salvert, D., Leger, L., Boissard, R., Salin, P., Peyron, C., Luppi, P.H. (2003): A role of melanin-concentrating hormone-producing neurons in the central regulation of paradoxical sleep. *BMC Neurosci.* **4**: 19.

Viale, A., Ortola, C., Hervieu, G., Furuta, M., Barbero, P., Steiner, D.F., Seidah, N.G., Nahon, J.L. (1999): Cellular localization and role of prohormone convertases in the processing of pro-melanin concentrating hormone in mammals. *J. Biol. Chem.* **274**: 6536–6545.

4.21
Melanocyte-stimulating Hormone

4.21.1
General Aspects

The melanocortins form a family of proopiomelanocortin-derived peptides which carry the melanocyte-stimulating hormone (MSH) core sequence, His-Phe-Arg-Trp (Fig. 4.33). Three different forms of MSH have been isolated and purified: α-, β- and γ-MSH.

The α-MSH of amphibians is identical to the mammalian neuropeptide, whereas teleosts produce two isoforms of α-MSH, one of which is identical to mammalian and amphibian α-MSH, except that it lacks acetylation at the amino-terminal sequence. The second MSH is a 15-amino-acid peptide, 13 residues of which are identical with mammalian ACTH (1–15).

Melanocortins have been described as exerting a variety of cardiovascular effects (hemorrhagic shock, hypotension and bradycardia), but they also affect behavior.

Different fragments and analogs of MSH have been exploited in behavioral experiments in order to study their influence on learning and memory. The results indicate that the various peptides reveal different potencies, depending on the site of application and the learning task.

Five melanocortin receptor subtypes have been cloned thus far, but only four of them seem to be expressed in the central nervous system. One of the cloned

Ac-Ser-Tyr-Ser-Met-Glu-His-Phe-Arg-Trp-Gly-Lys-Pro-Val-NH$_2$

Fig. 4.33 Sequence of melanocyte-stimulating hormone.

receptors (the MC2 receptor) has been shown to be an ACTH receptor. Thus, only three receptors exist in the central nervous system, which interact specifically with MSH.

4.21.2
Localization Within the Central Nervous System

The distribution of the a-MSH precursor proopiomelanocortin (POMC) has been investigated in more detail than that of a-MSH.

POMC is not only the precursor of MSH, but is also the source of some other neuroactive substances. Because of its importance in neuropeptide generation, the action of POMC and its function as a neuropeptide precursor are considered separately (Section 4.28). POMC is expressed in several tissues, including the pituitary, brain, skin, pancreas and testis; and its posttranslational processing varies in a tissue- and species-specific manner. a-MSH-containing neurons are found in different areas of the hypothalamus, such as in the preoptic area, the ventromedial nucleus of the hypothalamus and in the arcuate nucleus. In addition, MSH immunoreactivity occurs in the amygdala, the septum, the hippocampus, the caudate-putamen and the cortex.

The arcuate a-MSH immunoreactive axons project to the paraventricular nucleus of the hypothalamus, the dorsomedial hypothalamic nucleus and the lateral hypothalamic area. Interestingly, all these hypothalamic nuclei appear to be important in appetite regulation and energy homeostasis.

Neurons in the lateral arcuate nucleus exhibit immunoreactivity for a-MSH projections to pre-ganglionic sympathetic neurons in the spinal cord. Aside from this, a-MSH occurs in within the anterior and intermediate lobe of the pituitary.

4.21.3
Biosynthesis and Degradation

a-MSH is a peptide of 13 amino acids. Like ACTH and endorphins, a-MSH results from proteolytic cleavage of the larger proopiomelanocortin precursor (POMC) molecule. The biosynthesis of different neuromodulators from the common precursor is described separately (see Section 4.28). Alpha-MSH is cleaved by endopeptidases at positions seven and eight (Phe-Arg) and then further degraded by exopeptidases.

4.21.4
Receptors and Signal Transduction

The melanotrophic receptors belong to the family of G protein-coupled receptors with seven transmembrane domains. They form a unique subfamily of G protein-coupled receptors and are highly homologous to each other. One of the features of the melanotropic receptors is their relatively small size (297–360 amino acids). They appear to be more closely related to the cannabinoid receptors than to other receptors.

In primary cultures of melanoma cells, binding of a-MSH to its receptor stimulates the activation of adenylate cyclase and the phosphorylation of a 34-kDa molecule. This step is blocked by phorbolesters, suggesting that protein kinase C might be involved in the signal transduction of a-MSH.

In neurons, a-MSH stimulates the activation of adenylate cyclase and decreases the phosphorylation of a protein, called B-50, by inhibition of protein kinase C. Activation of the melanocortin receptors depends on extracellular calcium, indicating that these receptors belong to a unique class of calcium-requiring receptors.

Between 1992 and 1994, five distinct melanocortin receptors (MC1–MC5) were been identified, of which three (MC3-R, MC4-R and MC5-R) are expressed in the brain and selectively bind a-MSH.

MC-1 receptor

The MC-1 receptor binds a-MSH with high selectivity and seems to be expressed in melanocytes, in melanoma cells and, to a limited extent, in the testis. Thus, pigmentation through melanocytes is mediated by the melanocortin MC-1 receptor.

MC-2 receptor

The melanocortin MC-2 receptor binds ACTH but none of the other melanocortins. The human, murine and bovine melanocortin MC-2 receptors have been cloned. MC-2 receptors are expressed in the adrenal cortex, especially in the zona fasciculata (site of glucocorticoid expression) and zona glomerulosa, but also to some extent in the zona reticularis.

MC-3 receptors

MC-3 receptors are said to be selective for a-MSH, γ-MSH and ACTH, depending on the species. In the 1990s, the melanocortin MC-3 receptor was cloned (Roselli-Rehfuss et al. 1993).

The MC-3 receptor is expressed in various parts of the central nervous system. Melanocortin MC-3 receptor mRNA was found to be expressed in the cortex, thalamus, septum, hippocampus, olfactory cortex, amygdala, periaqueductal gray, ventral tegmental area, interfascicular nuclei, central linear raphe nucleus and the hypothalamus.

In addition, MC-3 receptor mRNA is also expressed in the periventricular nucleus and the posterior hypothalamic area, regions which have been implicated in the neural control of cardiovascular and thermo-regulatory functions.

MC-3 receptors are detectable in various peripheral tissues, such as the stomach, intestine, placenta and heart.

MC-4 receptors

The MC-4 receptors selectively bind α-MSH, β-MSH and ACTH. The mRNA of the melanocortin MC-4 receptor is ubiquitous throughout the central nervous system. In particular, MC-4 receptor mRNA has been detected in the cortex, the hippocampus, the amygdala, the septum, the striatum, the nucleus accumbens, the hypothalamus, the nucleus of the tractus solitarius and the dorsal horn of the spinal cord.

It is assumed that the MC-4 receptor plays a role in glucagon and insulin metabolism, since knockout mice with a deletion of the MC-4 receptor gene display a pronounced hyperglycemia.

MC-5 receptor

The MC-5 receptors preferentially bind α-MSH. The melanocortin MC-5 receptor has been cloned from human, rat, mouse and sheep.

Expression sites of the receptor are the cortex, the hippocampus, the hypothalamus, the substantia nigra, the cerebellum and the pituitary gland.

In addition, MC-5 receptors are located in various peripheral tissues, such as skeletal muscle, adrenal gland, stomach and lacrimal gland.

4.21.5
Biological Effects

The general function of MSH is the regulation of epidermal pigmentation. Activation of MC-1 receptors preferentially found in melanocytes and melanoma cells affects the distribution of melanin. The functional significance of this action centers on protective color changes in lower vertebrates – amphibians, reptiles and fishes. These species are able to change the number and size of the secretory granula of their epidermal melanocytes to adapt their body color to the background. MSH has also been identified in the skin of mammals (including humans). However, the involvement of MSH in pigmentation of the skin in higher vertebrates has not been demonstrated, although some evidence indicates that MSH in higher vertebrates, including humans, may affect hair color (rodents) and skin pigmentation. An example of a feasible relationship is given by the inverse correlation between susceptibility to sunburn, photo-ageing and skin cancer and the ability of an individual to tan after sun exposure. Healey and coworkers (2000) provided evidence of an association between the degree of tanning after repeated sun exposure and the number of variant MC1R alleles in individuals from Ireland and the UK. They suggest that MC1R gene status determines sun sensitivity.

In higher vertebrates, MSH is involved in a variety of additional functions. Intraventricular administration of α-MSH causes a stereotypic behavior in rats. MSH, ACTH and some of their derivates enhance aggression and influence learning and memory.

Dopamine (DA) is one of the most important transmitters, regulating MSH secretion from melanotrophs. Dopamine, acting via D2 receptors, has an inhibitory influence upon MSH secretion. Thus, lesions which destroy the DAergic fibers running to the intermediate lobe cause a rise in circulating a-MSH. Conversely, treatment with dopamine agonists can induce inhibition of a-MSH release.

Melanocortins participate in the regulation of multiple physiological functions. They are involved in grooming behavior, food intake and thermoregulation processes and can also modulate the response of the immune system in inflammatory states.

Animals have developed highly adaptive and redundant mechanisms to maintain energy balance by matching caloric intake to caloric expenditure. A role for a-MSH in the regulation of energy homeostasis is suggested by several lines of evidence. Regulation of signaling by melanocortin 3 and melanocortin 4 receptors in the CNS are controlled via neuronal cell bodies in the arcuate nucleus that produce a-MSH. Increased melanocortin signaling via pharmacological or genetic means in the CNS causes potent reductions in food intake and weight loss, whereas decreased melanocortin signaling results in increased food intake and weight gain. Injection of a-MSH into the lateral ventricle of rodents leads to suppression of food intake in a dose-dependent manner. In contrast, intraventricular (icv) administration of the melanocortin agonist MTII inhibits food intake in fasted mice. Moreover, the icv injection of a pharmacological antagonist of a-MSH (SHU9119) at sites containing MC-4 receptors results in increased feeding, supporting a role for endogenous melanocortins in appetite regulation. Alpha-MSH does not antagonize the appetite-stimulating effect of NPY after i.c.v. injection, although the melanocortin agonist MTII appears to suppress NPY-induced feeding. The idea that the melanocortin receptor system may be particularly important in modulating food intake is strengthened by the fact that transgenic mice, which lack MC-4 receptors, are massively overweight. Null mutations of the MC-4 receptor are associated with hyperphagia, obesity, hyperinsulinemia and hyperglycemia (Huszar et al. 1997).

In contrast to MC4-R knockout mice, MC3-R knockouts exhibit an exclusively metabolic syndrome. Homozygous MCR3-R $-/-$ mice are not significantly overweight, but they exhibit an approximately 50–60% increase in fat mass and also exhibit an unusual increase in respiratory quotient when transferred onto high-fat diet, suggesting a reduced ratio of fat/carbohydrate oxidation. Furthermore, these knockout mice also exhibit an approximately 50% reduction in locomotor behavior, indicating reduced energy expenditure. Additionally, MSH seems to play an important role during early neurogenesis and exhibits a peak around parturition. The latter has been interpreted to mean that MSH plays a signaling role in the initiation of parturition.

4.21.6
Neurological Disorders and Neurodegenerative Diseases

The limited data currently available are consistent with the view that melanocortins act as regulative factors in controlling food intake in humans. Autosomal recessive mutations that interfere with POMC expression have been reported to result in early onset of obesity, together with red hair color (suggesting a role for melanocortins in hair pigmentation in humans), and in deficiency of adrenal corticoid synthesis as a result of defective ACTH biosynthesis. A decrease in the activity of prohormone convertase 1 (PC 1; an important enzyme in POMC processing) has been reported in a woman with childhood obesity and a mutation of the PC 1 gene. This patient exhibited several endocrine abnormalities, including adrenal insufficiency, hypogonadotrophic hypogonadism, impaired glucose tolerance and reactive hypoglycemia, presumably due to impaired processing of POMC and proinsulin.

In addition, systemic application of MSH produces anti-inflammatory and antipyretic effects. Among the melanocortins, a-MSH is one of the most potent endogenous antipyretic agents inhibiting IL-1-induced fever. Centrally administered alpha-MSH can induce antipyretic effects.

It has been shown that MSH exerts strong anti-inflammatory activity via direct action on peripheral host cells:
- inhibition of descending neurogenic anti-inflammatory pathways arising from central nervous melanocortin receptors sites;
- local actions on receptors that control inflammation within the brain.

Since glia can secrete alpha-MSH and express melanocortin receptors, the effects of alpha-MSH on both fever and inflammation in the brain might be associated with the activation of glial cells.

To date, there are no clear indications for a role of MSH in neurodegenerative diseases or in neurological disorders. However, there is pharmaceutical evidence that MC-4 receptor antagonists could have a role in depression.

Further Reading

Beckwith, B.E., Sandman, C.A., Hothersall, D., Kastin, A.J. (1977): Influence of neonatal injections of a-MSH on learning, memory and attention in rats. *Physiol. Behav.* **18**: 63–71.

Benoit, S., Schwartz, M., Baskin, D., Woods, S.C., Seeley, R.J. (2000): CNS melanocortin system involvement in the regulation of food intake. *Hormone Behav.* **37**: 299–305.

Butler, A.A., Kesterson, R.A., Khong, K., Cullen, M.J., Pelleymounter, M.A., Dekoning, J., Baetscher, M., Cone, R.D. (2000): A unique metabolic syndrome causes obesity in the melanocortin-3 receptor-deficient mouse. *Endocrinology* **141**: 3518–3521.

Catania, A., Lipton, J.M. (1998): Peptide modulation of fever and inflammation within the brain. *Ann. N.Y. Acad. Sci.* **856**: 62–68.

Chaki, S., Hirota, S., Funakoshi, T., Suzuki, Y., Suetake, S., Okubo, T., Ishii, T., Nakazato, A., Okuyama, S. (2003): Anxiolytic-like and antidepressant-like activities of MCL0129 {1-[(S)-2-(4-fluorophenyl)-2-(4-isopropylpiperadin-1-yl)ethyl]-4-[4-(2-methoxynaphthalen-1-yl)butyl]piperazine}, a novel and potent nonpeptide antagonist of the melanocortin-4 receptor. *J. Pharmacol. Exp. Ther.* **304**: 818–826.

Healey, E., Flannagan, N., Ray, A., Todd, C., Jackson, I.J., Matthews, J.N., Birch-Machin, M.A., Rees, J.L. (2000): Melanocortin-1-receptor gene and sun sensitivity in individuals without red hair. *Lancet* **355**: 1072–1073.

Huszar, D., Lynch, C.A., Fairchild-Huntress, V., Dunmore, J.H., Fang, Q., Berkemeier, L.R., Gu, W., Kesterson, R.A., Boston, B.A., Cone, R.D., Smith, F.J., Campfield, L.A., Burn, P., Lee, F. (1997): Targeted disruption of the melanocortin-4 receptor results in obesity in mice. *Cell* **88**: 131–141.

Kastin, A. J., Nissen, C., Nikolics, K., Medzihradszky, K., Coy, D. H., Teplan, I., Schally, A. V. (1976): Distribution of ^{3}H-a-MSH in the rat brain. *Brain Res. Bull.* **1**: 19–26.

Luger, T. A. (1997): The skin immune system: role of a-melanocyte stimulating hormone. *Int. J. Immunopathol. Pharm.* **10**: 47–48.

Medina, F., Siddiqui, A., Scimonelli, T., Fenske, C., Wilson, C. A., Celis, M. E. (1998): The interrelationship between gonadal steroids and POMC peptides, beta-endorphin and alpha-MSH, in the control of sexual behavior in the female rat. *Peptides* **19**: 1309–1316.

Miller, C. L., Hruby, V. J., Matsunga, T. O., Bickford, P. C. (1993): Alpha-MSH and MCH are functional antagonists in a CNS auditory gating paradigm. *Peptides* **14**: 431–440.

Mountjoy, K. G., Robbins, L. S., Mortud, M. T., Cone, R. D. (1992): The cloning of a family of genes that encode the melanocortin receptors. *Science* **257**: 1248–1251.

Mountjoy, K. G., Wong, J. (1997): Obesity, diabetes and functions for proopiomelanocortin-derived peptides. *Mol. Cell Endocrinol.* **128**: 171–177.

Papadopoulos, A. D., Wardlaw, S. L. (1999): Endogenous alpha-MSH modulates the hypothalami—pituitary–adrenal response to the cytokine interleukin-1beta. *J. Neuroendocrinol.* **11**: 315–319.

Roselli-Rehfuss, L., Mountjoy, K. G., Robbins, L. S., Mortrud, M. T., Low, M. J., Tatro, J. B., Entwistle, M. L., Simerly, R. B., Cone, R. D. (1993): Identification of a receptor for gamma melanotropin and other proopiomelanocortin peptides in the hypothalamus and limbic system. *Proc. Natl Acad. Sci. USA* **90**: 8856–8860.

Seeley, R. J., Drazen, D. L., Clegg, D. J. (2004): The critical role of the melanocortin system in the control of energy balance. *Annu. Rev. Nutr.* **24**: 133–149.

Starowicz, K., Przewlocka, B. (2003): The role of melanocortins and their receptors in inflammatory processes, nerve regeneration and nociception. *Life Sci.* **73**: 823–847.

Taherzadeh, S., Sharma, S., Chhajlani, V., Gantz, I., Rajora, N., Demitri, M. T., Kelly, L., Zhao, H., Ichiyama, T., Catania, A., Lipton, J. M. (1999): alpha-MSH and its receptors in regulation of tumor necrosis factor-alpha production by human monocyte/macrophages. *Am. J. Physiol.* **276**: R1289–R13294.

Tritos, N. A., Maratos-Flier, E. (1999): Two important systems in energy homeostasis: melanocortins and melanin-concentrating hormone. *Neuropeptides* **33**: 339–349.

Vaudry, H., Eberle, A. N. (1993): The melanotropic peptides. *Ann. N.Y. Acad. Sci.* **680**: 1–687.

4.22
Neuropeptide Y

4.22.1
General Aspects and History

It was not until 1981 that pancreatic polypeptide-like immunoreactivity in mammalian brain was definitely demonstrated. Tatemoto and coworkers (1981) successfully isolated a 36-amino-acid peptide from porcine brain extracts. Sequencing of this peptide revealed the presence of a tyrosine amino acid residue at both its amino- and carboxy-terminals. The peptide was therefore named neuropeptide Y (NPY, Y being the amino acid one-letter code for tyrosine).

Earlier, the same group had successfully isolated a very similar peptide from porcine intestine extracts (Tatemoto and Mutt 1980). This peptide was labeled peptide YY (PYY). NPY and PYY have an identical length (36 amino acids), but vary slightly in their amino acid composition. Both peptides are related to pancreatic polypeptide (PP). They are all encoded by different genes with high structural similarity.

Each of these genes encodes a propeptide, comprising a signal peptide, the neuroactive peptide itself and a carboxy-terminal flanking peptide.

PP is expressed exclusively in the pancreatic islets, PYY is found in the endocrine cells of the pancreas and in the duodenum, while NPY is widely distributed in the entire neuronal system with a large amount in the enteric nervous system.

4.22.2
Localization Within the Central Nervous System

NPY is found in remarkably high concentrations within the central nervous tissues of mammals. With the exception of the cerebellum, NPY is synthesized in large amounts in the entire brain. The hypothalamus (paraventricular nucleus, arcuate nucleus, suprachiasmatic nucleus) contains the highest concentrations of NPY. In the striatum, the basal forebrain, the hippocampus and the amygdala, NPY is concentrated in interneurons and is often colocalized with GABA or somatostatin. Colocalization of GABA and NPY has also been described in the hypothalamus.

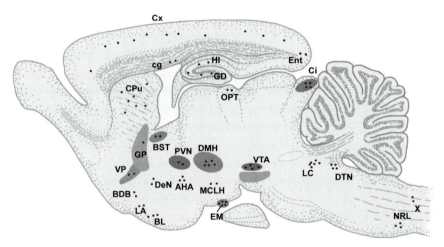

Fig. 4.34 Distribution of NPY-expressing neurons in the brain. In addition to regions already mentioned in the text, NPY-synthesizing cells are also present in some additional areas of the rat brain.
Abbreviations: AHA=anterior hypothalamic area; BDB=diagonal band of Broca; BL=basolateral nucleus of the amygdala; BST=bed nucleus of the *stria terminalis*; cg=cingulum; Ci=inferior colliculus; Cx=cortex; CPu=caudate putamen; DeN=dorsoendopiriform nucleus; DMH=dorsomedial nucleus of the hypothalamus; DTN=dorsal tegmental nucleus; EM=*eminentia mediana*; Ent=entorhinal cortex; GD=*gyrus dentatus*; GP=*globus pallidus*; Hi=hippocampus; LA=lateral nucleus of the amygdala; LC=*locus coeruleus*; MCLH=*nucleus magnocellularis* of the lateral hypothalamus; NRL=lateral *nucleus reticularis*; OPT=olivary pretectal nucleus; PVN=paraventricular nucleus of the hypothalamus; VP=ventral pallidum; VTA=ventral tegmental area; X=nucleus of the *nervus vagus*.

4 Neuromodulators

In addition, NPY can be detected in monoaminergic neurons of the ventro-lateral regions of the olfactory bulb, within the nucleus of the tractus solitarius and in the locus coeruleus. These neurons project to various cerebral areas, including the cerebral cortex, the hypothalamus and the spinal cord. NPY-containing neurons seem to be involved in a variety of neuroendocrine, neurovegetative and behavioral mechanisms.

The distribution of neurons expressing NPY is illustrated schematically in Fig. 4.34.

PYY is mainly concentrated in the endocrine pancreas. However, PYY has also been found in neurons of the myenteric plexus and in some areas of the brain stem, in contrast to PP, which is not known to occur in nervous tissues.

4.22.3
Biosynthesis and Degradation

The NPY gene is composed of four exons and three introns and covers a length of 7.2 kb. The human NPY gene is located on chromosome 7.

The large amount of NPY in pheochromocytoma cells has allowed the isolation of the corresponding mRNA, which encodes for the precursor of NPY. This precursor has a total length of 97 amino acids and consists of a 28-amino-acid peptide, the 36-amino-acid long NPY, and a Gly-Lys-Arg sequence followed by a 30-amino-acid peptide, which is referred as the C-flanking peptide of neuropeptide Y (CPON). The steps involved in the generation of NPY are illustrated in Fig. 4.35.

The enzymatic hydrolysis of NPY leads to the generation of peptidergic fragments, which may also have biological activity. These fragments are inactive at

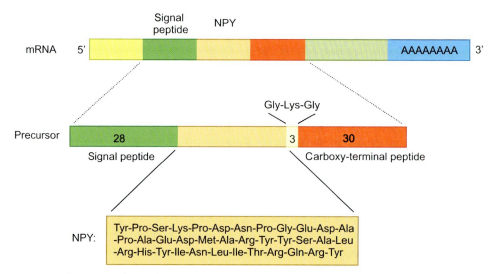

Fig. 4.35 Synthesis of NPY from its precursor.

the NPY receptor Y1, but some of them interact with the second NPY receptor type Y2.

4.22.4
Receptors and Signal Transduction

The various biological effects of NPY and its homolog are mediated by the activation of at least five classes of receptor, known as the Y1, Y2, Y4, Y5 and y6 subtypes. All of these receptors have been cloned and most are expressed in a range of species. However, the y6 receptor subtype has only been shown to be functionally expressed in mouse and rabbit but not in rat and primates. The genuine existence of a Y3 receptor subtype remains to be established.

Y1 receptors
The cloning and functional expression of a human NPY receptor of the Y1 type was achieved by Larhammar and coworkers in 1992.

Further studies revealed that this receptor is expressed not only in mammals, but also in non-mammalian species and that it has been well conserved during evolution.

The Y1 receptor is a receptor with seven transmembrane regions and is coupled to G proteins. This receptor has a length of about 380 amino acids.

Y2 receptors
Screening of cDNA libraries led to the isolation of a human cDNA receptor clone, which upon transfection in COS cells possessed a pharmacological binding profile similar to that of the Y2 receptor. The Y2 receptor has been cloned from different species and reveals high interspecies homology.

The Y2 receptor has a putative length of 381 amino acids and shares a homology of about 31% with the Y1 receptor. The Y2 receptor recognizes the complete NPY molecule as well as some short fragments of the carboxy-terminal sequence of NPY. In contrast, the Y1 receptor needs the complete NPY molecule for its activation.

Y4 receptors
Sequence homology screening with an Y1 receptor probe led to the isolation of a new human NPY receptor cDNA. This receptor was originally designated as either PP1 or Y4, and is now recognized as the Y4 receptor.

The Y4 receptor displays a high affinity for PP-related peptides. This feature allows Y4 to be distinguished from Y1 and Y2, since both have low affinities for the pancreatic polypeptide.

Y5 receptors
The Y5 receptor has also been cloned from mouse and dog. Short carboxy-terminal fragments of NPY are much less potent at the Y5 receptor than NPY, PYY or long carboxy-terminal fragments such as NPY(2–36) and PYY(3–36).

The Y5 receptor seems to exist as two subtypes. One subtype has been cloned in rats, while the other has been cloned in mice.

The activation of NPY receptors induces several of intracellular events, such as inhibition of adenylate cyclase and activation of phospholipase C.

It is assumed that NPY-activated G proteins act directly upon ion channels. In some cell types (for example smooth muscle cells), NPY induces the intracellular accumulation of calcium by an inositol-trisphosphate-dependent mechanism.

In neurons NPY seems to suppress the influx of calcium ions through voltage-gated ion channels.

Since no selective NPY-antagonists are available, there is no definite data as to action of NPY during signal transduction.

4.22.5
Biological Effects

NPY is one of the most abundant neuropeptides found in the central nervous system and, since it is also one of the most conserved peptides in evolution, it is likely to play an important role in the regulation of basic physiological functions.

NPY is a pleiotropic factor participating to the control of physiological processes, such as reproductive and cardiovascular functions, neuroendocrine mechanisms and eating behavior. NPY is also involved in the control of circadian rhythms and in the release of reproductive hormones. In addition, NPY seems to be involved in mechanisms associated with learning and memory; and it appears to interfere with complex behavior, for example fear and anxiety.

Injection of NPY into the central nucleus of the amygdala or into the ventricle results in an anxiolytic reaction. Furthermore, it has been demonstrated that injection of NPY into the rostral hippocampus and the septal area enhances memory retention, whereas NPY injection into the amygdala and the caudal hippocampus induces amnesia. Further support for a physiological role for NPY in cognitive behavior comes from passive immunization studies with NPY antibodies which, when injected into the hippocampal regions, induced amnesia.

Additional evidence suggests that NPY, together with LHRH, is involved in the regulation of gonadotropin secretion. *In vitro* experiments have shown that NPY potentiates the effect of LHRH on the release of gonadotropins from gonadotropic cells of the pituitary.

When released from the paraventricular nucleus of the hypothalamus, NPY enhances the release of corticotropin-releasing factor (CRF), which in turn leads to an increase in ACTH levels.

Furthermore, it is known that NPY inhibits the release of vasopressin from neurons of the paraventricular nucleus of the hypothalamus and seems to stimulate the secretion of growth hormones.

NPY coexists with other neurotransmitters or neuromodulators, especially with norepinephrine. Presynaptically, NPY inhibits the release of norepinephrine from sympathetic neurons as well as the release of acetylcholine from parasympathetic neurons through activation of Y2 receptors.

NPY and NPY receptors appear to be involved in the control of eating behavior. The expression of both NPY and its receptors is modulated by changes of food intake and energy balance. Food deprivation increases levels of NPY and its corresponding mRNA in the paraventricular nucleus of the hypothalamus; and food intake shows a reverse effect. These data are supported by antisense oligonucleotide experiments, which showed a reduction of food intake when applied intracerebrally. Along this line, acute administration of NPY into the hypothalamus or into the brain ventricles leads to increased food intake. In the case of chronic administration, the hyperphagic effects of NPY are prolonged, leading to the development of an obese state.

The highest NPY concentration in the paraventricular nucleus can be measured prior to the onset of natural feeding. NPY concentrations in the paraventricular nucleus are found to be elevated before the intake of food and, thereafter, levels decrease significantly during the postprandial phase. Administration of NPY into the hypothalamus triggers food intake, the paraventricular nucleus being most sensitive in this respect. In essence, these data strongly indicate that NPY plays an important role in feeding behavior.

Intravenously applied NPY induces an increase in blood pressure, resulting from a Y1 receptor-mediated vasoconstriction. In addition, very low concentrations of NPY potentiate the vasoconstricitive effects induced by histamine or norepinephrine. In contrast to the fast- and short-lasting action of norepinephrine on vasoconstriction, the NPY-induced effects occur slowly and are more prolonged.

4.22.6
Neurological Disorders and Neurodegenerative Diseases

Investigations to date have implicated NPY in the pathophysiology of several diseases, including feeding disorders, seizures and depression. Moreover, changes to NPY-containing neurons have been demonstrated in some neurological diseases, such as Alzheimer's and Huntington's chorea disease.

The hippocampal formation is severely affected in Alzheimer's disease. Some studies have reported a significant decrease in NPY-like immunoreactivity in cortical, amygdaloid and hippocampal areas under Alzheimer conditions. NPY-like immunoreactivity has also been detected within neuritic plaques present in the brains of patients with Alzheimer's disease. These data suggest that the degenerative processes associated with this disease may involve changes in NPY-related innervation.

Hippocampal NPY-positive neurons increase their NPY expression in some models of epilepsy, such as kindling. Since NPY inhibits glutamatergic excitation under seizure conditions in human, Patrylo and coworkers (1999) speculated that the increase in NPY in the hippocampus plays a beneficial role in reducing cell excitability in chronically epileptic tissue and subsequently limits severity of the seizure. Thus, it has been suggested that NPY may act as an endogenous anticonvulsant. Indeed, administration of NPY attenuated epileptiform-like activity in various animal models of epilepsy and reduced seizure activity following kainic acid administration to NPY-deficient mice.

There might be also an involvement of the NPY system in other neurological diseases, since reduced NPY mRNA levels have been detected in the cortex of patients with schizophrenia and bipolar disorders.

Several brain structures are involved in mediating anti-stress actions of NPY; and anti-stress actions of NPY are mimicked by Y1 receptor agonists and blocked by Y1 antagonists. Thus, the NPY system might be a useful target for novel pharmacological treatments of stress-related disorders.

Further Reading

Adrian, T.E., Allen, J.M., Bloom, S.R., Ghatei, M.A., Rossor, M.N., Roberts, G.W., Crow, T.J., Tatemoto, K., Polak, J.M. (1983): Neuropeptide Y distribution in human brain. *Nature* **306**: 584–586.

Allen, Y.S., Adrian, T.E., Allen, J.M., Tatemoto, K., Crow, T.J., Bloom, S.R., Polak, J.M. (1983): Neuropeptide Y distribution in the rat brain. *Science* **221**: 877–879.

Baraban, S.C. (2004): Neuropeptide Y and epilepsy: recent progress, prospects and controversies. *Neuropeptides* **38**: 261–265.

Beal, M.F., Mazurek, M.F., Chattha, G.K., Svendsen, C.N., Bird, E.D., Martin, J.B. (1986): Neuropeptide Y immunoreactivity is reduced in cerebral cortex in Alzheimer's disease. *Ann. Neurol.* **20**: 282–288.

Beck-Sickinger, A.G., Jung, G. (1995): Structure–activity relationships of neuropeptide Y analogues with respect to Y1 and Y2 receptors. *Biopolymers* **37**: 123–142.

Chen, G., van den Pol, A.N. (1996): Multiple NPY receptor coexist in pre- and postsynaptic sites: Inhibition of GABA release and isolated self-innervating SCN neurons. *J. Neurosci.* **16**: 7711–7724.

Colmers, W.F., Bleakman, D. (1994): The effects of neuropeptide Y on the electrical properties of neurons. *Trends Neurosci.* **17**: 373–379.

Heilig, M. (2004): The NPY system in stress, anxiety and depression. *Neuropeptides* **38**: 213–224.

Kuromitsu, J., Yokoi, A., Kawai, T., Nagasu, T., Aizawa, T., Haga, S., Ikeda, K. (2001): Reduced neuropeptide Y mRNA levels in the frontal cortex of people with schizophrenia and bipolar disorder. *Gene Expr. Patterns* **1**: 17–21.

Larhammar, D. (1996): Evolution of neuropeptide Y, peptide YY and pancreatic polypeptide. *Regul. Peptides* **62**: 1–11.

Larhammar, D., Blomqvist, A.G., Yee, F., Jazin, E., Yoo, H., Wahlested, C. (1992): Cloning and functional expression of a human neuropeptide Y/peptide YY receptor of the Y1 type. *J. Biol. Chem.* **267**: 10935–10938.

Levens, N.R., Della-Zuana, O. (2003): Neuropeptide Y Y5 receptor antagonists as anti-obesity drugs. *Curr. Opin. Investig. Drugs* **4**: 1198–1204.

Magni, P. (2003): Hormonal control of the neuropeptide Y system. *Curr. Protein Peptide Sci.* **4**: 45–57.

Michel, M.C. (1991): Receptors for neuropeptide Y: multiple subtypes and multiple second messengers. *Trends Neurosci.* **12**: 389–394.

Patrylo, P.R., van den Pol, A.N., Spencer, D.D., Williamson, A. (1999): NPY inhibits glutamatergic excitation in the epileptic human dentate gyrus. *J. Neurophysiol.* **82**: 478–483.

Pelletier, G. (1990): Ultrastructural localization of neuropeptide Y in the hypothalamus. *Ann. N.Y. Acad. Sci.* **611**: 232–246.

Redrobe, J.P., Dumont, Y., St-Pierre, J.-A., Quirion, R. (1999): Multiple receptors for neuropeptide Y in the hippocampus: putative roles in seizures and cognition. *Brain Res.* **848**: 153–166.

Tatemoto, K., Mutt, V. (1980): Isolation of two novel candidate hormones using a chemical method for finding naturally occurring polypeptides. *Nature* **285**: 417–418.

Tatemoto, K., Carlquist, M., Mutt, V. (1981): Neuropeptide Y – a novel brain peptide with structural similarities to peptide YY and pancreatic polypeptide. *Nature* **296**: 659–660.

Toth, P. T., Bindokas, V. P., Bleakman, D., Colmers, W. F., Miller, R. J. (1993): Presynaptic inhibitor by neuropeptide Y is mediated by reduced Ca^{2+} influx at sympathetic nerve terminals. *Nature* **364**: 635–639.

Walker, P., Grouzmann, E., Burnier, M., Waeber, B. (1991): The role of neuropeptide Y in cardiovascular regulation. *Trends Pharmacol. Sci.* **12**: 111–115.

Weinberg, D. H., Sirinathsinghji, D. J., Tan, C. P., Shiao, L. L., Morin, N., Rigby, M. R., Heavens, R. H., Rapoport, D. R., Bayne, M. L., Cascieri, M. A., Strader, C. D., Linemeyer, D. L., MacNeil, D. J. (1996): Cloning and expression of a novel neuropeptide Y receptor. *J. Biol. Chem.* **271**: 16435–16438.

Woods, S. C., Figlewicz, D. P., Lisa Madden, L., Porte, D., Sipols, A. J., Seeley, R. J. (1998): NPY and food intake: discrepancies in the model. *Regul. Peptides* **75**: 403–408.

4.23
Neurotensin

4.23.1
General Aspects and History

Neurotensin (NT) was first isolated by Carraway and Leeman in 1973 from extracts of bovine hypothalamus. It is a peptide which consists of 13 amino acids with the primary sequence shown in Fig. 4.36.

This primary structure of neurotensin is highly conserved in mammals. Immunohistochemical studies have demonstrated that neurotensin is widely expressed in vertebrates and invertebrates and has consistently been found down to Protozoa. This high degree of evolutionary conservation and its neuronal expression is suggestive of an essential role for this neuropeptide in signal transmission.

The neurotensin precursor as well as the neurotensin receptors have been identified and cloned. In addition, pharmacologically applicable neurotensin antagonists have been developed.

The first potent non-peptidergic neurotensin receptor antagonist (SR48692) was developed by Gully and coworkers (1993). This compound proved capable of inhibiting a number of the central and peripheral effects of neurotensin. The second generation of antagonists (e.g. SR142948A) was found to be more potent than SR48692 and to exhibit a wider spectrum of activities since it antagonizes hypothermia and analgesia induced by intraventricular injection of neurotensin, effects not observed with SR48692.

pGlu-Leu-Tyr-Glu-Asn-Lys-Pro-Arg-Arg-Pro-Tyr-Ile-Leu-OH

Fig. 4.36 Sequence of neurotensin.

4.23.2
Localization Within the Central Nervous System

In the central nervous system, neurotensin is expressed exclusively in neurons. Its distribution and localization have been studied by autoradiography, immunohistochemistry and by *in situ* hybridization.

Neurotensin-containing neurons have been identified in anterior regions of the hypothalamus, in the preoptic area and in the premammillary area, as well as in the central and medial nucleus of the amygdala and in the endopiriform nucleus. Neurotensin-containing neurons are also localized in the septum, the stria terminalis, the median eminence, the zona incerta, the ventral tegmental area, the substantia nigra, the nucleus of the tractus solitarius and in the area postrema.

Neurotensin-containing neurons have also been identified in smaller quantities in the nucleus accumbens, the caudate-putamen, the periaqueductal gray, the lateral hypothalamus, the locus coeruleus, the thalamus and the spinal cord. In the anterior pituitary gland, a population of neurotensin-containing neurons has also been demonstrated.

Figure 4.37 depicts some of the neuronal neurotensin populations, as well as their projections.

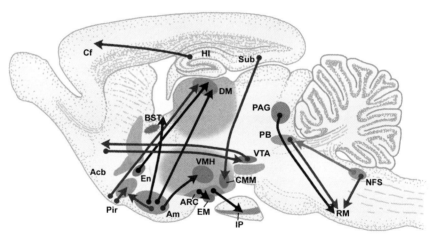

Fig. 4.37 Schematic overview of some neuronal neurotensin-containing nuclei and their projections. *Abbreviations*: Acb = *nucleus accumbens*; Am = amygdala; ARC = arcuate nucleus; BDB = diagonal band of broca; BST = bed nucleus of the *stria terminalis*; Cf = frontal cortex; CMM medial mammillary complex; DM = dorsomedial nucleus of the thalamus; EM = *eminentia mediana*; En = endopiriform nucleus; HI = hippocampus; IP = intermediate pituitary; NFS = nucleus of the *fasciculus solitarius*; PAG = periaqueductal gray; PB = parabrachial nucleus; Pir = piriform cortex; RM = magnocellular raphe nucleus; Sub = subiculum; VMH = ventromedial nucleus of the hypothalamus; VTA = ventral tegmental area.

Neurotensin coexists with other neurotransmitters and neuromodulators, including dopamine, norepinephrine, epinephrine, enkephalin, GABA, cholecystokinin and corticotropin-releasing hormone (CRH).

4.23.3
Biosynthesis and Degradation

The gene which encodes for the neurotensin precursor (pro-neurotensin) is composed of four exons and three introns (total base length 10.2 kb). The primary structure of the precursor has a putative length of 170 amino acids (Fig. 4.38); and the fragment which corresponds to neurotensin is located in the carboxy-terminal region of the precursor.

Neurotensin and its structurally related analog neuromedin N are synthesized from the common precursor, pro-neurotensin. Expression of the precursor mRNA has been observed for example in the lateral septal nucleus, hippocampus, nucleus accumbens, medial preoptic area, bed nucleus of the stria terminalis, lateral hypothalamus, central amygdaloid nucleus, subthalamic nucleus, geniculate nucleus and substantia nigra.

Neuromedin N is a hexapeptide with the following amino acid sequence: Lys-Ile-Pro-Tyr-Ile-Leu-OH. This sequence is also conserved in the carboxy-terminal of neurotensin. Neurotensin and neuromedin N are located in the carboxy-terminal domain of the precursor, where they are flanked and separated by three Lys-Arg sequences. A fourth dibasic sequence preceding a neuromedin N-like sequence is present in the amino-terminal domain.

The Lys-Arg sites provide consensus sequences which are used by endoproteases belonging to the recently described family of proprotein convertases (PC).

While neuromedin N shows rapid inactivation by aminopeptidases, neurotensin degradation proceeds much more slowly. The degradation of neurotensin involves the activity of membranous and cytosolic peptidases. These include the endopeptidases 24.11, 24.15 and 24.16.

The endopeptidase 24.16 is a metalloendopeptidase of 70 kDa, which cleaves the peptide in its carboxy-terminal domain.

Co-application of neurotensin and inhibitors of endopeptidase 24.11 have been shown to potentiate central neurotensin-related effects.

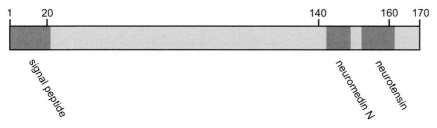

Fig. 4.38 Schematic representation of the common precursor (pro-neurotensin) of neurotensin and neuromedin N.

4.23.4
Receptors and Signal Transduction

Neurotensin is released by depolarizing stimuli in a calcium-dependent manner and binds to specific receptors. Three subtypes of neurotensin receptors have been cloned. All three subtypes recognize the same carboxy-terminal sequence and display a structure–function relationship similar to that of neurotensin. Two of the receptors belong to the family of G protein-coupled receptors, whereas the third is an entirely new type of neuropeptide receptor, consisting of a 100-kDa protein with a single transmembrane domain.

The initial evidence for the existence of different neurotensin receptor subtypes was based on binding studies, which showed two classes of neurotensin-binding site. A high-affinity site (site H), which proved susceptible for Na^+ and GTP, and a low-affinity site (site L), which was less susceptible to Na^+ and insensitive to GTP. The antagonist of the histamine receptor H1 levocabastine, which is devoid of any neurotensin-like properties, selectively blocks neurotensin binding to the L-sites without changing the binding properties of the H-sites.

NTS1 receptors

In 1990 Nakanishi and his group cloned the first NT receptor (NTS1, also known as NTR-1) and showed that it belongs to the family of G protein-coupled receptors. Screening of cDNA expression libraries from rat brain allowed the cloning of a 3633-kb cDNA. The cDNA encodes for a putative 424-amino-acid protein with seven transmembrane domains, indicating that the NTS1 receptor belongs to the family of G protein-coupled receptors. When expressed in COS cells, the receptor binds selectively to neurotensin with high affinity and is insensitive to levocabastine.

The human NTS1 receptor has also been cloned. The human receptor consists of 418 amino acids and the amino acid sequence shares 84% homology with the rat homolog. The human NTSR1 gene is localized on chromosome 20 (20q13).

In situ hybridization studies have been used to clarify the distribution of NTS1 receptors in the central nervous system. High levels of NTS1 mRNAs are found in neurons of the hypothalamus (suprachiasmatic nucleus, supramammillary area), as well as in neurons of the diagonal band of Broca, the medial septal nucleus, the substantia nigra, the ventral tegmental area and in the some neurons of the spinal cord.

The signal transduction of the NTS1 receptor seems to be coupled to phospholipase C. Activation of these receptors leads to an increase in intracellular calcium as well as an increase in inositol-1.4.5-trisphosphate.

NTS2 receptors

Binding studies indicated the existence of a further receptor with low affinity for neurotensin. Indeed, a second neurotensin receptor has been cloned. The rodent NTS2 receptor shows a putative length of 416 amino acids as compared to 410 amino acids in humans. The NTS2 receptor exhibits the same seven transmembrane-spanning motif and thus belongs to the G protein-coupled receptor family. The rat NTS2 and NTS1 receptors share an amino acid homology of about 64%.

Within the central nervous system, NTS2 receptors have been localized to the olfactory bulb, cerebral cortex, hippocampus, bed nucleus of the stria terminalis, magnocellular preoptic nucleus, amygdaloid complex, anterodorsal thalamic nucleus, substantia nigra, ventral tegmental area, cerebellum, periaqueductal gray and raphe nuclei.

NTS3 receptors
In the central nervous system, a further high-affinity neurotensin-binding site is present which is distinct from the NTS1 and NTS2 receptors. One essential feature of this binding site is that it retains its activity after dissolving in the detergent 3-[(3-cholamidopropyl) dimethylammino]-1-propane sulfonate (CHAPS). Sequence analysis of this receptor reveals 100% homology with a previously cloned human gp95/sortilin 60 cDNA.

This nts3/gp95/sortilin cDNA encodes a protein of 833 amino acids. However, the gene product does not constitute the receptor, and subsequent investigations have revealed that the NTR3/sortilin is synthesized as a precursor, which is processed into a 5-kDa peptide fragment and a 95-kDa membranous protein.

Currently available pharmacological data indicate that both neurotensin and neuromedin N bind to the three neurotensin receptor subtypes (Neuromedin N seems to be slightly less potent than neurotensin at the NTS1 and NTS3 receptors and nearly equipotent on the NTS2 receptors).

NTS3 receptors have been mapped to the neocortex, the piriform cortex, the hippocampal formation, islands of Calleja, medial and lateral septal nuclei, amygdaloid nuclei, thalamic nuclei, supraoptic nucleus, substantia nigra, Purkinje cells of the cerebellar cortex and cranial motor nuclei.

4.23.5
Biological Effects

Intravenous administration of neurotensin produces endocrine and cardiovascular effects, which include hyperglycemia, hypotension and vasodilatation.

Intracerebroventricular or intracisternal administration of neurotensin produces physiological and behavioral alterations, including potentiation of barbiturate-mediated effects and effects mediated by ethanol. In addition, it induces hypothermia, reduces food intake and prevents the development of stress-induced gastric ulcers in rats.

Furthermore, it has been demonstrated that neurotensin inhibits locomotor hyperactivity induced by dopamine agonists. Neurotensin has been shown to modulate dopaminergic transmission and to modulate dopaminergic effects in cells which possess both dopamine and neurotensin receptors. For example, the activation of neurotensin receptors located on the soma or on nerve terminals of dopaminergic neurons can facilitate dopaminergic transmission. The action of dopamine itself is to suppress dopamine release via dopaminergic autoreceptors.

In contrast, the activation of neurotensin receptors antagonizes dopamine release at the postsynaptic level.

Neurotensin elicits hypothermic and naloxone-insensitive analgesic responses after brain injection. The effects are independent of the activation of NT-1 receptors; and recent studies (Dubuc et al. 1999) suggest that they are mediated via NT-2 receptors. An interesting point in this context is that the analgesic effect of neurotensin is not be blocked by opioid antagonists. In addition, by comparing mice deficient in the receptors NTS1 or NTS2, an involvement of NTS2 receptors in thermal nociception has been demonstrated. Thus, in a hot-plate test, a significant alteration in jump latency was observed in NTS2-deficient mice compared to NTS1-deficient or wild-type mice.

Food intake, especially the intake of lipids, induces a release of neurotensin; and increased levels of neurotensin can be detected in the circulating blood. A likely interpretation of this relationship is that neurotensin is involved in digestive processes.

Neurotensin also inhibits the secretion of gastric acid. Besides its endocrine functions, neurotensin induces paracrine effects in the periphery: these are evident as locally restricted effects on the contraction of smooth muscles in the gasto-intestinal tract and on the transepithelial transport of electrolytes.

4.23.6
Neurological Disorders and Neurodegenerative Diseases

Parkinson's disease is characterized by a significant destruction of nigrostriatal dopaminergic neurons concomitantly with a loss of neurotensin receptors in neurons of the substantia nigra and the striatum.

Neurotensin is a neuropeptide implicated in schizophrenia that specifically modulates neurotransmitter systems known to be dysregulated in this disorder. Modulation of neurotensin in the mesolimbic system is thought to underlie the mechanism of action of antipsychotic drugs. Indeed, several lines of evidence support this notion, since an association of neurotensins with neural circuits involved in the pathophysiology of schizophrenia and the therapeutic effects of antipsychotic drugs has been demonstrated. Thus, administration of antipsychotic drugs can increase neurotensin neurotransmission. Moreover, in some patients with schizophrenia, low concentrations of neurotensins have been measured in the cerebrospinal fluid, which are normalized after treatment with antipsychotic drugs. Along this line, the central administration of neurotensin results in a variety of neurobehavioral effects which, depending upon the administration site, resemble the effects of antipsychotic drugs and psychostimulants.

Further Reading

Adachi, D.K., Kalivas, P.W., Schenk, J.O. (1990): Neurotensin binding to dopamine. *J. Neurochem.* **54**: 1321–1328.

Binder, E.B., Kinkead, B., Owens, M.J., Nemeroff, C.B. (2001): The role of neurotensin in the pathophysiology of schizophrenia and the mechanism of action of antipsychotic drugs. *Biol. Psychiatry* **50**: 856–872.

Caceda, R., Kinkead, B., Nemeroff, C.B. (2003): Do neurotensin receptor agonists represent a novel class of antipsychotic drugs? *Semin. Clin. Neuropsychiatry* **8**: 94–108.

Carraway, R.E., Mitra, P., Spaulding, G. (1992): Post-translational processing of the neurotensin/neuromedin-N precursor. *Ann. N.Y. Acad. Sci.* **668**: 1–6.

Dobner, P.R., Deutch, A.Y., Fadel, J. (2003): Neurotensin: dual roles in psychostimulant and antipsychotic drug responses. *Life Sci.* **73**: 801–811.

Dubuc, I., Sarret, P., Labbe-Jullie, C., Botto, J.M., Honore, E., Bourdel, E., Martinez, J., Costentin, J., Vincent, J.P., Kitabgi, P., Mazella, J. (1999): Identification of the receptor subtype involved in the analgesic effect of neurotensin. *J. Neurosci.* **19**: 503–510.

Farkas, R.H., Chien, P.-H., Nakajima, S., Nakajima, S. (1997): Neurotensin and dopamine D2 activation oppositely regulate the same K^+ conductance in rat midbrain dopaminergic neurons. *Neurosci. Lett.* **231**: 21–24.

Gully, D., Canton, M., Boigegrain, R., Jeanjean, F., Molimard, J.C., Poncelet, M., Gueudet, C., Heaulme, M., Leyris, R., Brouard, A., et al. (1993): Biochemical and pharmacological profile of a potent and selective nonpeptide antagonist of the neurotensin receptor. *Proc. Natl Acad. Sci. USA* **90**: 65–69.

Henry, J.L. (1982): Electrophysiological studies on the neuroactive properties of neurotensin. *Ann. N.Y. Acad. Sci.* **440**: 216–227.

Jann, M.W. (2004): Implications for atypical antipsychotics in the treatment of schizophrenia: neurocognition effects and a neuroprotective hypothesis. *Pharmacotherapy* **24**: 1759–1783.

Kitabgi, P. (1989): Neurotensin modulates dopamine neurotransmission at several levels along brain dopaminergic pathways. *Neurochem. Int.* **14**: 111–119.

Kitabgi, P., Checler, F., Mazella, J., Vincent, J.P. (1985): Pharmacology and biochemistry of neurotensin receptors. *Rev. Clin. Basic Pharm.* **5**: 397–486.

Levant, B., Merchant, K.M., Dorsa, D.M., Nemeroff, C.B. (1992): BMY 14802, a potential antipsychotic drug, increases expression of proneurotensin mRNA in the rat striatum. *Mol. Brain Res.* **12**: 279–284.

Maeno, H., Yamada, K., Santo-Yamada, Y., Aoki, K., Sun, Y.J., Sato, E., Fukushima, T., Ogura, H., Araki, T., Kamichi, S., Kimura, I., Yamano, M., Maeno-Hikichi, Y., Watase, K., Aoki, S., Kiyama, H., Wada, E., Wada, K. (2004): Comparison of mice deficient in the high- or low-affinity neurotensin receptors, Ntsr1 or Ntsr2, reveals a novel function for Ntsr2 in thermal nociception. *Brain Res.* **998**: 122–129.

Rosas-Arellano, M.P., Solano-Flores, L.P., Ciriello, J. (1996): Neurotensin projections to the subfornical organ from arcuate nucleus. *Brain Res.* **706**: 323–327.

Sarret, P., Perron, A., Stroh, T., Beaudet, A. (2003a): Immunohistochemical distribution of NTS2 neurotensin receptors in the rat central nervous system. *J. Comp. Neurol.* **461**: 520–538.

Sarret, P., Krzywkowski, P., Segal, L., Nielsen, M.S., Petersen, C.M., Mazella, J., Stroh, T., Beaudet, A. (2003b): Distribution of NTS3 receptor/sortilin mRNA and protein in the rat central nervous system. *J. Comp. Neurol.* **461**: 483–505.

Seutin, V., Massotte, L., Dresse, A. (1989): Electrophysiological effects of neurotensin on dopaminergic neurones of the ventral tegmental area of the rat *in vitro*. *Neuropharmacology* **28**: 949–954.

Smits, S.M., Terwisscha van Scheltinga, A.F., van der Linden, A.J., Burbach, J.P., Smidt, M.P. (2004): Species differences in brain pre-pro-neurotensin/neuromedin N mRNA distribution: the expression pattern in mice resembles more closely that of primates than rats. *Brain Res. Mol. Brain Res.* **125**: 22–28.

Tanaka, K., Masu, M., Nakanislu, S. (1990): Structure and functional expression of the cloned rat neurotensin receptor. *Neuron* **4**: 847–854.

Uhl, G. R., Kuhar, M. J., Snyder, S. H. (1977): Neurotensin: immunohistochemical localization in rat central nervous system. *Proc. Natl Acad. Sci. USA* **74**: 4096–4063.

Vincent, J. P., Mazella, J., Kitabgi., P. (1999): Neurotensin and neurotensin receptors. *Trends Pharmacol. Sci.* **20**: 302–309.

4.24
Neurotrophins

4.24.1
General Aspects and History

Neurotrophins are relatively small, homodimeric polypeptides (120 amino acid residues) with a varied repertoire of biological activities influencing the generation, differentiation, survival and regeneration of vertebrate neurons. The neurotrophins belong to the heterogeneous family of growth factors which regulate growth and differentation of various cell types. Figure 4.39 indicates some of the most important members of this group of bioactive substances.

The family of neurotrophins consists of nerve growth factor (NGF), brain-derived neurotrophic factor (BDNF), neurotrophin-3 (NT-3), NT-4/5, NT-6, NT-7 and a newly cloned member from lamprey considered to be the ancestral form of neurotrophins.

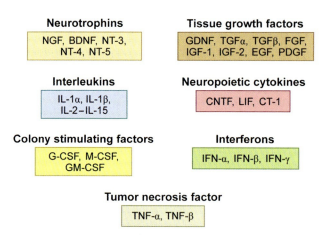

Fig. 4.39 The growth factor family.

4.24.2
Localization Within the Central Nervous System

BDNF
BDNF mRNA and protein has been found in neurons of the anterior olfactory nucleus, the cortex (orbital, insular, frontal, cingulate, piriform, parietal, perirhinal, occipital, temporal, and entorhinal cortices), the endopiriform nucleus, the hippocampus (areas CA1-CA3, dentate gyrus, and subiculum) and in the lateral and medial septal nuclei. Within the amygdala, BDNF has been found in the basolateral complex of the amygdala, including the lateral, basolateral and basomedial nucleus of the amygdala. Furthermore, BDNF has been detected in the amygdala within the medial amygdaloid nucleus and cortical amygdaloid nucleus as well as in the anterior amygdaloid area and amygdalohippocampal area. In addition, BDNF mRNA and protein have been detected in the substantia inominata, corpus callosum, caudate-putamen and the nucleus of the lateral olfactory tract.

Both, BDNF mRNA and protein have been found in a huge number of thalamic nuclei, including the paraventricular thalamic nucleus, central medial and central lateral thalamic nuclei, anterodorsal and intermediodorsal thalamic nuclei, rhomboid thalamic nucleus, posterior intralaminar thalamic nucleus, medial geniculate nucleus and suprageniculate thalamic nucleus. Within the hypothalamus, BDNF mRNA and protein are found in the lateral and medial preoptic nuclei, magnocellular preoptic nucleus, lateroanterior hypothalamic nucleus, lateral hypothalamic area, the parvocellular portion of the paraventricular hypothalamic nucleus, the medial and lateral mammillary nuclei and the posterior hypothalamic area.

Furthermore, BDNF protein and mRNA can be found in the substantia nigra (pars compacta), ventral tegmental area, raphe nuclei, locus coeruleus, pedunculopontine tegmental nucleus, pontine nuclei, inferior colliculus, nucleus of the tractus solitarius and inferior olive.

NGF
NGF mRNA has been located in the frontal, cingulate, entorhinal, piriform and perirhinal cortices, the hippocampus (areas CA1-CA3 and dentate gyrus), the amygdala, the striatum, the diagonal band nucleus, the ventral pallidum, the mediodorsal thalamic nucleus, the supramammillary complex and the mammillary nucleus. Furthermore, NGF mRNA has been detected in the superior and inferior olive, the nucleus raphe magnus, the nucleus of the tractus solitarius, the motor nucleus of the trigeminal nerve and the trochlear, facial, hypoglossal, dorsal vagal motor nuclei.

NT-3
Strong mRNA signals and protein immunoreactivity for neurotrophin-3 have been detected in the neocortex, the entorhinal cortex, the hippocampus and the molecular layer of the dentate gyrus. In addition, NT-3 is present in neurons of

the nucleus of the lateral olfactory tract, anterior olfactory nucleus, anteromedial thalamic nucleus, neostriatum, amygdala, substantia nigra, dorsal root ganglia, central gray of spinal cord and the granular layer of the cerebellum.

NT-4

Neurotrophin-4 has been identified within the neocortex, hippocampus, thalamus, hypothalamus, the locus coeruleus, the inferior olive, the cerebellum, the brain stem, the dorsal root ganglia and in the choroid plexus.

4.24.3
Biosynthesis and Degradation

The neurotrophins are closely structurally related: NGF, BDNF, NT-3 and NT-4 share approximately 50% sequence identity and all neurotrophins occur in different splice variants.

The structure of rat brain-derived neurotrophic factor (BDNF) gene is complex: four 5′ exons are linked to separate promoters and one 3′ exon encodes the BDNF protein. Alternative promoter usage, differential splicing and the use of two different polyadenylation sites within each of the four transcription units generate eight different BDNF mRNAs. The promoters I–HI are used predominantly in the brain, while promoter IV is more active in peripheral tissues. Moreover, the levels of BDNF mRNAs transcribed from promoters I–HI are markedly elevated in promoter-specific subsets of hippocampal and neocortical neurons in different models of neuronal activation, while only a modest increase was seen for promoter IV specific mRNA.

BDNF is generated from a higher molecular precursor. Normally, pro-BDNF (32 kDa) undergoes amino-terminal cleavage within the trans-Golgi network and/or immature secretory vesicles to generate mature BDNF (14 kDa). In addition, small amounts of a truncated form of the precursor (28 kDa) are generated in the endoplasmic reticulum through amino-terminal cleavage of pro-BDNF.

Since blocking of the generation of the 28-kDa BDNF has no effect on the level of mature BDNF and since blocking the generation of mature BDNF does not lead to accumulation of the 28-kDa form, it is suggested that the 28-kDa pro-BDNF is not an obligatory intermediate in the formation of the 14-kDa form in the constitutive secretory pathway. Interestingly, extracellularly released pro-BDNF is biologically active, since it is able to mediate TrkB phosphorylation.

There are two major alternatively spliced mRNAs for NGF, giving rise to different precursor proteins. These may be processed by proteases within the secretory pathway, leading to two major secreted proteins, "mature" fully processed NGF and several "pro-NGF" forms. These NGF isoforms differ in their affinity for the receptors p57^{NTR} and TrkA.

In the case of NGF, it is speculated that neurons release cleaved NGF, acting on TrkA receptors, while a non-neuronal population in the brain (presumably microglial cells) is able to release pro-NGF. The pro-NGF, like NGF, is able to bind to TrkA and p57^NTR receptors, but with a lower affinity. In this context, it has been shown that pro-NGF, via p57 receptors, can induce death of axotomized cortical neurons.

NFG, BDNF and NT-3 are sorted to dense-core vesicles. NT-3 and BDNF appeared to be processed in the regulated secretory pathway of brain neurons and secreted in an activity-dependent manner. In the case of NGF, it has been proposed that it is released through the constitutive secretory pathway from cells in peripheral tissues and nerves, where it can act as a target-derived survival factor. However, it has also been reported that NGF might be released via the regulated secretory pathway.

4.24.4
Receptors and Signal Transduction

The neurotrophins interact with two classes of receptors on responsive cells: a protein tyrosine kinase-type of receptor (members of the Trk family) and a smaller receptor, distantly related to the TNF and CD40 receptors, containing a short cytoplasmic tail of unknown function (the p75 low-affinity NGF receptor, p75^{NTR}). This receptor is a member of the tumor necrosis receptor superfamily. The function of the low-affinity receptor p75^{NTR}, which binds all neurotrophins and thus acts as a pan-neurotrophin receptor, is less well understood than that of the Trk receptors.

The Trk receptors mediate specificity and several of the neurotrophic functions of the neurotrophins. There are three main types of Trk receptors: TrkA is a receptor for NGF, TrkB a receptor for BDNF and NT-4/5, and TrkC is a receptor for NT-3.

Trk receptors
All the neurotrophins bind with high affinity to members of a family of receptor tyrosine kinases, the Trk family. The events leading to signal transduction include:
- binding of neurotrophin to the appropriate Trk receptor (in general, NGF binds to TrkA, BDNF and NT4 to TrkB, and NT3 to TrkC);
- receptor homodimerization;
- autophosphorylation of the tyrosine kinase domains of the bound receptors;
- activation of various intracellular signaling molecules that are associated with the receptor.

The cascade of intracellular signal transduction of neurotrophins follows the following pathway:
1. binding to their specific Trk receptor and promotion of the formation of a Trk-Shc-Grb2-Sos complex;
2. activation of the low-molecular-weight G protein Ras;
3. Ras activation then allows transmission of signals to other downstream pathways.

The most prominent is the MAP kinases cascade following the steps:
1. Ras activates Raf kinase, which phosphorylates and activates MEK.
2. MEK in turn phosphorylates the MAP kinases.
3. MAP, when activated by MEK, becomes proline-directed, which means that the kinase phosphorylates serine or threonine residues neighboring prolines. One direct substrate is the protein kinase p90rsk.

Importantly, activation of MAP kinases and rsk causes their translocation to the nucleus where a number of transcription factors can be phosphrylated, such as the immediate early genes c-fos and c-jun as well as the delayed-response genes like CREB (cAMP-response element-binding protein (see Fig. 4.40).

The biological effects are long-lasting and induce changes in gene expression. Two further pathways are activated:
- The phospolipase C (PLCγ) pathway, which generates DAG and IP3. The consequences of PLC activation are increase in intracellular calcium, pH, cytoskeletal responses and transcriptional changes.
- Phospatidylinositol-3 kinase (PI-3K), which catalyzes the production of phosphoinositides. They in turn activate the protein kinase Akt, which leads to the activation of pathways necessary for growth factor-mediated survival.

Thus, there are many substrates of Trk activation in both the cytoplasm and nucleus, including other protein kinases, cytoskeletal elements, tyrosine hydroxylase and transcription factors.

Two TrkA isoforms are known, which differ in their extracellular domain. Both isoforms appear to have similar biological properties, but they differ in their expression pattern. One isoform is expressed in cells of non-neuronal origin and the other (containing a VSFSPV sequence in the extracellular domain) is primarily expressed in neuronal cells. TrkA, which serves as a receptor for NGF, has been found in NGF-responsive cells, including sympathetic neurons, small spinal sensory neurons of the dorsal root ganglion (DRG) and basal forebrain cholinergic neurons.

TrkB is a receptor for BDNF and NT-4; and it is widely expressed in the peripheral and central nervous systems. Beside the full-length TrkB receptor, TrkB isoforms are known, having the same extracellular and transmembrane domains but lacking a tyrosine kinase catalytic domain.

At least four different isoforms of the TrkC receptor have been described. These isoforms differ from the canonical TrkC tyrosine kinase by the presence of 14, 25, or 39 additional amino acid residues in the middle of their kinase catalytic domains. Expression of TrkC, the main receptor for NT-3, has been documented in cells responsive to this neurotrophin including, among others, large spinal sensory neurons, motor neurons, noradrenergic neurons of the locus coeruleus and neurons of the substantia nigra.

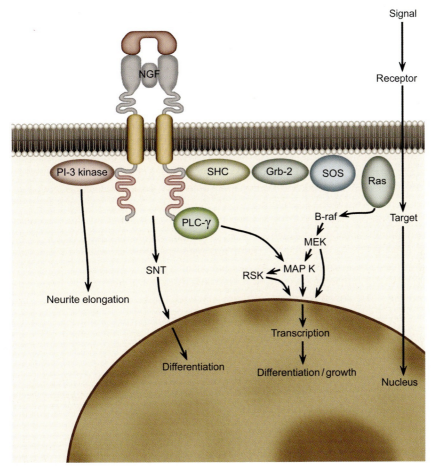

Fig. 4.40 Schematic drawing of the most important pathways of trk signaling. Upon NGF binding to two trk molecules, homodimers are formed which allow each trk molecule to phosphorylate tyrosine residues on its partners. This phosphorylation creates specific binding sites for PI-3, PLC-γ and Shc. Recruitment of these proteins into the complex initiates the different indicated signaling cascades.

p75^{NTR}, the p75 low-affinity NGF receptor

p57^{NTR} exists in two splice variants, which differ in their extracellular domains and ligand preference. Thus, TrkA exists in a short and a long isoform, with the short isoform lacking domains known to be important for neurotrophin-binding. The interaction of NGF with p75^{NTR} is mediated by a positively charged region that contains at least three lysine residues located in two spatially adjacent loops (I and IV). These positively charged residues are conserved in NT-3 and NT-4, but not in BDNF, which contains three consecutive positively charged residues in loop V that may compensate for the absence of the to lysines present in loop I in the other neurotrophins.

The p75 neurotrophin receptor was formerly thought to be a low-affinity binding receptor that interacted with the signal-producing Trk receptors to enable high-affinity binding. Binding and functional assays have indicated that $p75^{NTR}$ can cooperate with Trk receptors to increase the affinity of neurotrophin binding and/or the signaling efficiency. Although $p75^{NTR}$ can modulate Trk receptor activity, published reports have revealed unique functions of $p75^{NTR}$ that are independent of Trk receptors.

$p75^{NTR}$ binds each of the neurotrophins with approximately equal affinity. Recent work has also shown that $p75^{NTR}$ interacts with the Nogo receptor as a signal-transmitting subunit that mediates inhibitory effects on axon growth of three myelin-associated glycoproteins (Nogo, MAG and Omgp). Neurotrophin binding to $p75^{NTR}$ promotes survival of some cells and apoptosis of others as well as affecting axon outgrowth. $p75^{NTR}$ exerts these diverse actions through a set of signaling pathways largely distinct from those activated by Trk receptors. Pro-survival pathways activated by $p75^{NTR}$ include NFκB and Akt. Ligand binding to $p75^{NTR}$ also stimulates several proapototic pathways, which include the jun-kinase signaling cascade, sphingolipid turnover and association with several adaptors (e.g. NRAGE and NADE) that directly promote cell cycle arrest and apoptosis. The activation of $p75^{NTR}$ also activates the small G proteins Rac and Rho that directly affect growth cone motility.

4.24.5
Biological Effects

The neurotrophins play a critical role in the development, differentiation, maintenance and survival of specific neuronal populations in both the central and peripheral nervous systems during development.

Neurons exibit trophic requirements, either for survival or for maintenance of some aspect of differentiation. NGF acts on various classes of central and peripheral neurons to promote cell survival, to stimulate neurite outgrowth and to modulate differentiation. Thus, NGF is known as a potential target-derived survival factor. It has been demonstrated, for example, that sympathetic ganglion neurons require NGF for their survival during embryonic development and that NGF applied exogenously during this phase of development enhances sympathetic ganglion cell survival.

NGF is synthesized as a precursor, proNGF, which undergoes processing to generate mature NGF. It has been shown that proNGF is abundant in the central nervous system, whereas mature NGF is undetectable. Interestingly, it has been demonstrated that even proNGF has neurotrophic activity. However, not only NGF possesses trophic properties. For example, both BDNF and NT-3 promote the survival of dopaminergic neurons *in vitro* and *in vivo*.

Neurotrophins are also required for migration, morphological and biochemical differentiation of neurons and even for the maintenance of, at least, cortical dendrites.

Aside from their trophic action, neurotrophins are involved in processes attributed to learning and memory. The relatively high expression of NGF, BDNF and NT-3 as well as TrkB and TrkC receptors in the hippocampus coupled with the findings that neurotrophin levels are influenced by neuronal activity led to the suggestion that neurotrophic factors may play a role in synaptic plasticity. Indeed, it has been found that induction of LTP in the hippocampus increases BDNF and NT-3 mRNA in this structure. Hippocampal slices treated with BDNF or NT-3, but not NGF, show persistent increases in EPSP amplitude, which are similar to tetanically induced LTP (Kang and Schuman 1995). BDNF, via activation of the receptor tyrosin kinase TrkB, is critically involved in hippocampal LTP. Interestingly, both TrkB amd BDNF knockout mice showed impaired hippocampal LTP and deficits in hippocampal-dependent learning.

4.24.6
Neurological Disorders and Neurodegenerative Diseases

Parkinson's disease
Postmortem analyses of Parkinson's disease (PD)-diseased human substantia nigra have revealed a dramatic reduction in both NGF and BDNF mRNA and protein, raising the possibility that there may be a link between reduced levels of neurotrophins and Parkinson's disease. Recently, it has been reported that, in a Japanese population, homozygosity for a polymorphism of the BDNF gene occurs more frequently in patients with PD than in unaffected controls and, in addition, two single nucleotide polymorphisms at position C270T of the BDNF gene have been identified in patients with familial Parkinson, suggesting that BDNF may play a role in the development of Parkinson's disease.

Alzheimer' disease
After NGF, pro-NGF is the predominant form of NGF in mouse, rat and human brain tissues. Not only is NGF protein increased in Alzheimer diseased brains, but also pro-NGF is increased in Alzheimer's diseased parietal cortex, indicating that it is the precursor form, pro-NGF, that accumulates in Alzheimer's disease. This increase may reflect either a role for biologically active pro-NGF or posttranslational disturbances in NGF biosynthesis that decrease the processing of pro-NGF to mature NGF in Alzheimer's disease.

In addition, the neurotrophins act on neurons affected by other neurological and psychiatric pathologies, including ischemia, epilepsy, or depression.

Depression
Concerning depression, it is thought that a lack of trophic support can contribute to depression. It has been demonstrated in rodents, for example, that exposure to stress decreases BDNF levels in brain regions associated with depression. Interestingly, antidepressant treatment produces the opposite effect and blocks the effects of stress on BDNF. The increased neurotrophin signaling stimulated by the application of antidepressants may induce the formation and

stabilization of synaptic connectivity, which gradually leads to antidepressive effects and mood recovery (Saarelainen et al. 2003).

Epilepsy

Seizure activity increases the expression of BDNF mRNA and protein; and recent studies have shown that BDNF-mediated signal transduction inhibits the development of the epileptic state *in vivo*. Along that line, amygdala stimulation of kindled animals produces a great increase in hippocampal and cortical BDNF mRNA levels and – in addition – it has been shown that epileptogenic stimuli induce a dramatic accumulation of BDNF mRNA and protein in the dendrites of hippocampal neurons *in vivo*, suggesting that BDNF contributes to epileptogenesis.

Further Reading

Altar, C. A., Criden, M. R., Lindsay, R. M., DiStefano, P. S. (1993): Characterization and topography of high-affinity 125I-neurotrophin-3 binding to mammalian brain. *J. Neurosci.* **13**: 733–743.

Barbacid, M. (1994): The Trk family of neurotrophin receptors. *J. Neurobiol.* **25**: 1386–1403.

Barde, Y. A. (2004): Death of injured neurons caused by the precursor of nerve growth factor. *Proc. Natl Acad. Sci. USA* **101**: 5703–5704.

Castren, E. (2004): Neurotrophic effects of antidepressant drugs. *Curr. Opin. Pharmacol.* **4**: 58–64.

Dechant, G., Neumann, H. (2002): Neurotrophins. *Adv. Exp. Med. Biol.* **513**: 303–334.

Fahnestock, M., Yu, G., Coughlin, M. D. (2004): ProNGF: a neurotrophic or an apoptotic molecule? *Prog. Brain Res.* **146**: 107–110.

Friedman, W. J., Black, I. B., Kaplan, D. R. (1998): Distribution of the neurotrophins brain-derived neurotrophic factor, neurotrophin-3, and neurotrophin-4/5 in the postnatal rat brain: an immunocytochemical study. *Neuroscience* **84**: 101–114.

Friedman, W. J., Greene L. A. (1999): Neurotrophin signaling via Trks and p75. *Exp. Cell Res.* **253**: 131–142.

Gomez-Pinilla, F., Cotman, C. W., Nieto-Sampedro, M. R. (1987): NGF receptor immunoreactivity in rat brain: topographic distribution and response to entorhinal ablation. *Neurosci. Lett.* **82**: 260–266.

Hofer, M., Pagliusi, S. R., Hohn, A., Leibrock, J., Barde, Y. A. (1990): Regional distribution of brain-derived neurotrophic factor mRNA in the adult mouse brain. *EMBO J.* **9**: 2459–2464.

Huang, E. J., Reichardt, L. F. (2003): Trk receptors: roles in neuronal signal transduction. *Annu. Rev. Biochem.* **72**: 609–642.

Kang, H., Schuman, E. M. (1995): Long-lasting neurotrophin-induced enhancement of synaptic transmission in the adult hippocampus. *Science* **267**: 1658–1662.

Katoh-Semba, R., Ichisaka, S., Hata, Y., Tsumoto, T., Eguchi, K., Miyazaki, N., Matsuda, M., Takeuchi, I. K., Kato, K. (2003): NT-4 protein is localized in neuronal cells in the brain stem as well as the dorsal root ganglion of embryonic and adult rats. *J. Neurochem.* **86**: 660–668.

Minichiello, L., Korte, M., Wolfer, D., Kuhn, R., Unsicker, K., Cestari, V., Rossi-Arnaud, C., Lipp, H. P., Bonhoeffer, T., Klein, R. (1999): Essential role for TrkB receptors in hippocampus-mediated learning. *Neuron* **24**: 401–414.

Momose, Y., Murata, M., Kobayashi, K., Tachikawa, M., Nakabayashi, Y., Kanazawa, I., Toda, T. (2002): Association studies of multiple candidate genes for Parkinson's disease using single nucleotide polymorphisms. *Ann. Neurol.* **51**: 133–136.

Nagatsu, T., Mogi, M., Ichinose, H., Togari, A. (2000): Changes in cytokines and neurotrophins in Parkinson's disease. *J. Neural Transm. Suppl.* **60**: 277–290.

Saarelainen, T., Hendolin, P., Lucas, G., Koponen, E., Sairanen, M., MacDonald, E., Agerman, K., Haapasalo, A., Nawa, H., Aloyz, R., Ernfors, P., Castren, E. (2003): Activation of the TrkB neurotrophin receptor is induced by antidepressant drugs and is required for antidepressant-induced behavioral effects. *J. Neurosci.* **23**: 349–357.

Schinder, A. F., Poo, M. (2000): The neurotrophin hypothesis for synaptic plasticity. *Trends Neurosci.* **23**: 639–645.

Shelton, D. L., Reichardt, L. F. (1986): Studies on the expression of the beta nerve growth factor (NGF) gene in the central nervous system: level and regional distribution of NGF mRNA suggest that NGF functions as a trophic factor for several distinct populations of neurons. *Proc. Natl Acad. Sci. USA* **83**: 2714–2718.

Simonato, M., Molteni, R., Bregola, G., Muzzolini, A., Piffanelli, M., Beani, L., Racagni G., Riva M. (1998). Different patterns of induction of FGF-2, FGF-1 and BDNF mRNAs during kindling epileptogenesis in the rat. *Eur. J. Neurosci.* **10**: 955–963.

Tongiorgi, E., Armellin, M., Giulianini, P., Bregola, G., Zucchini, S., Paradiso, B., Steward, O., Cattaneo, A., Simonato, M. (2004): Brain-derived neurotrophic factor mRNA and protein are targeted to discrete dendritic laminas by events that trigger epileptogenesis. *J. Neurosci.* **24**: 6842–6852.

Vaidya, V. A., Duman, R. S. (2001): Depression – emerging insights from neurobiology. *Br. Med. Bull.* **57**: 61–79.

4.25
Nitric Oxide and Carbon Monoxide

4.25.1
General Aspects and History

Nitric oxide (NO) was the first gaseous moiety to be identified as a messenger molecule with highly unorthodox properties. The major sources of NO are endothelial cells, macrophages and neurons. In contrast to neurotransmitters and classic neuromodulators, nitric oxide – because of its gaseous nature – cannot be stored in vesicles. Conventional neurotransmitters are released by exocytosis from the presynaptic site, whereas nitric oxide is synthesized by nitric oxide synthase (NOS) and simply diffuses from the nerve terminals into the surrounding tissue.

Conventional neurotransmitters and neuromodulators undergo reversible interactions with cell surface receptors; and their lifetime is terminated by presynaptic re-uptake or by enzymatic degradation. Nitric oxide does not need such mechanisms. It reaches its targets simply by diffusion; and a specific re-uptake system seems not to exist.

Nitric oxide is a paracrine messenger, which was first described in the vascular system by Furchgott and Zawadski in 1980. In blood vessels, NO plays a crucial role. The well known vasodilatatory properties of acetylcholine and bradykinin are coupled to the release of NO, which in smooth muscle cells triggers an increase in the intracellular concentration of cyclic guanosine monophosphate (cGMP). This increase in cGMP is essential for the mediation of muscle relaxation. In addition, NO mediates some of the bactericidal and tumoricidal effects of macrophages.

The idea that the gaseous substance NO plays a role as an intracellular second messenger in the central nervous system was developed in the late 1980s. Garthwaite and coworkers (1989) demonstrated that nitric oxide as well as NMDA increases intraneuronal concentrations of cGMP. In addition, they showed that, in the presence of an inhibitor of nitric oxide synthase (NOS), NMDA fails to increase cGMP levels. Nitric oxide is not exclusively an intracellular second messenger, but is also a diffusable retrograde messenger. When synthesized in neurons, NO appears to modulate several physiological and pathophysiological processes. In addition to its involvement in various neuronal mechanisms, including regulation of cerebrovascular perfusion, modulation of wakefulness, mediation of nociception, olfaction food intake and drinking, NO seems also to contribute to mechanisms attributed to learning and memory.

Since NO lacks a vesicular storage site and is exclusively produced on demand by the activity of NOS, attempts to gain information on the distribution of NO in living systems have failed. However, by purification and cloning neuronal NOS (nNOS), it has been possible to generate monospecific antibodies against nNOS; and these have been extensively utilized as markers for NO-synthesizing cells.

In living cells, nitric oxide is produced in very low concentrations and has an extremely short lifetime, making direct measurements difficult. Physiological attempts to monitor local concentrations of nitric oxide *in vivo* or *in vitro* have been hampered by the lack of a suitable means to identify it directly.

To date, many issues related to nitric oxide metabolism remain unanswered because most common techniques for nitric oxide determination depend on indirect methods: i.e. the so-called Griess reaction, which utilizes a NADPH-diaphorase technique, and immunocytochemical approaches, which depend on NOS localization. Substances suitable for direct labeling of NO-producing cells, analogous for example to fura-2, which changes its fluorescent behavior with changes in Ca^{2+} concentrations, would be beneficial for further NO research. Two fluorochromes are regarded as promising candidates: 1,2-diaminoanthraquinone (DAQ) and 4,5-diaminofluorescein diacetate (DAF-2 DA). DAQ reacts specifically with NO by forming a fluorescent triazole. DAF-2 DA does not directly interact with nitric oxide, because the ester bonds of the dye must be hydrolyzed by intracellular esterase to generate DAF-2, which then reacts with NO to form the corresponding triazole ring (DAF-2 T). Suitable protocols for the application of these substances will ultimately further cell biological NO research.

Aside from NO, there is another small gaseous molecule that can act as a neuromodulator. This gaseous molecule is carbon monoxide (CO). Carbon monoxide is produced by heme oxygenase which cleaves the heme ring into CO and biliverdin, which is rapidly reduced to bilirubin. Since CO is a gaseous substance, the distribution of this molecule, like in case of NO, cannot be visualized directly. Since CO is produced by hemeoxygenase – the distribution of this enzyme can be examined.

4.25.2
Localization Within the Central Nervous System

Within the brain, three different forms of NOS (neuronal, endothelial and inducible) have been characterized.

The general distribution of NOS in the central nervous system can be demonstrated by using autoradiography, immunohistochemistry or *in situ* hybridization.

Immunohistochemical mapping of nNOS in the brain has revealed an association of this enzyme with specific neuronal populations and shows a remarkable coincidence of the catalytic activity of nNOS with NADPH diaphorase staining.

Thomas and Pearse (1964) were the first to show that NADPH diaphorase neurons possess the ability to stain dark blue in the presence of nitroblue tetrazolium (while NADH does not). These cells display a certain resistance to neurotoxic events.

The neuronal form of NOS has been detected mainly in neurons of the olfactory bulb, the pons and the thalamus (supraoptic nucleus). In addition, nNOS-expressing neurons have been found in the inferior and superior colliculus, the caudate-putamen, the hippocampus and the cerebellum.

Additionally, the endothelial isoform is present in some hippocampal neurons, while the inducible NOS is present in some glial cells within the central nervous system.

Colocalization of NOS with different neurotransmitters and neuromodulators has frequently been described; however, a uniform pattern seems not to exist. Colocalization of NOS with NPY and GABA as well as colocalizations with other neuropeptides has been found. In the brain stem NOS colocalizes with cholineacetyltransferase, the key enzyme necessary for the generation of acetylcholine.

Concerning carbon monoxide, the localization of this neuroactive substance in the brain has also been determined indirectly, by mapping the distribution of the heme oxygenases (HO). At least two different forms of HOs are known to exist. The first isoform, the inducible HO isoform (HO-1 or HSP32) seems to be limited to discrete populations of pituitary cells, to cells located in the the hilus of the dentate gyrus, the hypothalamus, cerebellum and brainstem. The other isoform, the constitutive isoform (HO-2) has a more widespread distribution. Thus, it can be found in mitral cells in the olfactory bulb and in pyramidal cells of the cortex and hippocampus. Furthermore, HO-2 has been detected in granule cells of the dentate gyrus as well as in many neurons of the thalamus, hypothalamus, cerebellum and caudal brainstem.

4.25.3
Biosynthesis and Degradation

The constitutively expressed neuronal and endothelial NOS synthesize nitric oxide in very small amounts (pmol min^{-1} mg^{-1} homogenized protein). The constitutive NO synthases catalyze the oxidation of one molecule of L-arginine in at least two steps, by forming N-ω-hydroxy-L-arginine as an intermediate metabolite which then serves as the substrate to deliver one molecule of citrulline and one molecule of nitric oxide (Fig. 4.41).

Nitric oxide production in neurons is triggered by activation of NMDA receptors. This activation augments intracellular calcium concentrations which, in response to the rise of cytosolic calcium, leads to a transient increase in NO production (Fig. 4.42).

Nitric oxide serves as a second messenger with very specific properties of a gaseous substance, exhibiting no interaction with a membrane-bound receptor, and no specific re-uptake or degradation system seems to exist. Instead, it is assumed that degradation results from chemical interactions between nitric oxide and oxygen or superoxide anions which reduce the half-life of NO to a few seconds.

Nitric oxide is a chemically highly reactive substance and can rapidly form reaction products, e.g. peroxynitrite (OONO$^-$). These reaction products have strong neurotoxic properties.

Carbon monoxide is produced by heme oxygenase, which cleaves the heme ring into CO and biliverdin, which is rapidly reduced to bilirubin.

There are different isoforms of heme oxygenases. HO-1 is an inducible isoform and HO-2 is a constitutive isoform. These isozymes are encoded by different genes and are related only by their catalytic activities and regions of homologous primary sequences. In addition, a third isoform (HO-3) has been discovered using rat brain tissue. The predicted amino acid structure of HO-3 seems to be closely related to HO-2.

The heme oxygenase and nitric oxide synthase systems display some similarities, since both HO and NOS are both oxidative enzymes using NADPH as an electron donor. However, the constitutive forms of these enzymes are differen-

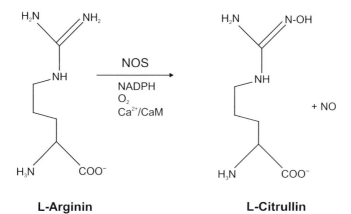

Fig. 4.41 Production of NO through constitutive NO synthase (NOS). One molecule of L-arginine is oxidized in at least two steps to L-citrullin and one molecule of NO.

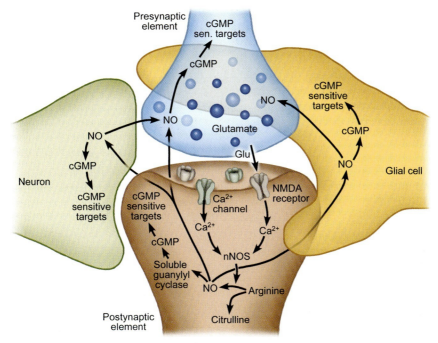

Fig. 4.42 Sketch of a nitric oxide containing neuron (nNOS) with the different neuronal and glial pathways of NO synthesis. Formation of NO follows activation of neuronal NO synthase by receptor- or ion channel-mediated increase of Ca^{2+}. NO freely diffuses across the postsynaptic membrane to the presynaptic neuron (as indicated here in the case of a glutamatergic neuron) serving a retrograde messenger function or diffusing to neighboring cells which are not directly apposed to the nNOS containing neuron (adapted from Byrne and Roberts 2004).

tially activated, since calcium entry activates protein kinase C to phosphorylate and activate HO-2, whereas calcium entry stimulates NOS by binding to calmodulin. Moreover, unlike NOS, HO utilizes a separate cytochrome P_{450} reductase protein to donate electrons for its oxidative cleavage of the heme ring and, in contrast to the multiple intracellular localizations of nNOS, HO-2 is thought to be localized exclusively to the endoplasmic reticulum.

4.25.4
Receptors and Signal Transduction

As already indicated, NO is not released by exocytosis like a classic neurotransmitter, but diffuses from the nerve terminals into the extracellular compartment and penetrates cells in the close neighborhood. Evidently, NO does not bind to specific receptors on the cell membrane, but binds with high affinity to specific intracellular enzymes. In the central nervous system, the main target for NO seems to be the heme moiety of guanylate cyclase. Nitric oxide is able to alter the conformation of this enzyme which is then activated and increases the level of cyclic GMP (cGMP), followed by activation of cGMP-dependent protein kinase activity (PKG). However, the role of the NO second messenger system in neurotransmission remains enigmatic.

Nitric oxide increases the production of prostaglandin E2 by activation of cyclo-oxygenase (an enzyme which also contains a heme group). In addition, numerous enzymes which contain non-heme iron are also regulated via NO. These enzymes include: cis-aconitase, NADH, NADH ubiquinone oxireductase and succinate oxireductase (Table 4.4).

A further chemical feature of NO is its capacity to nitrate or nitrosylate a variety of proteins and thereby alter their biological functions. In this way, nitric oxide can interact with the NMDA receptor. This interaction inactivates the NMDA receptor and down-regulates glutamatergic neurotransmission.

An important aspect with respect to the physicochemical property of NO is its diffusion coefficient, which allows the gas to diffuse over a distance of nearly 100 μm within 5 min, with an estimated radius of activity of about 300–350 μm.

Table 4.4 Nitric oxide is involved in various physiological processes: some potential cellular targets of nitric oxide have been identified.

Heme-carrying proteins	Cyclooxygenase 1 and 2
	Cytochrome P-450
	Guanylate cyclase
	Hemoglobin and myoglobin
	NO synthetase
Proteins with iron–sulfur clusters	Cis-aconitase (IRE-BP)
	CREB
	Ferritin and transferrin
	Mitochondrial aconitase
	Mitochondrial complex I
	Mitochondrial complex II
	Ribonucleotide reductase
Reaction with thiol compounds	Adenylatate cyclase
	Albumin
	GAP-43
	GAPDH
	Glutathione
	G Protein/p21^{ras}
	NADPH oxidase
	NMDA receptors
	Plasminogen activator
	Protein kinase C
	SNAP-25
Reaction with free radicals	NO
	O_2

For this reason, NO is considered to be an ideal endogenous substance for non-synaptic interactions and for the long-distance transport of information, extending the locally restricted effects of neurotransmitters like glutamate.

Like NO, CO is not released by exocytosis like classic neurotransmitters or neuromodulators. After generation, CO diffuses and penetrates neighboring cells. The main target for CO seems to be the heme moiety of guanylate cyclase to activate cGMP. Aside from this, there seems to exist alternative signaling pathways for CO in the brain, such as the activation of cyclooxygenase.

4.25.5
Biological Functions

Nitric oxide
Experimental data indicate that NO regulates cerebral blood flow through the activity-dependent activation of nNOS. Through activation of cGMP-dependent protein phosphorylation cascades, NO may also regulate transmitter and neuromodulator release.

The diffusion properties and the short half-life of nitric oxide make this substance a good candidate for playing a role in neurogenesis and synaptic plasticity.

Nitric oxide is suggested to exhibit a regulatory function in developmental processes of the central nervous system, since significant increases in the concentration of NO during critical periods of neuronal development have been shown; for instance, during the maturation of spinal cord motor neurons and the initial patterning of connections in the developing visual system.

Long-term potentiation (LTP) in the hippocampus and the cerebral cortex as well as long-term depression (LTD) in the cerebellum are sensitive to inhibitors of NOS, whereas NO precursors can stimulate the expression of LTP. Arancio and coworkers (1996) showed that nitric oxide constitutes a retrograde messenger which is produced in the postsynaptic neuron, from which it diffuses through the extracellular space and operates in the presynaptic neuron to sustain LTP production.

Indirect evidence for a function for NO during LTP generation stems from behavioral studies, which showed that chronic systemic application of nitric oxide synthase inhibitors can induce deficits in the acquisition of spatial memory, a complex higher-order function of the hippocampus and the cerebellum.

In contrast to spatial memory, other forms of memory seem to be independent of nitric oxide. Therefore, two forms of LTP have been postulated: a nitric oxide-dependent and a nitric oxide-independent form.

Besides its regulatory properties, nitric oxide exhibits neuroprotective and neurotoxic actions. In the neostriatum, for example, cells containing NOS were found to be resistant to glutamate-induced neurotoxicity. Apparently, the effect of NO depends on its actual concentration in the tissue.

Under normal conditions, NOS-carrying neurons are resistant to the neurotoxic effects of glutamate. Excessive glutamatergic receptor stimulation and subsequent increase in the intracellular calcium concentrations are thought to initiate most forms of glutamate neurotoxicity. During ischemia, high concentrations of glutamate are liberated from neuronal sources, which causes an increase in the intracellular calcium, accompanied by an excessive increase in NO which concurrently can display neurotoxicity. Support for the neurotoxicity theory derives from tissue culture experiments and recent data on mice with ablated nNOS gene. Administration of NMDA to neuronal cell cultures for a short time (12–24 h) promptly elicits cell death. This toxicity depends to a high degree on calcium and can be blocked by several types of NOS inhibitors. In contrast, neuronal cell cultures from nNOS knockout mice are resistant to this NMDA-mediated form of neurotoxicity.

Not only can nitric oxide be neurotoxic, but also the reaction products of nitric oxide with other substances. It is known that nitric oxide contributes to the extracellular potassium ion concentration ($[K^+]_o$)-induced hydroxyl radical (*OH) generation; and cytotoxic free radicals such as peroxinitrite ($ONOO^-$) and *OH may also be implicated in NO-mediated cell injury.

Carbon monoxide
The function of CO in the central nervous system is less clear than that of NO. Carbon monoxide is generated by heme oxygenase, an enzyme that degrades heme in red blood cells. Thus, it is not surprising that heme oxygenase activity was first discovered in the spleen, which represents the prime organ for erythrocyte turnover. The best experimental evidence for a role of CO in neurotransmission comes from studies in the peripheral nervous system, particularly in the enteric nervous system. HO-2 is confined to neurons of the myenteric plexus in the gut and colocalizes with nNOS in about 25 to 50%. HO-2 also occurs in the so-called interstitial cells of Cajal (ICC). HO-2 knockout mice exhibit depolarized jejunal smooth muscle cells, which are innervated by ICC cells. The current concept is that HO-2 is responsible for setting the resting membrane potential of smooth muscle cells in the jejunum.

In the central nervous system, it has been shown that HO-2 knockout mice have normal levels of cGMP, with the exception of the olfactory bulb and cortex. The concept that CO plays a role in olfaction is strengthened by the finding that HO inhibitors largely reduce basal cGMP levels and block the generation of cGMP in response to odorants in cultured olfactory neurons. Since olfactory neurons lack NOS activity, it is thought that CO is responsible for the generation of cGMP through the activation of soluble guanylyl cyclase.

Several studies also show that CO, at the hypothalamic level, plays an important role in the modulation of the stress response, because CO is able to inhibit the release of antiinflammatory neuropeptides, such as corticotropin-releasing hormone and arginine–vasopressin. Moreover, CO has been found to increase body temperature in rodents exposed to psychological stressors (stress fever).

Furthermore, CO – like NO – seems to play a role in LTP generation. Inhibitors of either NOS synthase or HO are able to block the induction of LTP in the CA1 region of the hippocampus. Brief application of either NO or CO to slices produces a rapid and long-lasting increase in the size of synaptic potentials if the application occurs concurrently with weak tetanic stimulation of the presynaptic fibers. A feasible explanation of this phenomenon is that both NO and CO, either alone or in combination, serve as retrograde messengers that produce activity-dependent presynaptic enhancement, perhaps by stimulating soluble guanylyl cyclase and cGMP-dependent protein kinase, during LTP in the hippocampus (Arancio et al. 1996).

However, hippocampal LTP is normal in heme oxygenase-2 mutant mice (Poss et al. 1995). A plausible explanation is that heme oxygenase-1 or -3 in the brain might compensate for the loss of HO-2. In contrast, in a behavioral study, it was found that NO, but not CO, seems to be involved in spatial learning (a learning task that depends on the hippocampus) of mice.

4.25.6
Neurological Disorders and Neurodegenerative Diseases

Nitric oxide
Although the question of NO neurotoxicity versus neuroprotection and NMDA receptor activation is still a debatable issue, some evidence suggests that this gas may play a role in the pathogenesis of neurodegenerative diseases.

Excessive generation of nitric oxide and its toxic metabolite peroxynitrite ($ONOO^-$) can inhibit the mitochondrial respiratory chain, leading to energy failure and ultimately cell death. Moreover, evidence suggests that NO/$ONOO^-$ causes the release of neuronal glutamate, which leads to glutamate-induced activation of neuronal NO synthase and the generation of further damage by reactive oxygen species. Such forms of oxidative stress contribute to the cascade leading to dopamine cell degeneration in Parkinson's disease.

Mitochondrial dysfunction, apoptosis and overproduction of reactive oxygen species is a final common pathogenic mechanism not only in Parkinson's disease but also in other neurodegenerative disease, such as Alzheimer's disease and amyotrophic lateral sclerosis.

Ischemia leads to increased activity of nitric oxide synthase. Overproduction of nitric oxide in ischemia may represent an effort to reestablish normal blood flow. However, the overproduction of NO can induce reactions of NO with cellular proteins that can subsequently induce cell death and thereby contribute to cerebral damage, as observed under ischemic conditions.

Carbon monoxide
There are some indications that CO, via heme oxygenases, might be affected in several neurodegenerative diseases.

For example HO-1 immunoreactivity is augmented in neurons and astrocytes of the hippocampus and cerebral cortex in Alzheimer's disease, and colocalizes to senile plaques and neurofibrillary tangles.

In Parkinson's disease, HO-1 was found to colocalize to Lewy bodies of affected dopaminergic neurons and was found to be overexpressed in astrocytes residing within the substantia nigra.

Further Reading

Arancio, O., Kiebler, M., Lee, C. J., Lev-Ram, V, Tsien, R. Y., Kandel, E. R., Hawkins, R.D. (1996): Nitric oxide acts directly in the presynaptic neuron to produce long-term potentiation in cultured hippocampal neurons. *Cell* **87**: 1025–1035.

Boehning, D., Snyder, S. H. (2003): Novel neural modulators. *Annu. Rev. Neurosci.* **26**: 105–131.

Baranano, D. E., Snyder, S. H. (2001): Neural roles for heme oxygenase: contrasts to nitric oxide synthase. *Proc. Natl Acad. Sci. USA* **98**:10996–101002.

Böhme, G. A., Bon, C., Stutzmann, J. M., Doble, A., Blanchard, J. C. (1991): Possible involvement of nitric oxide in long-term potentiation. *Eur. J. Pharmacol.* **199**: 379–381.

Bredt, D. S., Snyder, H. (1992): Nitric oxide, a novel neuronal messenger. *Neuron* **8**: 3–11.

Crépel, F., Audinat, E., Daniel, H., Hémart, N., Jaillard, D., Rossier, J., Lambolez, B. (1994): Cellular locus of the nitric oxide-synthetase involved in cerebellar long-term depression induced by high external potassium concentrations. *Neuropharmacology* **33**: 1399–1405.

Dawson, T. M., Snyder, S. H. (1994): Gases as biologic messengers: nitric oxide and carbon monoxide in the brain. *J. Neurosci.* **14**: 5147–5159.

Garthwaite, J. (1991): Glutamate, nitric oxide and cell–cell signalling in the nervous system. *Trends Neurosci.* **14**: 60–67.

Garthwaite, J., Boulton, C. L. (1995): Nitric oxide signaling in the central nervous system. *Ann. Rev. Physiol.* **57**: 683–706.

Garthwaite, J., Garthwaite, G., Palmer, R. M., Moncada, S. (1989): NMDA receptor activation induces nitric oxide synthesis from arginine in rat brain slices. *Eur. J. Pharmacol.* **172**: 413–416.

Haley, J. E. (1998): Gases as neurotransmitters. *Essays Biochem.* **33**: 79–91.

Heiduschka, P., Thanos, S. (1998): NO production during neuronal cell death can be directly assessed by a chemical reaction *in vivo*. *NeuroReport* **9**: 4051–4057.

Jenner, P. (2003): Oxidative stress in Parkinson's disease. *Ann. Neurol.* **53**[Suppl 3]: S26–S36.

Kojima, H., Nakatsubo, N., Kikuchi, K., Urano, Y., Higuchi, T., Tanaka, J., Kudo, Y., Nagano, T. (1998): Direct evidence of NO production in rat hippocampus and cortex using a new fluorescent indicator: DAF-2 DA. *NeuroReport* **9**: 3345–3348.

Luo, D., Vincent, S. R. (1994): NMDA-dependent nitric oxide release in the hippocampus *in vivo*: interactions with norepinephrine. *Neuropharmacology* **33**: 1345–1350.

Mancuso, C. (2004): Heme oxygenase and its products in the nervous system. *Antioxid. Redox. Signal.* **6**: 878–887.

Marletta, M. A. (1994): Nitric oxide synthase: aspects concerning structure and catalysis. *Cell* **78**: 927–930.

McCoubrey, W. K. Jr, Huang, T. J., Maines, M. D. (1997): Isolation and characterization of a cDNA from the rat brain that encodes hemoprotein heme oxygenase-3. *Eur. J. Biochem.* **247**: 725–732.

Murphy, K. P. S. J., Williams, J. H., Bettache, N., Bliss, T. V. P. (1994): Photolytic release of nitric oxide modulates NMDA receptor-mediated transmission but does not induce long-term potentiation at hippocampal synapses. *Neuropharmacology* **33**: 1375–1385.

Obata, T. (2002): Nitric oxide and depolarization induce hydroxyl radical generation. *Jpn J. Pharmacol.* **88**: 1–5.

Poss, K. D., Thomas, M. J., Ebralidze, A. K., O'Dell, T. J., Tonegawa, S. (1995): Hippocampal long-term potentiation is normal in heme oxygenase-2 mutant mice. *Neuron* **15**: 867–873.

Rodrigo, J., Fernandez, A. P., Serrano, J., Peinado, M. A., Martinez, A. (2005): The role of free radicals in cerebral hypoxia and ischemia. *Free Radic. Biol. Med.* **39**:26–50.

Schipper, H. M. (2004): Heme oxygenase expression in human central nervous system disorders. *Free Radic. Biol. Med.* **37**: 1995–2011.

Schuchmann, S., Albrecht, D., Heinemann, U., von Bohlen und Halbach, O. (2002): Nitric oxide modulates low-Mg^{2+}-induced epileptiform activity in rat hippocampal–entorhinal cortex slices. *Neurobiol. Dis.* **11**: 96–105.

Stamler, J. S. (1994): Redox signaling: nitrosylation and related target interaction of nitric oxide. *Cell* **78**: 931–936.

Stewart, V. C., Heales, S. J. (2003): Nitric oxide-induced mitochondrial dysfunction: implications for neurodegeneration. *Free Radic. Biol. Med.* **34**: 287–303.

Thomas, E., Pearse, A. G. E. (1964): The solitary active cells. Histochemical demonstration of damage-resistant nerve cells with a TPH-diaphorase reaction. *Acta Neuropathol.* **3**: 238–249.

Tieu, K., Ischiropoulos, H., Przedborski, S. (2003): Nitric oxide and reactive oxygen species in Parkinson's disease. *Mol. Neurobiol.* **27**: 325–355.

Toyoda, M., Saito, H., Matsuki, N. (1996): Nitric oxide but not carbon monoxide is involved in spatial learning of mice. *Jpn J. Pharmacol.* **71**: 205–211.

Vizi, E. S. (2000): Role of high-affinity receptors and membrane transporters in nonsynaptic communication and drug action in the central nervous system. *Pharmacol. Rev.* **52**: 63–89.

Vincent, S. R., Hope, B. T. (1991): Neurons that say NO. *Trends Neurosci.* **15**: 108–113.

Vincent, S. R., Das, S., Maines, M. D. (1994): Brain heme oxygenase isoenzymes and nitric oxide synthase are co-localized in select neurons. *Neuroscience* **63**: 223–231.

von Bohlen und Halbach, O. (2003): Nitric oxide imaging in living neuronal tissues using fluorescent probes. *Nitric Oxide* **9**: 217–228.

4.26
Nociceptin (Orphanin FQ)

4.26.1
General Aspects and History

The heptadecapeptide nociceptin, also known as orphanin FQ, is a neuropeptide recently discovered by deorphanizing GPCRs via expression cloning (see Section 1.2).

By extending the screening of genomic and cDNA libraries, perhaps in an effort to identify putative subtypes of the classic opioid receptors, a novel receptor was isolated which showed a significant degree of homology with the classic opioid receptors. Mollereau and coworkers (1994) succeeded in cloning the hORL1 gene, which was found to encode for a putative protein of 370 amino acids. Although the ORL1 receptor is generally accepted as a member of the opioid receptor family on the basis of its structural homology, it lacks any pharmacological similarities to the other members. Even non-selective ligands that exhibit uniformly high affinity towards μ, κ and δ receptors have a very low affinity for the ORL-1 receptor. For this reason, as much as for the initial lack of an endogenous ligand, the receptor was called "orphan opioid receptor". The ORL-1 receptor was identified in rodents and man, where it shows a high degree of interspecies homology (>90%).

An endogenous ligand for this receptor was discovered independently by the groups of Meunier and Reinscheid in 1995. Both groups reported the same heptadecapeptidergic sequence of the ligand. Each group coined its own term for this new ligand: "nociceptin" (Meunier and coworkers) and "orphanin FQ" (Renscheid and coworkers 1995).

An interesting feature of the amino acid sequence of this peptide is that it shares consensus sequences with other endogenous ligands of opioidergic receptors (Fig. 4.43). Nociceptin has structural homologies to dynorphin A. Thus,

Nociceptin   Phe-*Gly-Gly-Phe*-Thr-Gly-Ala-Arg-Lys-Ser-Ala-Arg-*Lys*-Leu-Ala-*Asn-Gln*

Dynorphin A (1-17)   Tyr-*Gly-Gly-Phe*-Leu-Arg-Arg-Ile-Arg-Pro-Lys-Leu-*Lys*-Trp-Asp-*Asn-Gln*

Met-enkephalin   Tyr-*Gly-Gly-Phe*-Met

Fig. 4.43 Structure of nociceptin/orphanin FQ in comparison to the ligands of other opioid receptors (dynorphin, Met-enkephalin).

both peptides are composed of 17 amino acids bounded by pairs of basic amino acids important in their production from their precursors. Furthermore, both have internal pairs of basic amino acids, raising the possibility of further processing. Despite these similarities, the peptides are functionally quite distinct; and nociceptin has no high affinity with any of the opioid receptors.

4.26.2
Localization Within the Central Nervous System

Nociceptin-containing neurons have been identified by immunohistochemistry or *in situ* hybridization using probes for nociceptin or for its precursor pronociceptin.

Nociceptin-positive neurons are mainly located in the cortex, the anterior olfactory nucleus, the amygdala, the bed nucleus of the stria terminalis, the hippocampus, the periaqueductal gray, the hypothalamus (especially in the arcuate nucleus and in the ventromedial hypothalamus), the nucleus vestibularis and in various raphe nuclei (raphe magnus, medial and dorsal raphe) as well as in the nucleus of the tractus solitarius and the eminentia mediana.

4.26.3
Biosynthesis and Degradation

Detailed knowledge about the synthesis and the degradation process of nociceptin is still poor. Nociceptin, as is the case for most neuropeptides, derives from a high-molecular-weight precursor. The gene which encodes for the nociceptin precursor ("pro-nociceptin") is the PNOC gene. The locus of the human gene is 8p21.

The nociceptin precursor shares several properties with the precursors of other opioidergic neuropeptides; and it has been suggested that they all share a common evolutionary origin. For example, the nociceptin precursor shares 27% similarity with the dynorphin precursor, 25% with the enkaphalin precursor and 13% with proopiomelanocortin (POMC). The nociceptin residue occupies amino acids 135–151 in the precursor sequence. A pair of basic amino acids flanks the nociceptin sequence. These pairs of basic amino acids provide the recognition sites for different endopeptidases. Since additional restriction sites for endopeptidase have been identified in the nociceptin precursor, it seems possible that additional biologically active peptides can be generated from this precursor.

Downstream of the nociceptin sequence, there is a stretch of 28 amino acids which is completely conserved in several mammalian species. Neither this 28-amino-acid peptide nor two other shorter peptides, generated from the precursor, bind to the nociceptin receptor. The high degree of conservation of the sequences among several species argues strongly for biological significance.

The nociceptin sequence contains pairs of basic amino acids that might imply additional processing of the peptide to nociceptin(1–11) and/or nociceptin(1–7). Both of these truncated peptides are functionally active when administered *in vivo* (Rossi et al. 1997).

Precise data concerning the degradation of nociceptin are still missing. It seems likely that nociceptin is degraded by the activity of aminopeptidase N.

4.26.4
Receptors and Signal Transduction

The nomenclature of the nociceptin receptor varies according to the species from which the receptor is cloned. While the human receptor was designated ORL-1, the mouse ortholog was termed MOR-C and the rat ROR-C, oprl, LC132, XOR1 Hyp 8-1 or C3, respectively. The receptor is now named NOP_1. It is a typical G protein-coupled receptor with seven predicted transmembrane domains (Fig. 4.44) and is localized to murine chromosome 2. Like other members of the opioid receptor family, NOP_1 undergoes alternative splicing. Three NOP_1 variants (NOP_{1a}, NOP_{1b}, and NOP_{1c}) have been described that contain mini-exons located between the first and second coding exons (exons 2 and 5).

The overall amino acid similarity between the nociceptin receptor and the other opiate receptors from various species is about 65%. However, the sequence homology is more extensive in certain domains of the receptor. For example, the transmembrane domains collectively display about 80% similarity. In

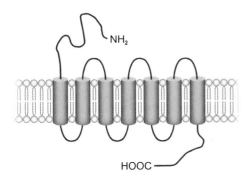

Fig. 4.44 Schematic drawing of the NOP_1 receptor.

contrast, the domains of lowest homology occur at the second and third extracellular loops and at the amino- and carboxy-terminals.

As already indicated, according to the structural homology, the receptor is accepted as a member of the family of opioid receptors, in spite of the lack of any pharmacological similarity.

The distribution of nociceptin receptors has been evaluated using *in situ* hybridization. The highest level of nociceptin receptor mRNA expression has been found in the paraventricular and ventromedial nuclei of the hypothalamus, the amygdala and the piriform cortex. Lower mRNA levels have been found in the dorsal raphe and in the locus coeruleus. Moreover, mRNA signals have been detected in the cortex, the hippocampus, the thalamus, the substantia nigra, the periaqueductal gray and in the spinal cord. In contrast to the presence of the other opiodergic receptors, no signal has been detected in the striatum.

In their study, Mathis and coworkers (1999) proposed the existence of a biologically active fragment of nociceptin (nociceptin (1–11) and evaluated the binding sites for this substance. Apparently, binding sites for nociceptin (1–11) are present at various locations within the central nervous system, but their distribution differs from that of nociceptin and other opioidergic receptors.

4.26.5
Biological Effects

The first behavioral effect of nociceptin to be described was a hyperalgesic response after intraventricular injection. In fact, this behavioral effect was the reason why Meunier and colleagues labeled this peptide "nociceptin".

Originally characterized as a nociception-producing peptide, nociceptin has been shown to have diverse effects on nociception, as well as effects on many other behaviors, such as locomotion, feeding, anxiety, spatial attention, reproductive behaviors and opiate tolerance. With regard to nociception, nociceptin has been reported to produce hyperalgesia, reversal of opioid-mediated analgesia, and allodynia.

However, there are also reports that nociceptin produces antinociceptive effects. For example, Xu and coworkers (1996) reported that intrathecal administration of nociceptin could produce dose-dependent depression of a spinal nociceptive flexor reflex in the rat and behavioral antinociception with no signs of sedation or motor impairment. Furthermore, Erb and coworkers (1997) observed a decrease in flinching behavior with intrathecally delivered nociceptin, from which the authors concluded that nociceptin possesses antinociceptive effect. Thus, nociceptin seems to have no influence on algesia and could even induce analgesia. The observed differences in nociceptin effects seem crucially to be related to the site of intracerebral application.

Moreover, the role of nociceptin as an antiopiate remains controversial. Initially, nociceptin was proposed to exhibit nociception-producing properties, rather than counteracting the opiate-mediated analgesic effect. Since nociceptin shows both properties in terms of an analgesic as well as an anti-analgesic substance, it is reasonable to conclude that it may well be an opiate-modulating rather than antiopiate substance.

The bivalent behavior of nociceptin has also been demonstrated for other antiopiates, like peptides of the Tyr-MIF family. Nociceptin seems to be involved in water regulation. Administration of nociceptin not only elicits a decrease in heart rate and blood pressure, but also results in dose-dependent diuresis and antinatriuresis. In the central nervous system, nociceptin can inhibit the release of glutamate and GABA from nerve terminals, depending on the cell type and the brain area. In contrast, the release of tachykinins from sensory neurons is enhanced by nociceptin. Nociceptin also shows effects on feeding behavior which depend on the fasting state of the animal: intraventricular injection of nociceptin increases feeding in satiated rats during the first hour after injection.

Additionally, it has been shown that injections of nociceptin into the nucleus accumbens or the ventromedial hypothalamus produce a significant increase in food intake in satiated rats. Along this line, nociceptin has been found to act as a functional antagonist of corticotropin-releasing factor to inhibit its anorectic effect.

Approximately 98% of projection neurons in the lateral nucleus of the amygdala respond to nociceptin. Given the preeminent function of the amygdala in fear avoidance mechanisms, it has been suggested that nociceptin plays a role in the reduction of fear responsiveness and stress. In addition, based on the effects of nociceptin in several behavioral test batteries of anxiety, nociceptin has been proposed to have anxiolytic properties and/or to act like a general modulator of anxiety.

Learning and memory seem also to be affected by nociceptin, since central administration of nociceptin impairs learning. Microinjection of nociceptin into the hippocampal CA3 region of rats, for example, markedly impairs spatial learning without affecting swimming ability in the Morris water maze. Nociceptin can also impair the induction of LTP; and this effect upon LTP seems to be due to postsynaptic mechanisms, including the inhibition of dentate gyrus granule cells and inhibition of NMDA receptor-mediated currents (Yu and Xie 1998). Further along this line, nociceptin knockout mice display increased anxiety and impaired adaptation to repeated stress, while mice lacking the cognate receptor show improved spatial attention and memory.

4.26.6
Neurological Disorders and Neurodegenerative Diseases

The prospects for clinical application of agonists or antagonists for the ORL1 receptor can only be the subject of speculation at the moment. The development of non-peptide selective agonists, and perhaps more specific antagonists, is an essential prerequisite to any clinical trials in which the role of the NOP_1 receptor system in pain control can be explored.

Nociceptin and its receptor are expressed in the substantia nigra, a brain area involved in Parkinson's disease. Pharmacological blockade of the nociceptin system in the substantia nigra has been described to attenuated Parkinsonian-like symptoms in 6-hydroxydopamine (6-OHDA) hemilesioned rats.

Moreover, an enhancement of nociceptin expression and release was detected in the lesioned compared with the unlesioned substantia nigra in 6-OHDA-treated animals. In addition, mice with a deletion of the nociceptin gene showed a partial protection against MPTP-induced loss of nigral dopamine neurons. Based on these data, it is proposed that nociceptin receptor antagonists may represent a novel approach for asymptomatic and neuroprotective therapy of Parkinson's disease (Marti et al. 2005).

Further Reading

Berzeteigurske, I.P., Schwartz, R.W., Toll, L. (1996): Determination of activity for nociceptin in the mouse vas deferens. *Eur. J. Pharmacol.* **302**: R1–R2.

Ciccocioppo, R., Cippitelli, A., Economidou, D., Fedeli, A., Massi, M. (2004): Nociceptin/orphanin FQ acts as a functional antagonist of corticotropin-releasing factor to inhibit its anorectic effect. *Physiol. Behav.* **82**: 63–68.

Civelli, O. (1998): Functional genomics: the search for novel transmitters and neuropeptides. *FEBS Lett.* **430**: 55–58.

Darland, T., Heinricher, M.M., Grandy, D.K. (1998): Orphanin FQ/nociceptin: a role in pain and analgesia, but so much more. *Trends Neurosci.* **21**: 215–221.

Erb, K., Liebel, J.T., Tegeder, I., Zeilhofer, H.U., Brune, K., Geisslinger, G. (1997): Spinally delivered nociceptin/orphanin FQ reduces flinching behaviour in the rat formalin test. *NeuroReport* **8**: 1967–1970.

Giuliani, S., Maggi, C.A. (1996): Inhibition of tachykinin release from peripheral endings of sensory nerves by nociceptin, a novel opioid peptide. *Br. J. Pharmacol.* **118**: 1567–1569.

Harrison, L.M., Grandy, D.K. (2000): Opiate modulating properties of nociceptin/orphanin FQ. *Peptides* **21**: 151–172.

Henderson, G., McKnight, A.T. (1997): The orphan opioid receptor and its endogenous ligand – nociceptin/orphanin FQ. *Trends Pharmacol. Sci.* **18**: 293–300.

Marti, M., Mela, F., Fantin, M., Zucchini, S., Brown, J.M., Witta, J., Di Benedetto, M., Buzas, B., Reinscheid, R.K., Salvadori, S., Guerrini, R., Romualdi, P., Candeletti, S., Simonato, M., Cox, B.M., Morari, M. (2005): Blockade of nociceptin/orphanin FQ transmission attenuates symptoms and neurodegeneration associated with Parkinson's disease. *J. Neurosci.* **25**: 9591–6901.

Mathis, J.P., Goldberg, I.E., Letchworth, S.R., Ryan-Moro, J.P., Pasternak, G.W. (1999): Identification of a high-affinity orphanin FQ/nociceptin (1–11) binding site in mouse brain. *Synapse* **34**: 181–186.

Meis, S., Pape, H.C. (1998): Postsynaptic mechanisms underlying responsiveness of amygdaloid neurons to nociceptin/orphanin FQ. *J. Neurosci.* **18**: 8133–8144.

Mollereau, C. Parmentier, M., Mailleux, P., Butour, J.L., Moisand, C., Chalon, P., Caput, D., Vassart, G., Meunier, J.C. (1994): ORL1, a novel member of the opioid receptor family. Cloning, functional expression and localization. *FEBS Lett.* **34**: 33–38.

New, D.C., Wong, Y.H. (2002): The ORL1 receptor: molecular pharmacology and signalling mechanisms. *Neurosignals* **11**: 197–212.

Reinscheid, R.K., Nothacker, H.P., Bourson, A., Ardati, A., Henningsen, R.A., Bunzow, J.R., Grandy, D.K., Langen, H., Monsma, F.J. Jr, Civelli, O. (1995): Orphanin FQ: a neuropeptide that activates an opioidlike G protein-coupled receptor. *Science* **270**: 792–794.

Reinscheid, R.K., Civelli, O. (2002): The orphanin FQ/nociceptin knockout mouse: a behavioral model for stress responses. *Neuropeptides* **36**: 72–76.

Rossi, G.C., Leventhal, L., Bolan, E.A., Pasternak, G.W. (1997): Pharmacological characterization of orphanin FQ/nociceptin and its fragments. *J. Pharmacol. Exp. Ther.* **282**: 858–865.

Xie, G.X., Meuser, T., Pietruck, C., Sharma, M., Palmer, P.P. (1999): Presence of opioid receptor-like (ORL1) receptor mRNA splice variants in peripheral sensory and sympathetic neuronal ganglia. *Life Sci.* **64**: 2029–2037.

Xu, X.J., Hao, J.X., Wiesenfeld, H.Z. (1996): Nociceptin or antinociceptin: potent spinal antinociceptive effect of orphanin FQ/nociceptin in the rat. *NeuroReport* **7**: 2092–2094.

Yu, T.P., Xie, C.W. (1998): Orphanin FQ/nociceptin inhibits synaptic transmission and long-term potentiation in rat dentate gyrus through postsynaptic mechanisms. *J. Neurophysiol.* **80**: 1277–1284.

4.27
Pituitary Adenylate Cyclase-activating Polypeptide

4.27.1
General Aspects and History

Pituitary adenylate cyclase activating polypeptide (PACAP) is a hypothalamic peptide which activates the adenylate cyclase of the pituitary gland. PACAP was first discovered in the hypothalamus in 1989. Arimura and Shioda (1989) demonstrated that, in the pituitary gland, there are two substances which exhibit adenylate cyclase-activating properties. Both substances were isolated, purified and sequenced; and it turned out that one protein consisted of 38 amino acids (PACAP 38), while the second proved to be a 27-amino-acid (PACAP 27) fragment of the former.

PACAP shows high homology (68%) with VIP (vasoactive intestinal polypeptide) and thus belongs to a superfamily of structurally and functionally closely related proteins, named the vasoactive intestinal polypeptide (VIP)-glucagon-growth hormone releasing factor-secretin superfamily (Fig. 4.45). Additional members of this family include corticotropin-releasing hormone (CRH), growth hormone-releasing hormone (GHRH), secretin and glucagon.

The structure of PACAP has been highly conserved through evolution among tunicata, amphibians and mammals, suggesting that PACAP is involved in the regulation of important biological functions.

PACAP is abundantly expressed in the nervous and endocrine systems where it has diverse functions. In the nervous system, PACAP peptides not only serve roles in neuronal communication and signaling but, because of their neurotrophic properties, they also promote neuronal survival, mitosis, proliferation and differentiation. PACAP can therefore be regarded as a hypothalamic hormone with neurotransmitter, neuromodulator and neurotrophic abilities.

^{1}His-Ser-Asp-Gly-**Ile**-Phe-Thr-Asp-Ser-Tyr-Ser-Arg-Tyr-Arg-Lys-Gln-
Met-Ala-Val-Lys-Lys-Tyr-Leu-Ala-Ala-Val-Leu-^{27}Gly-Lys-Arg-Tyr-Lys-Gln-
Arg-Lys-Asn-^{38}Lys-NH$_2$

Fig. 4.45 Structure of PACAP 38. PACAP 27 is identical to PACAP 38 in the sequence of the first 27 amino acids. Those amino acids which are identical in PACAP 38, PACAP 27 and in vasoactive intestinal polypeptide (VIP) are indicated in gray.

4.27.2
Localization Within the Central Nervous System

In mammals, PACAP-containing neurons were found originally in the pituitary gland, but they have since been detected in the hypothalamus, the median eminence, the thalamus, amygdala, hippocampus, cerebellum and pons. Furthermore, PACAP-containing neurons have been identified in the brain of birds, amphibians and fishes.

4.27.3
Biosynthesis and Degradation

The structure of the human PACAP gene was determined in 1992 (Hosoya et al. 1992). In all sub-mammalian species investigated so far, PACAP and a GHRH-like peptide are located on the same precursor. The cDNA encoding for the common precursor have been characterized in birds, frogs and fishes. However, in mammals, the PACAP precursor is encoded by a separate gene. The PACAP gene has the gene locus 18p11 in humans; and the cDNA encodes a 176-amino-acid prepro-protein, which comprises a 24-amino-acid signal peptide. The sequence of PACAP is located in the carboxy-terminal domain of the precursor.

The PACAP precursor is not exclusive to PACAP, but carries an additional 29-amino-acid peptide, which is designated as "PACAP-related protein" or PRP.

4.27.4
Receptors and Signal Transduction

At least three receptors for PACAP exist in mammals; and two of them (VPAC1 and VPAC2) share receptor properties for VIP. In 1993, the cloning of the cDNA encoding the PAC1 receptor (previously known as the PACAP type I or PVR1 receptor) was reported by six independent groups (Hashimoto et al. 1993; Hosoya et al. 1993; Morrow et al. 1993; Pisegna and Wank 1993; Spengler et al. 1993; Svoboda et al. 1993). In contrast to VPAC1 and VPAC2 receptors, which recognize VIP and PACAP with identical affinity, the PAC1 receptor is selective for PACAP. This receptor is expressed predominantly in the central nervous system.

The PACAP receptors are G protein-coupled and exhibit the seven transmembrane domain motif. They all belong to the VIP receptor family. At least eight subtypes of PAC1 receptors are generated through alternative splicing. Each subtype is coupled with specific signaling pathways; and its expression is tissue- and cell-specific. The PAC1 receptor is expressed predominantly in the central nervous system, where PACAP is thought to act as a neuropeptide, a neurotrophic factor and a neuroprotectant. The VPAC1 receptor is expressed in the central nervous system (most abundantly in the cerebral cortex and hippocampus) and in peripheral tissues, including liver, lung, intestine and T-lymphocytes. The VPAC2 receptor has been found at high levels in the thalamus and suprachiasmatic nucleus and at lower levels in the hippocampus, brain stem, spinal cord and sensory ganglia. Peripheral tissues which express VPAC2 receptors include pancreas, skeletal muscle, heart, kidney, adipose tissue, testis and stomach.

4.27.5
Biological Effects

The physiological spectrum of PACAP covers central nervous effects, which include secretory activity on pituitary, and peripheral effects, like enhanced release of insulin from the endocrine pancreas and enhanced catecholamine release. PACAP exerts a direct positive ionotropic action on heart and thus may serve an important role in cardiovascular control.

PACAP not only enhances the release of gonadotropins, but also modulates the GnRH-dependent release of LH. The major regulatory task of PACAP in pituitary cells appears to be the regulation of gene expression of pituitary hormones and/or regulatory proteins, which control growth and differentiation of pituitary cells. These effects appear to be mediated by direct and indirect paracrine or autocrine mechanism. One important endocrine effect of PACAP is as a potent secretagogue for epinephrine from the adrenal medulla. The stage-specific expression of PACAP in testicular germ cells during spermatogenesis suggests a regulatory role in the maturation of germ cells. In the ovary, PACAP is transiently expressed in the granulosa cells of the preovulatory follicles and appears to be involved in the LH-induced cellular events in the ovary, including the prevention of follicular apoptosis.

In the central nervous system, PACAP acts as a neuromodulator. Furthermore, PACAP has neurotrophic properties, some of which may play an important role during brain development. PACAP, together with VIP, seems to play an important role during development. Both neuropeptides have been shown to peak during periods of brain development when critical morphogenetic mechanisms take place, for example cell proliferation and cell differentiation. Furthermore, it is known that PACAP receptors are expressed in proliferative zones in the embryonic and postnatal nervous system. PACAP exerts a variety of growth factor-like actions depending on the developmental stage and origin of the cells, including regulation of proliferation, survival, maturation and neurite outgrowth.

In the adult brain, PACAP appears to act as a neuroprotective factor that attenuates neuronal damage resulting from central insult. PACAP is capable of protecting neurons from cell death and enhancing neuronal survival *in vitro*. In addition, PACAP can act *in vivo* as a neurotrophic factor controlling histogenesis of the cerebellar cortex (Vaudry et al. 2000).

4.27.6
Neurological Disorders and Neurodegenerative Diseases

PACAP is also produced by lymphoid cells and exerts a wide spectrum of immunological functions controlling homeostasis of the immune system. PACAP has been identified as a potent anti-inflammatory factor that exerts its function by regulating the production of both anti- and proinflammatory mediators.

The PAC1 receptor is induced in mouse brain following transient focal cerebral ischemia; and PACAP has been reported to prevent ischemia-induced death of hippocampal neurons in vivo. Furthermore, PACAP has strong anti-apoptotic effects in several neuronal cultures and in vivo, supporting the idea that PACAP might have protective effects in several neurodegenerative diseases. Indeed, protective effects of PACAP have been shown in various models of brain injuries, including cerebral ischemia, Parkinson's disease and nerve transections.

Further Reading

Anderson, S.T., Sawangjaroen, K., Curlewis, J.D. (1996): Pituitary adenylate cyclase-activating polypeptide acts within the medial basal hypothalamus to inhibit prolactin and luteinizing hormone secretion. *Endocrinology* **137**: 3424–3429.

Arimura, A. (1998): Perspectives on pituitary adenylate cyclase activating polypeptide (PACAP) in the neuroendocrine, endocrine, and nervous systems. *Jpn J. Physiol.* **48**: 301–331.

Arimura, A., Shioda, S. (1995): Pituitary adenylate cyclase-activating polypeptide (PACAP) and its receptors: neuroendocrine and endocrine interactions. *Front. Neuroendocrinol.* **16**: 53–88.

Culler, M.D., Paschall, C.S. (1992): Pituitary adenylate cyclase-activating polypeptide (PACAP) potentiates the gonadotropin-releasing activity of luteinizing hormone-releasing hormone. *Endocrinology* **129**: 2260–2261.

Dejda, A., Sokolowska, P., Nowak, J.Z. (2005): Neuroprotective potential of three neuropeptides PACAP, VIP and PHI. *Pharmacol. Rep.* **57**: 307–320.

Delgado, M., Abad, C., Martinez, C., Juarranz, M.G., Leceta, J., Ganea, D., Gomariz, R.P. (2003): PACAP in immunity and inflammation. *Ann. N.Y. Acad. Sci.* **992**: 141–157.

Harmar, T., Lutz, E. (1994): Multiple receptors for PACAP and VIP. *Trends Pharmacol. Sci.* **15**: 97–99.

Hashimoto, H., Ishihara, T., Shigemoto, R., Mori, K., Nagata, S. (1993): Molecular cloning and tissue distribution of a receptor for pituitary adenylate cyclase-activating polypeptide. *Neuron* **11**: 333–342.

Hosoya, M., Kimura, C., Ogi, K., Ohkubo, S., Miyamoto, Y., Kugoh, H., Shimizu, M., Onda, H., Oshimura, M., Arimura, A., et al. (1992): Structure of the human pituitary adenylate cyclase-activating polypeptide (PACAP) gene. *Biochim. Biophys. Acta* **1129**: 199–206.

Hosoya, M., Onda, H., Ogi, K., Masuda, Y., Miyamoto, Y., Ohtaki, T., Okazaki, H., Arimura, A., Fujino, M. (1993): Molecular cloning and functional expression of rat cDNAs encoding the receptor for pituitary adenylate cyclase-activating polypeptide (PACAP). *Biochem. Biophys. Res. Commun.* **194**: 133–143.

Laburthe, M., Couvineau, A., Marie, J.C. (2002): VPAC receptors for VIP and PACAP. *Receptors Channels* **8**: 137–153.

Morrow, J.A., Lutz, E.M., West, K.M., Fink, G., Harmar, A.J. (1993): Molecular cloning and expression of a cDNA encoding a receptor for pituitary adenylate cyclase-activating polypeptide (PACAP). *FEBS Lett.* **329**: 99–105.

Pisegna, J.R., Wank, S.A. (1993): Molecular cloning and functional expression of the pituitary adenylate cyclase-activating polypeptide type I receptor. *Proc. Natl Acad. Sci. USA* **90**: 6345–6349.

Rawlings, S.R., Hezareth, M. (1996): Pituitary adenylate cyclase-activating polypeptide (PACAP) and PACAP/vasoactive intestinal polypeptide receptors: action on the pituitary gland. *Endocrinol. Rev.* **17**: 24–46.

Somogyvari-Vigh, A., Reglodi, D. (2004): Pituitary adenylate cyclase activating polypeptide: a potential neuroprotective peptide. *Curr. Pharm. Des.* **10**: 2861–2889.

Spengler, D., Waeber, C., Pantaloni, C., Holsboer, F., Bockaert, J., Seeburg, P.H., Journot, L. (1993): Differential signal transduction by five splice variants of the PACAP receptor. *Nature* **365**: 170–175.

Svoboda, M., Tastenoy, M., Ciccarelli, E., Stievenart, M., Christophe, J. (1993): Cloning of a splice variant of the pituitary adenylate cyclase-activating polypeptide (PACAP) type I receptor. *Biochem. Biophys. Res. Commun.* **195**: 881–888.

Vaudry, D., Gonzalez, B.J., Basille, M., Pamantung, T.F., Fournier, A., Vaudry, H. (2000): PACAP acts as a neurotrophic factor during histogenesis of the rat cerebellar cortex. *Ann. N.Y. Acad. Sci.* **921**: 293–299.

Waschek, J.A. (2002): Multiple actions of pituitary adenylyl cyclase activating peptide in nervous system development and regeneration. *Dev. Neurosci.* **24**: 14–23.

Zhou, C.J., Shioda, S., Yada, T., Inagaki, N., Pleasure, S.J., Kikuyama, S. (2002): PACAP and its receptors exert pleiotropic effects in the nervous system by activating multiple signaling pathways. *Curr. Protein Pept. Sci.* **3**: 423–439.

4.28
Proopiomelanocortin

4.28.1
General Aspects and History

In 1977, a common precursor of adrenocorticotropic hormone (ACTH) and several endogenous opioids was identified. This precursor consists of a 241-amino-acid protein termed proopiomelanocortin (POMC).

Besides the endogenous morphins and ACTH, POMC is the precursor of the melanocyte-stimulating hormone (MSH). Evidently, proopiomelanocortin represents a potent precursor molecule which can be processed to a variety of neuroactive peptides. Among these peptides are β-endorphin and β-lipotropin, as well as the adrenocorticortopic hormone (ACTH) and different melanotrope peptides, including α-, β- and γ-melanocyte-stimulating hormone (Fig. 4.46). Other neuroactive peptides which derive from proopiomelanocortin are the corticotropin-like intermediary peptide (CLIP) and the joining peptide (JP).

The proopiomelanocortin gene is expressed in corticotropes and melanotropes of the pituitary gland, as well as in two additional cell populations within the brain.

Alpha-MSH:	Ser-Tyr-Ser-Met-Glu-His-Phe-Arg-Trp-Gly-Lys-Pro-Val-NH$_2$
ACTH:	Ser-Tyr-Ser-Met-Glu-His-Phe-Arg-Trp-Gly-Lys-Pro-Val-Gly-Lys-Lys-Arg-Arg-Pro-Val-Lys-Val-Tyr-Pro-Asn-Gly-Ala-Glu-Asp-Glu-Ser-Ala-Glu-Ala-Phe-Pro-Leu-Glu-Phe-OH
Beta-Endorphin:	Tyr-Gly-Gly-Phe-Met-Thr-Ser-Glu-Lys-Ser-Gln-Thr-Pro-Leu-Val-Thr-Leu-Phe-Lys-Asn-Ala-Ile-Ile-Lys-Asn-Ala-Tyr-Lys-Lys-Gly-Glu-OH

Fig. 4.46 Structure of the active neuropeptides which are generated from the precursor proopiomelanocortin.

4.28.2
Localization Within the Central Nervous System

POMC-synthesizing cells are frequently found in the pituitary gland. In brain tissues, two neuronal systems have been identified which synthesize proopiomelanocortin:

- One system is located in the nucleus of the solitary tract. These neurons project to the spinal cord.
- The second system is located in the arcuate nucleus of the hypothalamus in rats (corresponding to POMC localization in the nucleus infundibularis in humans). The neurons of the arcuate nucleus project ventrally to the median eminence, rostrally to the preoptic area and anterior hypothalamus and laterally to the amygdala. A further dorsal projection extends to the thalamus, as well as caudally to the substantia nigra, the locus coeruleus, the nucleus of the tractus solitarius and the spinal cord (Fig. 4.47).

In both systems, the primary structure of POMC is identical (241 amino acids), as is the structure and length of the POMC mRNA (1.1 kb). However, tissue-specific processing of the precursor POMC is common and gives rise to different site-specific endproducts in POMC-producing cells. This fact explains tissue-specific differences in PMOC function.

4.28.3
Biosynthesis and Degradation

Proopiomelanocortin, its mRNA and the POMC gene are all highly conserved and, not surprisingly, the human POMC gene and, for instance, the POMC gene of the frog *Xenopus laevis*, as well as the corresponding mRNA, reveals considerable similarities. On the evidence of statistic phylogenetic analysis, the structure of the POMC gene has not varied during the past 350 million years of vertebrate evolution. However, some minor differences among species are evident. For instance, two biologically active genes can be found in *X. laevis*, but

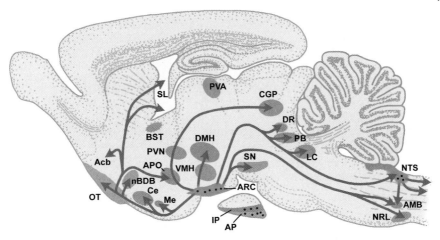

Fig. 4.47 Schematic overview of the distribution and projections of POMC-synthesizing cells in the brain. *Abbreviations*: Acb=*nucleus accumbens*; AMB=*nucleus ambiguus*; AP=anterior pituitary; APO=preoptic area; ARC=arcuate nucleus; BST=bed nucleus of the *stria terminalis*; Ce=central nucleus of the amygdala; CGP=central gray, periaqueductal; DMH=dorsomedial nucleus of the hypothalamus; DR=dorsal raphe; IP=intermediate pituitary; LC=*locus coeruleus*; Me=medial nucleus of the amygdala; nBDB=nucleus of the diagonal band of Broca; NRL=lateral *nucleus reticularis*; NTS=nucleus of the *tractus solitarius*; OT=olfactory tubercle; PB=parabrachial nuclei; PVA=paraventricular nucleus of the thalamus; PVN=paraventricular nucleus of the hypothalamus; SL=lateral septum; SN=*substantia nigra*; VMH=ventromedial nucleus of the hypothalamus.

only one active gene is present in humans. With the exception of mice, all mammals possess only one POMC gene. In some species, the existence of a pseudogene has been demonstrated.

The transcriptional product of the POMC gene is RNA containing three exons and two introns. Alternative splicing of the primary transcript leads to two mRNAs, which differ in a short sequence of 30 nucleotides close to the initial codon.

From these mRNAs, a 267-amino-acid protein is translated, which consists of a 26-amino-acid signal peptide. This protein is the pre-POMC. Maturation of this pre-POMC gives rise to POMC.

The POMC molecule has ten pairs of basic amino acids. These pairs provide potential recognition sites for specific endopeptidases; and the latter are known as prohormone convertases or proprotein convertases (PC).

Seven such convertases have been characterized, including PC1 (also known as PC3), PC2, PC4, PC5 (also known as PC6), PC7 (also known as LPC), PACE4 and furin. Differential cleavage by the convertases is responsible for the processing of POMC into different biologically active peptides. Consequently, the type of biologically active peptide secreted from a particular tissue depends largely on the site-specific enzymatic complement with which the tissue is equipped.

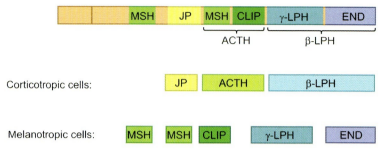

Fig. 4.48 Schematic overview of the different peptides which are generated from the common precursor POMC. *Abbreviations:* ACTH = adrenocorticotropic hormone; CLIP = corticotropin-like intermediate lobe peptide; END = endorphin; JP = joining peptide; LPH = lipotropin; MSH = melanocyte-stimulating hormone.

In corticotropes of the anterior pituitary, the cleavage products of POMC are the adrenocorticotropic hormone and β-LPH (Fig. 4.48), the production of which is under the control of corticotropin-releasing hormone (CRH).

The melanotropes of the intermediate lobe of the pituitary, in contrast, express α-MSH and β-endorphin.

These differences become explicable when the site-restricted expression of PCs is considered, in particular that of PC1 and PC2. For instance, corticotropes express high levels of PC1, which generates ACTH and β-LPH from POMC, while in melanotropes β-endorphin and α-MSH are processed because of the presence of both convertases.

ACTH and α-MSH are under control of the corticotropin-releasing factor (CRH), which stimulates secretion of ACTH from the anterior pituitary, and α-MSH from the intermediate lobe. However, CRH inhibits the secretion of POMC.

Epinephrine stimulates the release of α-MSH and β-endorphin from melanotropes, whereas dopamine is the most important inhibitor of these cells.

4.28.4
Receptors and Signal Transduction

The receptors responsible for β-endorphin or α-MSH binding and the subsequent signal transduction pathways are described in the sections on *endorphin* and *melanocyte-stimulating hormone* (Sections 4.11 and 4.21, respectively).

4.28.5
Biological Effects

A striking example of differential processing of POMC in the anterior pituitary occurs during development. In adults, the corticotropes produce ACTH but, during the first two postnatal weeks, some of these cells show high levels of PC2, which is essential for the generation of α-MSH (see above).

In mice, it has been shown that there is a transient presence of a-MSH-like immunoreactivity in the adenohypophysis during ontogeny. While PC1 expression in the anterior pituitary increased from E15 to adulthood, the PC2 mRNA expression peaked between postnatal day 1 (P1) and postnatal day 14 (P14) and then decreased to adult levels.

This variation during development in the relative ratio of PC1/PC2 expression rationalizes the observed plasticity of POMC processing in the adenohypophysis (Marcinkiewicz et al. 1993).

POMC of the pituitary also plays an essential role in stress-regulation:
- through the release of ACTH from corticotropes which affects directly the synthesis and secretion of steroids from the adrenal gland;
- by negative feedback control through adrenal steroids on the corticotropin-releasing hormone (CRH).

Thus, POMC is of central importance in the hypothalamo-hypophysial axis of stress regulation.

In vitro studies have shown that the activation of POMC-synthesizing neurons in the arcuate nucleus of the hypothalamus is controlled by dopamine, GABA, glutamate, serotonin and neuropeptide Y (NPY). The release of POMC metabolites in this area is stimulated by glutamate and inhibited by dopamine, GABA and NPY, as well as by benzodiazepines.

Further Reading

Castro, M.G., Morrison, E. (1997): Post-translational processing of proopiomelanocortin in the pituitary and in the brain. *Crit. Rev. Neurobiol.* **11**: 35–57.

Civelli, O., Birnberg, N., Herbert, E. (1982): Detection and quantification of pro-opiomelanocortin mRNA in pituitary and brain tissues from different species. *J. Biol. Chem.* **257**: 6783–6787.

Dalayeun J.F., Nores J.M., Bergal S. (1993): Physiology of beta-endorphins. A close-up view and a review of the literature. *Biomed. Pharmacother.* **47**: 311–320.

Herz, A. (ed.) (1993): *Opioids.* Springer-Verlag, Berlin.

Hughes, S., Smith, M.E. (1994): Upregulation of the pro-opiomelanocortin gene in motoneurons after nerve section in mice. *Mol. Brain Res.* **25**: 41–49.

Levin, N., Roberts, J.L. (1991): Positice regulation of proopiomelanocortin gene expression in corticotropes and melanotropes. *Front. Neuroendocrinol.* **12**: 1–22.

Mains, R.E., Eipper, B.A., Ling, N. (1977): Common precursor to corticotropins and endorphins. *Proc. Natl Acad. Sci. USA* **74**: 3014–3018.

Marcinkiewicz, M., Day, R., Chrétien, M., Seidah, N.G. (1993): Ontogeny of the prohormone convertases PC1 and PC2 in the mouse hypophysis and their colocalization with ACTH and a-MSH. *Proc. Natl Acad. Sci. USA* **90**: 4922–4926.

Schäfer, M.K.H., Day, R., Watson, S.J., Akil, H. (1990): Distribution of opioids in the brain and peripheral tissues. In: *Opioid Peptides and Receptors: Biochemistry, Physiology and Pharmacology.* ed. Almeida, O.F.X., Shippenberg, T.S., Springer-Verlag, Berlin.

4.29
Purines

4.29.1
General Aspects

The purine nucleoside adenosine is a component of nucleic acids and of the nucleotides ATP, ADP and cyclic AMP, all of which play important roles in cellular metabolism. Besides their general function in cell metabolism, ATP and adenosine (Fig. 4.49) can themselves act as neuroactive substances. While ATP is released non-specifically from the cytosol of damaged cells, it is also co-packaged in exocytotic vesicles/granules containing conventional neurotransmitters and hormones. The diverse biological responses to ATP appear to be mediated by a variety of so-called purinergic receptors which are activated upon binding to ATP and other nucleotides. Recent studies suggest that there are multiple ATP receptor subtypes. These include:
- G protein-coupled ATP receptors which stimulate inositol phospholipid hydrolysis, Ca^{2+} mobilization, and activation of protein kinase C;
- ATP receptors which directly activate non-selective cation channels in the plasma membranes of a variety of excitable cell types.

It is generally recognized that extracellular purines (ATP and adenosine) serve as ubiquitous signaling substances in neuronal and non-neuronal systems. ATP acts like a fast neurotransmitter in parts of the brain and also as a co-transmitter together with norepinephrine and acetylcholine.

Adenosine is a well known neuromodulator with neuroprotective capabilities under various pathological conditions, e.g. hypoxia and ischemia. Although individual functional effects mediated by ATP and adenosine can be well distinguished on the level of receptor action, both share common secretory and metabolic pathways.

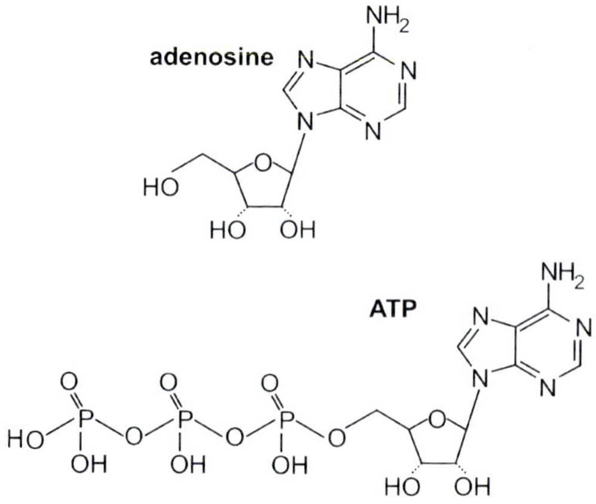

Fig. 4.49 Structure of adenosine and ATP.

ATP binds to so-called P2 purinoceptors, while its metabolic product adenosine binds to adenosine receptors, also referred to as P1 purinoceptors. Four different subtypes of P1 purinoceptors (A1, A2A, A2B and A3) are known. The A1 and A2A subtypes are expressed predominantly in the CNS.

4.29.2
Biosynthesis and Degradation

An enzymatic cascade located at the cell surface terminates nucleotide signaling. The final hydrolysis product is the nucleoside, adenosine. Enzymes involved in the processing of the nucleotides belong to the class of ecto-enzymes. The following enzymes participate in extracellular ATP processing:
- Ecto-ATPase preferentially hydrolyzes ATP to adenosine 5′-diphosphate (ADP).
- Ecto-apyrase converts both ATP and ADP to adenosine 5′-monophosphate (AMP).
- Ecto-5′-nucleotidase converts AMP to adenosine.

Each of the metabolites can activate its specific receptor complex: the ligands for the P2 receptor are ADP and ATP and, for the P1 receptors, AMP and adenosine. Extracellular concentrations of adenosine are regulated by bi-directional nucleoside transporters and by enzymatic activites. Inactivation of adenosine is achieved by deaminase (ADA, a purine catabolic enzyme) and adenosine kinase (AK), which both serve to remove adenosine from the extracellular space.

4.29.3
Receptors and Signal Transduction

P1 receptors
Four P1 receptors (designated A1, A2A, A2B and A3) have been cloned and pharmacologically characterized. They all belong to the GPCR superfamily. The A1 receptor is composed of 326 amino acids and has been cloned from several species, including human, rat, chicken, dog, cattle and guinea-pig with 90–95% amino acid homology. The A1 receptor is widely distributed in the central nervous system and functionally coupled to the inhibition of cAMP formation, stimulation of phospholipase C synthesis and modulation of nitric oxide production. The receptor is believed to mediate the inhibition of neurotransmitter release and reduction of neuronal activity.

Two molecularly and pharmacologically distinct subtypes of the A2 receptor have been identified, both of which are linked to activation of adenylate cyclase.

The A2A receptor is a high affinity receptor consisting of 409 amino acids with seven transmembrane domains and is coupled via the G_s pathway to the stimulation of adenylate cyclase. The human A2A receptor gene is located on chromosome 22q11.2 and contains two exons interrupted by a single intron of about 7 kb.

The A2A receptor is highly expressed in the striatum, nucleus accumbens, olfactory tubercles and subnuclei of the amygdala. Lower expressions have been found in the cortex, the hippocampus, the thalamus, the cerebellum and in portions of the hindbrain. The A2A receptor is not restricted to neurons, but is also expressed in microglia.

The low-affinity A2B receptor is more ubiquitously distributed throughout the central nervous system and peripheral tissues. The A2B receptor is composed of 332 amino acids and shares a 45% homology with A1 and A 2A receptors.

The A3 receptor is linked to inhibition of adenylate cyclase and elevation of cellular IP_3 levels and intracellular Ca^{2+}. This subtype is one of the smallest receptor ever cloned (between 316 and 320 amino acids, depending on the species investigated). The human A3 receptor is sensitive to xanthine blockade, while the rat receptor is not. The A3 receptor shows widespread distribution, with low levels in brain. The A3 receptor has been cloned from various species including human, rat, dog, rabbit and sheep.

P2 receptors

The purinoceptors through which ADP and ATP act are classified as P2 receptors and are quite distinct from the P1 purinoceptors.

The P2 purinoceptors were initially defined as a family of receptors responsive to extracellular adenine nucleotides. In the late 1980s, it became evident that extracellular uridine nucleotides also modulate cell function. The existence of a nucleotide receptor common to both ATP and UTP was suggested by indirect pharmacological evidence and later verified by cloning of the P2U receptor (which is believed to correspond to the P2Y2 receptor) which is equally responsive to ATP and UTP.

The P2 receptor family is divided into two major subclasses (the P2X receptor and the P2Y receptors) and a minor subclass (the P2Z receptors).

The P2X purinoceptors (P2X1 to P2X7) are ligand-gated ion channel receptors; and the P2Y purinoceptors receptors (P2Y1 to P2Y7) are G protein-coupled receptors.

Neuronal P2X receptors mediate fast synaptic responses, while P2Y receptors elicit slow synaptic responses or the release of Ca^{2+} from intracellular pools in response to an increase of ATP or UTP. ATP modulates transmitter release via presynaptic P2X purinoceptors, which are usually inhibitory in nature. It is therefore thought that presynaptic P2X purinoceptors mediate positive and G protein-coupled P2Y purinoceptors negative feedback mechanisms in transmitter release.

Glial cells (asterocytes, oligodendrocytes and microglia) also posses P2 purinoceptors. Activation of the receptor is believed to induce acute effects in these cells via an increase in intracellular Ca^{2+} concentration or chronic effects, e.g. proliferation, via mitogen-activated protein (MAP) kinases.

P2X purinoceptors
The P2X receptors share a common motif of two transmembrane-spanning regions linked by a long extracellular loop with the amino- and carboxy-terminals residing inside the cell.

The functional domain constitutes a non-selective pore, permeable to calcium, potassium and sodium, mediating rapid neurotransmission processes.

Seven functional members of the P2X receptor family (P2X1 to P2X7) have been cloned. The cloned receptors have been grouped into three major classes:
- Group 1 includes P2X1 and P2X3 receptors. They have high affinity for ATP and are rapidly activated and desensitized. P2X1 receptors are present in dorsal root ganglia, in the spinal cord and in the brain, while P2X3 receptors have been localized in a subset of sensory neurons in the dorsal root and trigeminal ganglia.
- Group 2 includes P2X2, P2X4, P2X5 and P2X6 receptors. They reveal lower affinity for ATP and show a slow desensitization. P2X2 receptors are present in the brain, the spinal cord, the superior cervical ganglia and in the adrenal medulla. P2X4 receptors are present in the hippocampus and in the spinal cord, but they have also been found in some additional areas of the central nervous system. P2X5 receptors occur in the central horn of the spinal cord and in the dorsal root and trigeminal ganglia. P2X6 receptors are present in Purkinje cells of the cerebellum, in neurons of lamina IX of ventral horn and in neurons of lamina II of the dorsal horn. Additionally, these receptors are expressed in neurons of the dorsal root and trigeminal ganglia.
- Group 3 is represented only by the P2X7 receptor. This receptor has very low affinity to ATP and shows little or no desensitization. In addition to its role as an ATP-gated ion channel it seems to function as a non-selective ion pore. P2X7 receptors have not yet been identified in neurons, but they are present in mast cells and macrophages.

P2Y purinoceptors
Over a dozen P2Y-like receptors have been cloned, but only five subtypes (P2Y1, P2Y2, P2Y4, P2Y6 and P2Y11) have been shown to possess functional activity. All five are coupled to G_q proteins and their activation results in activation of PLC and elevation of IP_3 levels, with subsequent release of calcium from intracellular stores.

P2Y1 receptors are abundant in the brain, but the other subtypes of P2Y purinoceptors have not yet been demonstrated in either the peripheral or central nervous system.

P2Z purinoceptors
In addition to the known P2X and P2Y purinoceptors, a third family of purinoceptors has been identified, the so-called P2Z receptors. These receptors bind to tetra-anionic $ATP4^-$.

P2Z purinoceptors are ligand-gated ion channels which are highly selective for the ATP4 species; and the addition of Mg^{2+} in excess over ATP closes the channel. Activation of P2Z purinoceptors stimulates phospholipase D activity and this has been suggested to be involved in membrane remodeling in some cell types. However, this receptor class has as yet not been identified in neuronal tissues.

4.29.4
Biological Effects

In addition to their well known roles within cells, purines are released into the extracellular space, where they act as intercellular signaling molecules. In the nervous system, they mediate both immediate effects, such as neurotransmission, and trophic effects which induce changes in cell metabolism, structure and function and therefore have a longer time-course. Some trophic effects of purines are mediated via purinergic cell surface receptors, whereas others require uptake of purines by the target cells.

ATP

After neuronal injury, microglia are engaged in remodeling processes called synaptic stripping. This process is considered to lead to the displacement of synaptic boutons from the neuronal surface. During such remodeling, microglial cells are directly exposed to elevated levels of extracellular ATP released from damaged neurons. Activation of the P2X, P2Z and P2Y purinoceptors results in an increase in the intracellular Ca^{2+} concentration in microglial cells and macrophages. Important functions of these phagocytic cells are regulated by intracellular Ca^{2+} (for example cytokine production and phagocytosis) and may therefore be modulated by changes in the extracellular concentrations of nucleotides. Thus, the purinoreceptor complement on microglia seems to provide a molecular link for functional cross-talk between both cellular elements.

Application of ATP to sensory terminals results in hyper-excitability and pain perception. ATP can also induce local nociceptive responses at sites of application and facilitates nociception to noxious stimuli. The pro-nociceptive effects of ATP are mediated via P2X receptors present on sensory afferents in the spinal cord. ATP is released from a number of cell types in response to trauma; and evidence exists that activation of P2X3 receptors initiates peripheral and central sensitization associated with visceral nociception.

An involvement of P2 purinoceptors in memory and learning has been suggested, since P2 receptors seem to be involved in the induction of fast synaptic currents in cultured hippocampal neurons and in long-lasting enhancement of population spikes. ATP release upon glutamate stimulation, which evokes an increase in intracellular Ca^{2+}, has also been shown in hippocampal slices; and the presence of mRNAs of P2X-purinoceptors in the hippocampus makes a contribution of ATP to the modulation of synaptic efficiency in the hippocampus likely (Inoue et al. 1996).

Adenosine

The endogenous metabolite adenosine reveals modulatory effects in the peripheral and central nervous system, which are mediated through specific cell surface-associated receptors. Adenosine is a potent inhibitor of dopamine, GABA, glutamate, acetylcholine, serotonin, and norepinephrine release via presynaptic A1 receptors.

In the brain, exogenously administered adenosine receptor agonists have been shown to exert a neuroprotective effect. Furthermore, adenosine has been demonstrated to be involved in pain, cognition, movement and sleep.

While ATP acts to facilitate nociceptive transmission, adenosine has the opposite effects, inhibiting nociceptive processes in the brain and spinal cord. The current view is that adenosine receptors of the A1 subtype are associated with a modulatory effect on pain transmission at spinal cord levels. Animal studies have demonstrated adenosine- and adenosine analog-mediated inhibitory influences on nociceptive reflex responses. It has also been proposed that endogenous adenosine formation is involved in opioid antinociception.

4.29.5
Neurological Disorders and Neurodegenerative Diseases

Modulation of adenosine level occurs under several pathological conditions. In hypoxia and focal ischemia, extracellular adenosine accumulates rapidly and adenosine receptor agonists can reduce damage from cerebral ischemia. These observations support a role for adenosine as a neuroprotective agent. Seizure activity is associated with a rapid and marked increase in adenosine concentrations in the brain; and adenosine agonists can reduce seizure via A1 receptors.

The adenosine A2 receptor has emerged as an attractive target for Parkinson's disease (PD) therapy, primarily because of its localized expression in striatum and its motor enhancement function. Recent studies indicate that A2A antagonists offer neuroprotective effects and may possibly modify chronic L-DOPA-induced maladaptive responses in animal models of PD.

Further Reading

Brundege, J. M., Dunwiddie, T. V. (1997): Role of adenosine as a modulator of synaptic activity in the central nervous system. *Adv. Pharm.* **39**: 353–391.

Burnstock, G. (1996): A unifying purinergic hypothesis for the initiation of pain. *Lancet* **347**: 1604–1605.

Chen, J. F. (2003): The adenosine A(2A) receptor as an attractive target for Parkinson's disease treatment. *Drug News Perspect.* **16**: 597–604.

De Mendonca, A., Ribeiro, J. A. (1997): Adenosine and neuronal plasticity. *Life Sci.* **60**: 245–251.

Fredholm, B. B. (1995): Purinoceptors in the nervous system. *Pharmacol. Toxicol.* **76**: 228–239.

Fredholm, B. B., Abbracchio, M. P., Burnstock, G., Daly, J. W., Harden, T. K. Jacobson, K. A., Leff, P., Williams, M. (1994): Nomenclature and classification of purinoceptors. *Pharmacol. Rev.* **46**: 143–156.

Hasko, G., Pacher, P., Vizi, E. S., Illes, P. (2005): Adenosine receptor signaling in the brain immune system. *Trends Pharmacol. Sci.* **26**: 511–516.

Henning, R.H. (1997): Purinoceptors in neuromuscular transmission. *Pharmacol. Ther.* **74**: 115–128.

Illes, P., Norenberg, W. (1993): Neuronal ATP receptors and their mechanism of action. *Trends Pharmacol. Sci.* **14**: 50–54.

Inoue, K., Koizumi, S., Ueno, S. (1996): Implication of ATP receptors in brain functions. *Prog. Neurobiol.* **50**: 483–492.

Jacobson, K.A., van Galen, P.J.M., Williams, M. (1992): Adenosine receptors: pharmacology, structure–activity relationships and therapeutic potential. *J. Med. Chem.* **35**: 407–422.

Moreau, J.L., Huber, G. (1999): Central adenosine A(2A) receptors: an overview. *Brain Res. Brain Res. Rev.* **31**: 65–82.

Norenberg, W., Illes, P. (2000): Neuronal P2X receptors: localisation and functional properties. *Naunyn Schmiedebergs Arch. Pharmacol.* **362**: 324–339.

Prince, D.A., Stevens, C.F. (1992): Adenosine decreases neurotransmitter release at central synapses. *Proc. Natl Acad. Sci. USA* **89**: 8586–8590.

Rathbone, M.P., Middlemiss, P.J., Gysbers, J.W., Andrew, C., Herman, M.A., Reed, J.K., Ciccarelli, R., Di Iorio, P., Caciagli, F. (1999): Trophic effects of purines in neurons and glial cells. *Prog. Neurobiol.* **59**: 663–690.

Rosin, D.L., Robeva, A., Woodard, R.L., Guyenet, P.G., Linden J. (1998): Immunohistochemical localization of adenosine A2A receptors in the rat central nervous system. *J. Comp. Neurol.* **401**: 163–186.

Shibuya, I., Tanaka, K., Hattori, Y., Uezono, Y., Harayama, N., Noguchi, J., Ueta, Y., Izumi, F., Yamashita, H. (1999): Evidence that multiple P2X purinoceptors are functionally expressed in rat supraoptic neurones. *J. Physiol. Lond.* **514**: 351–367.

Sollevi, A. (1997): Adenosine for pain control. *Acta Anaesthesiol. Scand. Suppl.* **110**: 135–136.

4.30
Somatostatin

4.30.1
General Aspects and History

In 1968, Kruhlich and coworkers were the first to describe a factor in hypothalamic extracts which blocked the effects of growth hormone (GH). In 1972, Guillemin's group isolated a substance with inhibiting features on GH secretion. This factor was termed somatostatin (SS) or somatotropin releasing-inhibiting factor (SRIF). The discovery of somatostatin confirmed the idea of a dual regulation of anterior pituitary peptides by hypothalamic releasing and release-inhibiting factors and the existence of "brain–gut" peptides, present in both the gastrointestinal tract and the central nervous system.

Later, it was shown that somatostatin is present in a number of different tissues and so participates in a wide range of different biological functions. The most prominent effect of somatostatin is its inhibitory action on the secretion of GH, which is mediated through somatostatin receptors on glandotropic cells of the anterior pituitary. Somatostatin is a neuropeptide which occurs in two different isoforms. The short form consists of 14 amino acids, while the second, longer form is composed of 28 amino acids.

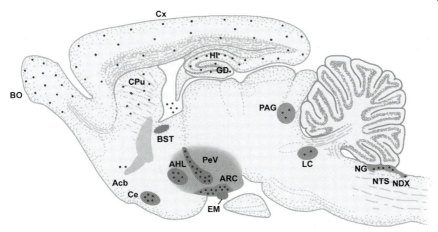

Fig. 4.50 Schematic representation of the distribution of somatostatin-containing neurons. Abbreviations: Acb=*nucleus accumbens*; AHL=lateral hypothalamic area; ARC=arcuate nucleus; BO=*bulbus olfactorius*; BST=bed nucleus of the *stria terminalis*; Ce=central nucleus of the amygdala; CPu=caudate-putamen; Cx=cortex; EM=*eminentia mediana*; GD=*gyrus dentatus*; HI=hippocampus; LC=*locus coeruleus*; NDX=dorsal nucleus of the vagal nerve; NG=*nucleus gracilis*; NTS=nucleus of the *tractus solitarius*; PAG=periaqueductal gray; PeV=periventricular nucleus of the hypothalamus.

4.30.2
Localization Within the Central Nervous System

The main sources of somatostatin in the periphery are the δ-cells of the endocrine pancreas and the entero-endocrine cells of the gut. Somatostatin is widely distributed throughout the central nervous system (Fig. 4.50), indicating that its function goes well beyond that originally described. Somatostatin-containing cell bodies have been identified in diverse brain regions, including the cortex, limbic structures, the hypothalamus, the caudate-putamen, the periaqueductal gray, the nucleus accumbens and the locus coeruleus.

While the highest local concentrations are found in the hypothalamus, the cortex accounts for approximately 50% of total brain somatostatin. Postmortem studies in humans indicated a distribution of somatostatin-containing neurons and fibers similar to that seen in rodents.

Somatostatin-containing nerve fibers have been found in various brain areas, including the median eminence (the source of anterior pituitary regulation), the posterior pituitary, the limbic system, the cortex, the hypothalamus and in hypothalamic projections to the brainstem and spinal cord.

Immunohistochemical studies have revealed two systems of somatostatinergic projections. One system consists of long-distance projections and the other of short-distance projections.

System of long-distance projections

This system includes the somatostatin-containing neurons of the hypothalamus, the amygdala and the nucleus of the tractus solitarius. The periventricular preoptic area is the primary source of somatostatinergic input to the median eminence, while the somatostatin-containing nerve terminals of the anterior hypothalamic area originate in the amygdala. Periventricular hypothalamic somatostatinergic fibers give rise to rostral and ascending fibers, which project to several limbic structures (e.g. the stria terminalis and the amygdala) and to caudal fibers running into the brainstem and spinal cord.

The somatostatinergic system within the amygdala is composed of intrinsic neurons (which project to the different nuclei of the amygdala) and of neurons (mainly located in the central nucleus of the amygdala) which project to limbic structures outside the amygdala.

Somatostatin-containing neurons of the nucleus of the tractus solitarius project to the nucleus ambiguus and to the magnocellular nuclei of the hypothalamus. Interestingly, this system seems to express exclusively somatostatin-28.

System of short-distance projections

The system of short-distance projections is for the most part composed of interneurons. They are located in layers II–III and layers V–VI of the cortex. Somatostatin-containing interneurons have also been shown in the hippocampus and in the striatum. They are believed to be involved in local and integrative regulation of neuronal activities in these brain areas.

Periventricular hypothalamic somatostatinergic fibers are known to form intrahypothalamic projections to different hypothalamic nuclei (for example, to the suprachiasmatic nucleus, the arcuate nucleus and the preoptic nuclei).

Colocalization of somatostatin with other neuromodulators has been shown in telencephalic neurons, where it coexists with NPY. Most of the somatostatin-containing neurons in the striatum also coexpress NPY. Furthermore, somatostatin-containing interneurons coexpress GABA, indicating a role for somatostatin in GABA circuits.

4.30.3
Biosynthesis and Degradation

Somatostatin is an evolutionary old peptide found even in Protozoa. While two separate genes code for somatostatin-14 and somatostatin-28 in fishes, a single gene encodes for both peptides in mammals.

In mammalian brains, somatostatin is mainly expressed in the form of a tetradecapeptide, the so-called somatostatin-14. The long form, somatostatin-28, has also been found in the brain to a minor extent. Somatostatin-28 is cleaved into somatostatin-14 and somatostatin-28 (1–12) (Fig. 4.51). In fishes additional forms of somatostatin have been described, as e.g. somatostatin-22, and somatostatin-25.

The different forms of somatostatin are cleavage products of a larger precursor, prepro-somatostatin. This precursor is composed of 116 amino acids and is

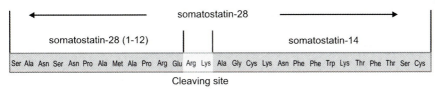

Fig. 4.51 The amino acid sequence of somatostatin-28 includes the sequences of somatostatin-28 (1–12) and somatostatin-14.

encoded by a single gene which consists of two exons (238 and 367 base pairs) separated by an intron (621 base pairs). Two 5′ upstream promoters flank an enhancer element, the cAMP-responsive element (CRE). Prepro-somatostatin is synthesized in the endoplasmatic reticulum. After cleavage of the signal sequence, a leader sequence is removed from the prepro-form and the 92-amino-acid prohormone (pro-somatostatin) is transported into the Golgi apparatus, where the final peptides maturate by proteolysis at monobasic and dibasic sites to yield the two final forms of somatostatin, namely somatostatin-14 and somatostatin-28. Somatostatin-14 is also processed by cleavage of somatostatin-28. The mature peptides are packed into granules which are anterogradely transported to the nerve endings for release.

Somatostatin-14 and somatostatin-28 are secreted in different amounts from somatostatin-containing cells. Somatostatin-14 is the dominant form in neuronal tissue and the exclusive form in the retina, peripheral nerves, pancreas and stomach. The diffuse endocrine system in the gut secretes mainly somatostatin-28.

Somatostatin-14 and somatostatin-28 have been detected in different quantities (in a ratio of about 3:1) within the central nervous system. Although the level of somatostatin-28 is lower than that of somatostatin-14, it has been found to be more effective in the inhibition of growth hormone secretion.

The somatostatins are degraded by several enzymes into proteolytic fragments. The identification of these fragments has made it possible to identify the enzymes involved in the degradation process. Lysomal peptides like cathepsin D and M are involved in the cleavage of somatostatin at Phe-Phe-sites. Furthermore, neutral endopeptidases (EC 24.11, EC 24.15 and EC 24.16) and aminopeptidases participate in the degradation.

4.30.4
Receptors and Signal Transduction

Five somatostatin-receptors (SST1–SST5) have been identified and cloned. They are coupled to G proteins (perhaps with the exception of the SST2 receptor) and to adenylate cyclase (Patel et al. 1994). The receptors show different molecular masses and variations in their amino acid sequences.

Most of them are encoded by single-copy genes, which reveal differential chromosomal localization. It has been found that the genes for the SST1, SST3,

SST4 and SST5 receptors possess no introns, while the gene coding for the SST2 receptor exhibits an intron.

Somatostatin-14 binds equally well to all receptors. This is not true of its analog octreotide, which does not bind to the SST1 and SST4 receptors. Pharmacological studies have failed to show clear differences in the binding capability of somatostatin-28 and somatostatin-14 to the SST1–SST4 receptors. Some evidence indicates that somatostatin-28 has a higher affinity to the SST5 receptor (this receptor is mainly found in the pituitary).

The distribution of the various types of somatostatin receptors in the central nervous system of mammals is summarized in Table 4.5.

The SST receptors use different signal transduction pathways, though few details are as yet available. Apparently, initial activation of the receptors requires G protein (perhaps with the exception of the SST2 receptor), followed by five different pathways, which involve suppression of cAMP formation, reduction of Ca^{2+} influx and K^+ efflux, decrease of tyrosine phosphatase activity and exocytosis.

None of the effects has been successfully related to a specific receptor subtype. A major difficulty in analyzing individual SST receptor signal transduction is that single cells express more than one receptor.

SST1 receptors

The SST1 receptor gene is located on human chromosome 14. It consists of 391 amino acids.

Table 4.5 Distribution of the different forms of somatostatin receptors in the rat brain. The number of asterisks indicates the relative density of SST receptors.

Brain area	Receptor subtype				
	SST1	SST2	SST3	SST4	SST5
Amygdala	++++	++	+++	+	++
Cerebellum	+	+	++++	–	–
Cortex	++++	++++	++	++	++
Hippocampus	++	++	++	++	++
Hypothalamus	++	++	+	+	++++
Pons	+	+	+	+	–
Nucleus accumbens	+	+	+	+	–
Olfactory bulbus	++	++	++	++	++
Olfactory tubercle	+	+	++	+	++
Preoptic area	+	+	+	+	++++
Striatum	+	+	++	+	+
Thalamus	++	+	+	+	–

SST2 receptors

The human SST2 receptor gene is located on chromosome 17. It is composed of two exons interrupted by an intron and it has been shown that this receptor type exists in two forms, a long (SST2A) and a short variant (SST2B).

SST3 receptors

The gene which encodes for the SST3 receptor is located on human chromosome 22. The SST3 receptor consists of 428 amino acids.

SST4 receptors

The receptors SST1, SST2 and SST3 bind somatostatin-14 and somatostatin-28 with similarly high affinity, the SST4 receptor has higher affinity for somatostatin-28. The human SST4 gene is located on chromosome 16 and consists of 388 amino acids.

SST5 receptors

The gene which encodes for the SST5 receptor is located on human chromosome 20. The SST5 receptor is somatostatin-28 selective and has a length of 364 amino acids.

4.30.5
Biological Effects

Somatostatin exhibits multiple biological activities. It blocks both insulin and glucagon secretion in the pancreas and inhibits growth hormone and TSH release in the pituitary gland. In the gut, somatostatin blocks the secretion of gas-

Table 4.6 Effects of somatostatin on neuroactive substances.
Abbreviations: CRF = Corticotropin-releasing factor;
GHRH = growth hormone-releasing hormone
(0 = without effect; + = excitatory effect; − = inhibitory effect).

Brain structure	Transmitter or neuromodulator	Effect
Cortex	Acetycholine	0
	Norepinephrine	+
	Serotonin	0
	Dopamine	+
	Histamine	−
Hippocampus	Acetycholine	+
	Histamine	−
Hypothalamus	Norepinephrine	0
	Histamine	+
	GHRH	−
	CRF	−

trointestinal hormones (e.g. gastrin) and reveals inhibitory effects on gastric acid and pepsin secretion. In gall bladder it blocks contraction.

Within the central nervous system, somatostatin shows neurohormonal and neuromodulatory functions, e.g. it modulates neurotransmitter efficiency like acetylcholine, dopamine or serotonin (Table 4.6).

Both excitatory and inhibitory effects of somatostatin have been observed in the cortex, the hippocampus and the hypothalamus (see Table 4.6).

Secretion of somatostatin is primarily stimulated by corticotropin-releasing factor (CRF) and, like CRF, somatostatin exerts inhibitory effects upon the immune system (for example by reducing T-cell proliferation).

Intraventricular application of somatostatin in rats induces various behavioral effects. Injection of small amounts of somatostatin elicits an increase in stereotypic behavior. Application of higher doses of somatostatin induces severe deficits in motor coordination.

4.30.6
Neurological Disorders and Neurodegenerative Diseases

Increased levels of somatostatin have been described in several inflammatory and degenerative neurological disorders, e.g. cerebral tumor, meningitis, nerve root compression, metabolic encephalopathies.

Human postmortem studies have shown decreased somatostatin concentrations in a variety of cortical and subcortical sites in Alzheimer's disease, accompanied by a significant reduction in somatostatin-containing neurons. Concentrations have been reported to be reduced to 20–60% of control values in the temporal cortex, with similar drastic reductions in the frontal and parietal cortex. Among different mental disorders, a significant reduction in somatostatin CSF levels has been found in depression. A significant negative correlation is also evident between somatostatin CSF levels and the incidence of depressive mood in Alzheimer's patients.

Not only are the somatostatin concentrations altered in Alzheimer's diseased cortex, but also the expression of somatostain receptors: neuronal expression of SST4 and SST5 has been found to be reduced, whereas SST3 expression has been found to be increased in Alzheimer's diseased cortex.

Furthermore, Parkinsonian patients who suffer from dementia (about 20–40%) show reduced levels of somatostatin in the frontal cortex, the entorhinal cortex and in the hippocampus, while patients without dementia do not exhibit somatostatin deficits.

Somatostatin seems also to be involved in epilepsy. Significant changes in the hippocampal SRIF system have been documented in experimental models of temporal lobe epilepsy, in particular in the kindling and kainate models. Somatostatin biosynthesis and release are increased in the kindled hippocampus, especially in the dentate gyrus. It has also been found that intraventricular or intracerebral injections of somatostatin are able to elicit seizures.

A.D. Hippocampus
↓   Mg
    K
    Glutamat

In patients with intractable temporal seizures, an increase in the level of somatostatin-14 has been found in the temporal lobe, which is accompanied by a loss of somatostatin-containing neurons in the dentate gyrus.

Further Reading

Bell, G. I., Reisine, T. (1993): Molecular biology of somatostatin receptors. *Trends Neurosci.* **16**: 34–38.

Binaschi, A., Bregola, G., Simonato, M. (2003): On the role of somatostatin in seizure control: clues from the hippocampus. *Rev. Neurosci.* **14**: 285–301.

Brazeau, P., Vale, W., Burgus, R., Ling, W., Butcher, M., Rivier, J., Guillemin, R. (1973): Hypothalamic peptide that inhibits the secretion of immunoreactive pituitary growth hormone. *Science* **179**: 77–79.

Epelbaum, J. (1986): Somatostatin in the central nervous system. *Prog. Neurobiol.* **27**: 63–100.

Epelbaum, J., Dournaud, P., Fodor, M., Viollet, C. (1994): The neurobiology of somatostatin. *Crit. Rev.Neurobiol.* **8**: 25–44.

Florio, T., Thellung, S., Schettini, G. (1996): Intracellular transducing mechanisms coupled to brain somatostatin receptors. *Pharmacol. Res.* **33**: 297–305.

Fries, J. L., Murphy, W. A., Sueiras-Diaz, J., Coy, D. H. (1982): Somatostatin antagonist analog increases GH, insulin and glucagon release in the rat. *Peptides* **3**: 811–814.

Hoyer, D., Lübbert, H., Bruns, C. (1994): Molecular pharmacology of somatostatin receptors. *Arch. Pharmacol.* **350**: 441–453.

Kang, T. C., Park, S. K., Do, S. G., Suh, J. G., Jo, S. M., Oh, Y. S., Jeong, Y. G., Won, M. H. (2000): The over-expression of somatostatin in the gerbil entorhinal cortex induced by seizure. *Brain Res.* **882**: 55–61.

Kumar, U. (2005): Expression of somatostatin receptor subtypes (SSTR1-5) in Alzheimer's disease brain: an immunohistochemical analysis. *Neuroscience* **134**: 525–538.

Li, X. J., Forte, M., North, R. A., Ross, C. A., Snyder, S. H. (1992): Cloning and expression of a rat somatostatin receptor enriched in brain. *J. Biol. Chem.* **267**: 21307–21312.

Olias, G., Viollet, C., Kusserow, H., Epelbaum, J., Meyerhof, W. (2004): Regulation and function of somatostatin receptors. *J. Neurochem.* **89**: 1057–1091.

Patel, Y. C., Greenwood, M. T., Warszynska, A., Panetta, R., Srikant, C. B. (1994): All five cloned human somatostatin receptors (hSSTR1-5) are functionally coupled to adenylyl cyclase. *Biochem. Biophys. Res. Comm.* **198**: 605–612.

Patel, Y. C., Greenwood, M. T., Panetta, R., Demchyshyn, L., Niznik, H., Srikant, C. B. (1995): The somatostatin receptor family. *Life Sci.* **57**: 1249–1265.

Reisine, T., Bell, G. I. (1993): Molecular biology of somatostatin receptors. *Trends Neurosci.* **16**: 34–38.

Rohrer, L., Raulf, F., Bruns, C., Buettner, R., Hofstaeter, F., Schule, R. (1993): Cloning and characterization of a forth human somatostatin receptor. *Proc. Natl Acad. Sci. USA* **90**: 4196–4200.

Selmer, I., Schindler, M., Allen, J. P., Humphrey, P. P., Emson, P. C. (2000): Advances in understanding neuronal somatostatin receptors. *Regul. Peptides* **90**: 1–18.

4.31
Substance P and Tachykinins

4.31.1
General Aspects and History

After its first description in 1931 by Euler and Gaddum, substance P (P was the laboratory code for "peparation") remained the only mammalian member of the family of tachykinins for several decades. Tachykinins were defined as peptides sharing the common carboxy-terminal amino acid sequence Gly-Leu-Met-NH$_2$ (Table 4.7).

In the 1970s, Chang and coworkers (1971) isolated and sequenced substance P. The family of mammalian tachykinins has subsequently grown with the isolation of four additional members. These peptides were named neurokinin A (NKA, neuromedin L or substance K), neuropeptide K (NPK), neurokinin B (NKB or neuromedin K). NKA is also present in two elongated forms, neuropeptide K and neuropeptide γ (NPγ).

In addition, some further tachykinins have been isolated from non-mammalian species (Table 4.7). These tachykinins include eledoisin, physalaemin and kassinin. The non-mammalian tachykinins will not be discribed here.

In parallel with the identification of the tachykinin family, several classes of tachykinin receptors were discovered. The receptors described to date are: tachy-

Table 4.7 Summary of mammalian and non-mammalian tachykinins. All these substances have a common N-terminal sequence (Gly-Leu-Met-NH$_2$; indicated in italics).

Name	Sequence
Mammalian tachykinins	
Neuromedin K (NMK) or neurokinin B (NKB)	Asp-Met-His-Asp-Phe-Phe-Val-*Gly-Leu-Met-NH$_2$*
Neuropeptide K (NPK)	Asp-Ala-Asp-Ser-Ser-Ile-Glu-Lys-Gln-Val-Ala-Leu-Leu-Lys-Ala-Leu-Tyr-Gly-His-Gly-Gln-Ile-Ser-His-Lys-Arg-His-Lys-Thr-Asp-Ser-*Gly-Leu-Met-NH$_2$*
Neuropeptide γ (NPγ)	Asp-Ala-Gly-His-Gly-Gln-Ile-Ser-His-Lys-Arg-His-Lys-Thr-Asp-Ser-*Gly-Leu-Met-NH$_2$*
Substance K (SK) or neurokinin A (NKA)	His-Lys-Thr-Asp-Ser-Phe-Val-*Gly-Leu-Met-NH$_2$*
Substance P	Arg-Pro-Lys-Pro-Gln-Gln-Phe-Phe-*Gly-Leu-Met-NH$_2$*
Non-mammalian tachykinins	
Eledoisin	<Glu-Pro-Ser-Lys-Asp-Ala-Phe-Ile-*Gly-Leu-Met-NH$_2$*
Kassinin	Asp-Val-Pro-Lys-Ser-Asp-Gln-Phe-Val-*Gly-Leu-Met-NH$_2$*
Physalaemin	<Glu-Ala-Asp-Pro-Asn-Lys-Phe-Tyr-*Gly-Leu-Met-NH$_2$*

kinin NK1, NK2 and NK3 receptors. The endogeous agonists for the receptors are: substance P for NK1, NKA for NK2 and NKB for NK3.

4.31.2
Localization Within the Central Nervous System

Substance P-like immunoreactivity has been demonstrated in various tissues, including the gastrointestinal tract and the central nervous system.

In the spinal cord substance P is found mainly in interneurons and in primary afferent fibers of the superficial layers of the dorsal horn (lamina I). High concentrations are also present in the septum, the striatum, the amygdala, the periaqueductal gray, the pons and in some nuclei of the brain stem. The highest concentration of substance P has been detected in the substantia nigra.

The distribution pattern and concentration of neurokinin A and neurokinin K are similar to that of substance P, while that of neurokinin B is different, with maximal concentrations in the cortex.

Two prominent substance P-positive tracts have been found in the brain: one consists of the habenulo-interpeduncular connection, which synapses upon the ventral tegmental area, while the other, the striato-nigratal pathway, descends from the striatum and projects to the substantia nigra.

Colocalizations of different tachykinins within individual cell have been described. Colocalization of substance P and neurokinin A is found in striato-nigratal neurons and in sensory afferent neurons.

In addition, tachykinins colocalize with neurotransmitters or other neuromodulators. For example, some cortical and hippocampal interneurons coexpress GABA and substance P. In sensory fibers, colocalization of substance P with CCK, CGRP, enkephalins and somatostatin has also been described. Additionally, substance P and acetylcholine seem to colocalize in several brain areas; and colocalization of substance P with serotonin has been demonstrated in neurons of the medulla oblongata.

4.31.3
Biosynthesis and Degradation

The biosynthesis and degradation of tachykinins are incompletely understood. It is known, however, that two genes encode for the precursors of tachykinins. The precursor encoded by one gene is named prepro-tachykinin A (PTT-A) and constitutes the source for three further precursors: PTT-Aα, PTT-Aβ and PTT-Aγ. All three precursors have in common the fact that they contain the sequence of substance P. Both, PTT-Aβ and PTT-Aγ, contain additional sequences for neurokinin A and neuropeptide K (a prolonged form of neurokinin A). NPγ is generated from the precursor PTT-Aγ. The second gene codes for a precursor named prepro-tachykinin B (PTT-B). Maturation of this precursor leads to the formation of neurokinin B.

The precursor PTT-Aγ, which is processed to substance P, neurokinin A and neuropeptide γ, seems to be the most abundant precursor in brain tissue, whereas the precursors PTT-Aα and PTT-Aβ have been found in only small quantities. Nerve growth factor (NGF) and dopamine seem to be involved in the transcriptional control of the PTT-A gene.

Extracellular tachykinins are degraded by local peptidases, which include endopeptidase 24.11 (known as enkephalinase), peptidyldipeptidase A [also known as angiotensin-converting enzyme (ACE)], and substance P-degrading enzyme for substance P and enkephalinase for neurokinin A and neurokinin B.

4.31.4
Receptors and Signal Transduction

Besides the three different receptors NK1, NK2 and NK3, a further receptor has been found. However, there is no consensus as to whether this receptor constitutes a new subtype (NK4 receptor) or just an isoform of the NK3 receptor. The endogenous ligands for all of these receptors are substance P, neurokinin A and neurokinin B.

Substance P is the most potent tachykinin for the NK1 receptor, whereas neurokinin A exhibits the highest affinity for the NK2 and neurokinin B for the NK3 receptor. NK1, NK2 and NK3 belong to the superfamily of G protein-coupled receptors.

The classification of the tachykinin receptors is somewhat confusing in the literature; e.g. the tachykinin NK1 receptor is also referred to as the SP-P receptor and the tachykinin NK2 receptor as SP-E, SP-K or NK-A. In addition, the NK3 receptor is known under SP-E, SP-N or NK-B receptor. For each tachykinin receptor, specific agonists and antagonists have been developed (Table 4.8).

NK1 receptors
Particularly rich in tachykinin NK1 receptors are the olfactory bulb, the cortex, the septum, the hippocampus, the striatum, the habenula, the nucleus accumbens, the lateral nucleus of the hypothalamus, the interpeduncular nucleus, the locus coeruleus, the superior colliculus, the nucleus of the tractus solitarius, the raphe nuclei and the medulla oblongata. NK1 receptors are not exclusively expressed in neurons, but also in glia.

Table 4.8 Agonists and antagonists of the three different tachykinin receptor subtypes.

Tachykinin receptor subtype	Agonists	Antagonists
NK1	SPOMe, GR 73632, Sar^{9}Met(O_2)^{11}SP, substance P	GR82334, RP67580, CP-99994
NK2	GR 64349, neurokinin A	L-659877, SR48968, MEN10207, GR94800, GR100679
NK3	Senktide, neurokinin B	GR138676

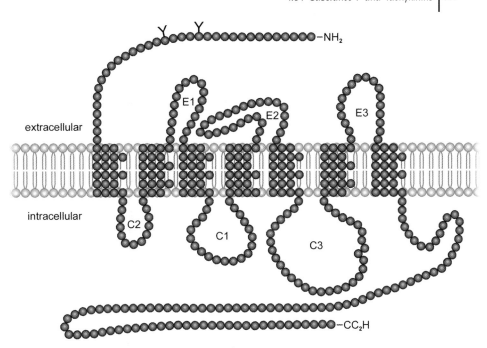

Fig. 4.52 Model of the tachykinin NK1 receptor. The receptor is composed of seven transmembrane-spanning domains and preferentially binds substance P.

The NK1 receptor has been cloned from several species. The primary sequence of the tachykinin NK1 receptor reveals some species-specific variations, but overall homology is high (human and rat NK1 receptors show about 95% homology).

The NK1 receptor (Fig. 4.52) is a protein of about 47 kDa and has seven putative transmembrane-spanning domains. The amino terminus possesses phosphorylation sites while the carboxyl terminus is believed to participate in desensitization mechanisms.

Binding of the ligand to tachykinin NK1 receptors activates several second-messenger systems: these start with activation of phosphatidyl-inositol turnover via phospholipase C, arachidonic acid formation via phospholipase A_2 and cyclic adenosine monophosphate accumulation via adenylate cyclase.

The expression of NK1 mRNA undergoes marked postnatal changes in the developing brain, supporting the idea of a role for tachykinins in synaptogenesis associated with morphological and functional development of the central nervous system.

NK2 receptors

NK2 receptors have been found at low density in the striatum and in the spinal cord. Activation of the NK2 receptor depends on its selectivity for the tachykinin. Its affinity is highest for neurokinin A and neuropeptide K followed in descending order by neuropeptide γ, neurokinin B and substance P.

NK3 receptors

NK3 receptors are expressed in the cortex and in the amygdalo-hippocampal area, in the hippocampus, in some hypothalamic nuclei, in the medial habenula and in the interpeduncular nucleus.

The affinity profile of NK3 is as follows (in descending order): neurokinin B and neurokinin A > neuropeptide K > neuropeptide γ > substance P.

4.31.5
Biological Effects

Tachykinins show prominent physiological effects in the periphery and in the central nervous system (CNS). In the CNS, tachykinins occur particularly in areas involved in the central control of several peripheral autonomic functions (blood pressure, respiration, micturition, gastrointestinal motility, etc.) and essential functions (e.g. drinking behavior).

The best known effects of substance P are profuse salivation after intravenous injection and a profound lowering of arterial blood pressure as a result of vasodilatation. Substance P also stimulates smooth muscle contraction in a variety of peripheral organs. The first effect described on central neurons was slowly developing long-lasting depolarization in spinal cord motor neurons. This effect was thought to result from stimulation of primary afferent inputs to spinal cord motor neurons. The characteristic responses of substance P on dorsal horn afferent neurons were described as delayed in latency, slow in rising and prolonged in after-discharge. These features qualify substance P as a neuromodulator.

Since the responses excited by substance P are among those that are excited by noxious cutaneous stimulation, it is considered that substance P plays a role in sensory transmission within the spinal cord and that this role is related specifically to pain.

Further arguments for an important role of tachykinins in pain transmission derive from the receptor localization:
- Tachykinin NK1 receptors are expressed at anatomical positions, which are considered to be appropriate to their involvement in the processing of afferent noxious input into the spinal cord.
- NK1 receptor expression in the spinal cord undergoes regulation after noxious manipulation.
- NK1 receptor antagonists act synergistically to inhibit NMDA-mediated nociceptive transmission.

Behavioral studies implicate tachykinins in functions connected with the control of hunger and thirst. These effects seem to depend on the stimulation of NK2 and NK3 receptors. Tachykinins are involved in cardiovascular control, control of respiration and in the release of several hypophyseal hormones.

Substance P, as indicated above, acts primarily as an excitatory neuromodulator, but it also exhibits inhibitory properties. For example, substance P inhibits the cholinergic excitation of Renshaw cells in the spinal cord.

Tachykinins seem to be involved in the regulation of the aminergic system, especially of the dopaminergic system. Neurokinin A and substance P excite neurons and modify the release of dopamine in the substantia nigra. Application of substance P or neurokinin A to the ventral tegmental area or injection into the substantia nigra stimulates locomotor activity.

Under experimental conditions, substance P exerts neurotrophic functions, since it supports the regeneration of lesioned neuronal tissue.

4.31.6
Neurological Disorders and Neurodegenerative Diseases

In addition to functioning as a neuromodulator or an endocrine or paracrine factor, substance P seems to participate in inflammatory responses, since it stimulates the release of histamine from mast cells and the proliferation of T-lymphocytes and increases phagocytosis. It also augments the secretion of cytokines from macrophages and the secretion of TNF-α from astroglial cells. Interestingly, substance P/NK-1 receptor interactions elicit activation of signal transduction pathways which can initiate, or augment, inflammatory responses by astrocytes and microglia (Marriott 2004).

Pathological alterations of tachykinin concentrations – especially substance P – have been found *post mortem* in several brain structures of patients with neurodegenerative diseases. In the normal human cerebral cortex, a subset of aspiny local circuit neurons in deep cortical layers contains preprotachykinin mRNA and substance P immunoreactivity. These neurons are depleted in Alzheimer's disease – in contrast to other local circuit neurons – suggesting that they provide an early target of the degenerative process.

A decrease in the density of tachykinin receptors has been described under diverse neurodegenerative conditions. In the brain of Alzheimer's patients and in individuals suffering from amyotrophic lateral sclerosis (ALS), a marked reduction of NK1 receptors has been found *post mortem*. The reduction occurred in the nucleus of Meynert (Alzheimer's disease) and in the motor neurons of the ventral horn of the spinal cord (ALS). In both diseases, the effects are accompanied by a degeneration of cholinergic fibers, which are known to possess NK1 receptors. Thus, the degeneration of cholinergic neurons is considered to elicit deficiencies in the effects of synaptic transmission causally related to substance P.

In Huntington's chorea disease, substance P is depleted in the striatum concomitantly with the characteristic dorsoventral gradient of neuronal loss and its depletion in the pallidum and substantia nigra. In Parkinsonism, a reduction of substance P is also obvious in the substantia nigra and striatum.

In addition, there are reports that NK1 receptor antagonists have antidepressant activity comparable to that of the selective serotonin re-uptake inhibitor, paroxetine. Although preliminary studies showed some therapeutic efficacy for

NK1 antagonists, the first compound developed has been discontinued from Phase III trials because it was no more effective than a placebo in the treatment of depression (Adell 2004).

Further Reading

Adell, A. (2004): Antidepressant properties of substance P antagonists: relationship to monoaminergic mechanisms? *Curr. Drug Targets CNS Neurol. Disord.* **3**: 113–121.

Atzori, M., Nistri, A. (1996): Effects of thyrotropin-releasing hormone on GABAergic synaptic transmission of the rat hippocampus. *Eur. J. Neurosci.* **8**: 1299–1305.

Chang, M. M., Leeman, S. E., Niall, H. D. (1971): Amino-acid sequence of substance P. *Nat. New Biol.* **232**: 86–87.

Chen, L. W., Yung, K. K., Chan. Y. S. (2004): Neurokinin peptides and neurokinin receptors as potential therapeutic intervention targets of basal ganglia in the prevention and treatment of Parkinson's disease. *Curr. Drug. Targets* **5**: 197–206.

Danks, J. A., Rothman, R. B., Cascieri, M. A., Chicchi, G. G., Liang, T., Herkenham, M. (1986): A comparative autoradiographic study of the distributions of substance P and eledoisin binding sites in rat brain. *Brain Res.* **385**: 273–281.

Gether, U., Johansen, T. E., Snider, R. M., Lowe, J. A., Edmonds-Alt, X., Yokoda, Y., Nakanishi, S., Schwartz, T. (1993): Binding epitopes for peptide and non-peptide ligands on the NK1 receptor. *Regul. Peptides* **46**: 49–58.

Guard, S., Watson, S. P. (1991): Tachykinin receptor types: classification and membrane signalling mechanisms. *Neurochem. Int.* **18**: 149–165.

Iritani, S., Fujii, M., Satoh, K. (1989): The distribution of substance P in the cerebral cortex and hippocampal formation: an immunohistochemical study in the monkey and rat. *Brain Res. Bull.* **22**: 295–303.

Krause, J. E., Hershey, A. D., Dykema, P. E., Takeda, Y. (1990): Molecular biological studies on the diversity of chemical signalling in tachykinin peptidergic neurons. *Ann. N. Y. Acad. Sci.* **579**: 254–272.

Lee, C.-M., Iversen, L. L., Hanley, M. R., Sandberg, B. E. B. (1982): The possible existence of multiple receptors for substance P. *Arch. Pharmacol.* **318**: 281–287.

Luber-Narod, N., Kage, R., Leeman, S. E. (1994): Substance P enhances the secretion of tumor necrosis factor-a from neuroglia cells stimulated with lipopolysaccharide. *J. Immunol.* **152**: 819–824.

Maggi, C. A. (1995): The mammalian tachykinin receptors. *Gen. Pharmacol.* **26**: 911–944.

Maggio J. E. (1988): Tachykinins. *Annu. Rev. Neurosci.* **11**: 13–28.

Marriott, I. (2004): The role of tachykinins in central nervous system inflammatory responses. *Front. Biosci.* **9**: 2153–2165.

McLean S. (2005): Do substance P and the NK1 receptor have a role in depression and anxiety? *Curr. Pharm. Des.* **11**: 1529–1547.

Nicoll, R. A., Schenker, C., Leeman, S. E. (1980): Substance P as a transmitter candidate. *Annu. Rev. Neurosci.* **3**: 227–268.

Ohkubo, H., Nakanoshi, S. (1991): Molecular characterization of the three tachykinin receptors. *Ann. N. Y. Acad. Sci.* **632**: 53–62.

Otsuka, M., Yoshioka, K. (1993): Neurotransmitter functions of mammalian tachykinins. *Physiol. Rev.* **73**: 229–267.

Pernow, B. (1983): Substance P. *Pharmacol. Rev.* **35**: 85–141.

Regoli, R., Boudon, A., Fauchère, J. L. (1994): Receptors and antagonists for substance P and related peptides. *Pharmacol. Rev.* **46**: 551–599.

Sandberg, B. E. B., Iversen, L.L. (1982): Substance P. *J. Med. Chem.* **25**: 1009–1015.

Saria, S. (1999): The tachykinin NK_1 receptor in the brain: pharmacology and putative functions. *Eur. J. Pharmacol.* **375**: 51–60.

4.32
Thyrotropin-releasing Hormone

4.32.1
General Aspects and History

In 1969, two groups (Guillemin and colleagues and Schally and coworkers) succeeded in isolating a hypothalamic substrate that causes the anterior pituitary gland to release thyrotropin.

The substrate was the first biochemically identified hypothalamic factor and, for the first time, it was demonstrated that the secretory activity of the pituitary gland depends on peptidergic hypothalamic hormones. Besides its stimulatory action on the secretion of the thyroid-stimulating hormone (TSH) and some other hormones, this hypothalamic peptide showed additional effects on the central nervous system.

The peptide was designated thyreoliberin, protirelin or thyrotropin-releasing hormone (TRH). TRH is a small neuropeptide composed of three amino acids (L-pyroglutamyl-L-histidyl-L-prolineamide; L-pGlu-L-His-L-ProNH$_2$) and belongs to the class of tripeptidergic hormones (Fig. 4.53).

TRH is present in the central nervous system (thalamus, cerebral cortex and spinal cord) as well as in peripheral organs (pancreas, gastrointestinal tract and placenta).

When secreted into the portal hypophyseal circulation, it has access to the pituitary gland where it stimulates the secretion of the thyroid-stimulating hormone (TSH). This in turn enhances the production of tri- and tetrajodthyronin (thyroxin) in the thyroid gland.

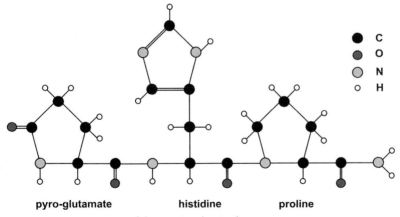

Fig. 4.53 Primary structure of thyrotropin-releasing hormone.

4.32.2
Localization Within the Central Nervous System

TRH-immunoreactive cell bodies are distributed throughout the central nervous system. They are predominantly clustered in the olfactory bulbs, the cortex, the hippocampus, the amygdala and in the paraventricular nucleus of the hypothalamus. TRH-containing cells are also scattered in some further brain areas, including the neostriatum, the arcuate nucleus of the hypothalamus, the tuberomammillary bodies, the reticular nucleus of the thalamus, the substantia nigra, the nucleus cuneatus and the nuclei of the nervus vagus (Fig. 4.54).

Furthermore, TRH-positive cells have been identified in the bed nucleus of the stria terminalis, the nucleus of the diagonal band of Broca and in several hypothalamic areas. Numerous TRH-containing cell bodies occur in the periaqueductal gray and the roots of the trigeminal nerve. In the medulla oblongata, TRH-containing cells are present in some raphe nuclei, in the area postrema and in the dorsal horn of the spinal cord. Immunohistochemical studies have demonstrated the coexistence in neurons of TRH with one or more neuroactive substances, including CCK, dopamine, enkephalins, GABA, growth hormone, histamine, NPY, serotonin and substance P.

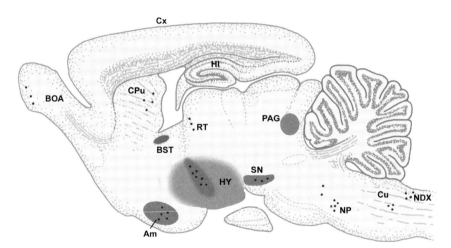

Fig. 4.54 Schematic representation of brain areas with TRH-synthesizing neurons
Abbreviations: BOa = accessory olfactory bulb; BST = bed nucleus of the *stria terminalis*; CPu = caudate-putamen; Cu = *nucleus cunealis*; Cx = cortex; HI = hippocampus; HY = hypothalamus; NDX = dorsal nucleus of the vagal nerve; NP = nuclei of the pons; RT = *nucleus reticularis* of the thalamus; SN = *substantia nigra*.

4.32.3
Biosynthesis and Degradation

TRH is synthesized from a precursor polypeptide whose sequence contains five copies of the TRH protein. The corresponding gene of the TRH precursor of rats was cloned in 1989 (Lee et al. 1989). The precursor consists of a putative 26-kDa protein which is composed of 255 amino acids. The rat gene contains five copies of the amino acid sequence of the TRH precursor, while the human pro-TRH contains six copies.

Translation of the mRNA gives rise to pro-TRH. Posttranslational processing of TRH starts with the excision of the precursor peptide by cleavage through a carboxy-peptidase. This is followed by amidation of proline by peptidyl-glycine a-amidating monoxygenase (PAMase), the amide moiety being donated by the carboxy-terminal glycine. Finally, cyclization of the amino-terminal glutamine is accomplished by glutaminyl cyclase. Posttranslational processing of TRH appears to be restricted to the neuronal cell, as indicated by lack of TRH progenitor immunoreactivity in axons or terminals.

The posttranslational processing of pro-TRH also gives rise to a number of other peptides, e.g. the peptides Ps4 and Ps5, which may have additional physiological properties.

It has been demonstrated that Ps4 and Ps5 are released in the hypothalamus in response to calcium-mediated neuronal depolarization. Ps4 seems to potentiate the thyreotrophic response as well as prolactin-mediated effects.

The peptide Ps4 has a stimulatory effect upon the expression of the prolactin gene and stimulates the transcription of β-TSH. In addition, in the pituitary, receptors have been found which bind Ps4 and which are different from the TRH receptors.

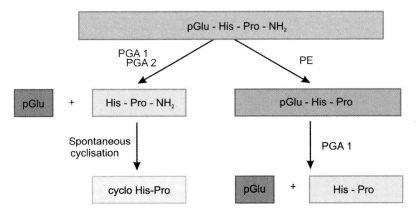

Fig. 4.55 Deactivation of TRH by enzymatic degradation. *Abbreviations*: His = histidine; PE = proline-endopeptidase; PGA 1 = pyroglutamate-aminopeptidase I; PGA 2 = pyroglutamate-aminopeptidase II; pGlu = pyroglutamate; Pro = proline.

Several TRH-degrading enzymes have been identified, including histidylproline imidopeptidase, prolylendopeptidase, pyroglutamyl aminopeptidase I (PGA 1, EC 3.4.19.3), pyroglutamyl aminopeptidase 2 (PGA 2, EC 3.4.19) and thyroliberinase. TRH released into the hypophyseal portal circulation is degraded rapidly to His-Pro-NH$_2$ by thyroliberinase, while TRH released into the synaptic clefts of the central nervous system is degraded by pyroglutamyl aminopeptidase 1 and 2 or by proline endopeptidase (PE, EC 3.4.21.26) and PGA 1 (Fig. 4.55). Developmental studies indicate differences in the degradation mechanisms during development and adulthood. In juveniles, PGA 1 and PE are expressed at high levels, whereas in adults the expression of both enzymes declines concomitantly with an increase in PGA 2 expression. PGA 2 has been characterized biochemically as well as by cloning. The cDNA codes for a putative protein of 1025 amino acids. PGA 2 is a dimeric glycoprotein composed of two identical subunits and possesses some characteristics of a metallopeptidase. To date, no specific inhibitor of PGA 2 is known, though some inhibitors of PGA 1 and PE have been characterized.

4.32.4
Receptors and Signal Transduction

In 1981, high-affinity binding sites for TRH ($K_d = 5.9 \times 10^{-9}$ M) have been described for the monkey brain (Ogawa et al. 1981). Significant progress has been made in investigating the TRH receptor complement, including cloning of TRH receptor cDNA and its corresponding gene, which in humans is located on chromosome 8q23.

Two isoforms of the rat TRH receptor have been shown to be generated from a single gene by alternative splicing. These isoforms are 387 and 412 amino acids in length and differ in their carboxy-terminal cytoplasmic tail.

TRH receptors belong to the class of G protein-coupled receptors with seven membrane-spanning domains and an extracellular amino-terminal region containing N-glycosylation sites. Receptors for TRH are found in cells of the anterior pituitary gland and in diverse neurons throughout the central nervous system. The structural and functional properties of both pituitary and central nervous TRH receptors have been characterized in detail and shown to be similar in structure and signal transduction.

The TRH receptors are coupled to the phospholipase C pathway via a pertussis toxin-insensitive G protein resulting in the liberation of calcium. TRH also seems to stimulate MAP kinases by proteinkinase C and tyrosine kinase-dependent mechanisms. TRH receptors reveal a negative autoregulation in the presence of high levels of TRH. This down-regulation is mediated by a reduction of TRH receptors, without any change in their affinity.

TRH receptors are widely distributed throughout the central nervous system. High densities have been detected in the anterior part of the pituitary, the cerebral cortex, the posterior mammillary nucleus, the arcuate nucleus, the hippocampus, the anterior cortical nucleus and in the lateral nucleus of the amygdala.

Kohortenstudie

Mit Antihypertensiva das Risiko senken?

Morbus Parkinson basiert auf einem neurodegenerativen Prozess, der zum Untergang dopaminerger Neuronen führt. Gibt es modifizierbare Risikofaktoren?

In eine Kohortenstudie aus Taiwan wurden nur Patienten mit Hypertonie (n = 65 001) eingeschlossen. Unter den Antihypertensiva wurden Betablocker als Referenzsubstanz gewählt.

Die Einnahme von Kalziumantagonisten des Dihydropyridin-Typs war mit einem verminderten Risiko für eine Erkrankung assoziiert (adjustierte Hazard Ratio 0,71). Dabei ergab sich ein Dosis-Wirkungs-Effekt. Dies galt nicht für Nicht-Dihydropyridin-Kalziumantagonisten. Man unterschied weiter zwischen ZNS-wirksamen (lipophilen) und nicht zentral wirkenden Kalziumblockern. Nur der Einsatz zentral effektiver Präparate war mit einer signifikanten Reduktion des Parkinson-Risikos verbunden (aOR: 0,69). Bei hohen kumulativen Dosen traf diese Unterscheidung allerdings nicht durchgängig zu.

Die Einnahme von ACE-Hemmern und AT$_1$-Blockern war insgesamt mit geringeren Reduktionen des Parkinson-Risikos verbunden, jedoch nicht statistisch signifikant. Betrachtete man nur die Patienten mit höheren kumulativen Dosen, ergab sich für beide Substanzklassen jedoch wiederum eine signifikante Assoziation.

Dass in dieser Untersuchung ZNS-gängige Kalziumblocker protektiv waren, ist biologisch plausibel. Man spekuliert, dass sie bestimmte Kalziumkanäle dopaminerger Neuronen hemmen und so die Degeneration aufhalten. Allerdings hatte auch Amlodipin in höheren Dosen einen protektiven Effekt. Demnach müsste es noch andere Mechanismen geben, die vor Parkinson schützen. Der Effekt von ACE-Hemmern und Sartanen in höheren Dosen könnte mit deren antioxidativen Eigenschaften zusammenhängen.

Weitere Langzeitstudien sind notwendig, um die Rolle von Antihypertensiva für das Parkinson-Risiko genauer abzustecken. **WE**

K Lee Y-C et al.: Antihypertensive agents and risk of Parkinson's disease: a nationwide cohort study. PLoS One 2014; 9(6) e98961 (Epub)
Mehr Infos: www.neuro-depesche.de/141389

Medikinet® adult

Das erste zugelassene Methylphenidat-Präparat für ADHS im Erwachsenenalter

0 mg. Wirkst.: Methylphenidathydrochlorid. **Zus.setzg.:** 1 Hartkps. enthält Methylphenidathydrochlorid 5 mg/10 mg/20 m isstärke, Methacrylsäure-ethylacrylat-Copolymer, Talkum, Trietylcitrat, Poly(vinylalkohol), Macrogol 3350, Polysorbat 80, Na-hyd Methylcellul., Sorbinsäure, Indigocarmin, Al-hydroxid. Kps.hülle: Gelatine, Ti-dioxid, Na-dodecylsulfat, ger. Wasser; zusätzl. b. 10 r ⟶. 40 mg: Erythrosin, Fe(II,III)-oxid, Indigocarmin. **Anw.:** Im Rahmen einer therap. Ges.strategie zur Behandl. einer seit Kindesal (ADHS) bei Erw. ab 18 J., wenn sich and. therap. Maßn. allein als unzureich. erwiesen haben. Die Behandl. muss unter Aufsic Diagn. erfolgt angelehnt an DSM-IV Krit. o. Richt. in ICD-10 u. basiert auf e. vollst. Anamn. u. Unters. d. Pat. Diese schließen akt. Sympt. d. Pat. ein. Die retrosp. Erf. d. Vorbest. e. ADHS im Kindesalt. muss anhand val. Instr. erfolgen. Die Diagn. darf si t. stützen. **Gegenanz.:** Bek. Überempfindlichkeit gg. Methylphenidat o. einen der sonst. Bestandt.; Glaukom; Phäochromozyto nmlern; Hyperthyreose o. Thyreotoxikose; Diagn. o. Anamn. v. schw. Depr.; Anorexia nerv./anorekt. Störg.; Suizidneig.; psych path./Borderline-Pers.k.störg.; Diagn. o. Anamn. v. schw. u. episod. (Typ I) bipol. affekt. Störg.; vorbest. Herz-Kreislauf-Erkr. einsc ., Angina pec., hämodyn. signifik., angeb. Herzfehler, Kardiomyopathien, Myokardinf., Arrhythmien u. Kanalopathien; vorbest. ze mit pH-Wert > 5,5, bei H2-Rezeptorenblocker- o. Antazidatherapie. **Nebenw.:** Sehr häufig: Schlaflosigk., Nervos., Kopfschm. Hä Gewichts- u. Größenzunahme b. längerer Anw. b. Kindern; Affektlab., Aggression, Unruhe, Angst, Depression, Reizbark., anor nnolenz; Arrhythmie, Tachykardie, Palpitationen; Hypertonie; Husten, Rachen- u. Kehlkopfschm.; Bauchschm., Durchfall, Übelke all, Pruritus, Hautausschl., Urtikaria; Arthralgie; Fieber; Veränd. v. Blutdr. u. Herzfreq. Gelegentlich: Überempf.keitsreakt., wie z bullöse u. exfol. Hauterkrank., Juckreiz, Eruptionen; psychot. Störg.; akust., opt. u. takt. Halluzinationen; Wut, Selbstmordgedan chk.; Tics, Verschlecht. best. Tics o. Tourette-Syndrom, erh. Wachheit, Schlafstörg.; Sedierung, Tremor; Diplopie, verschw. Sehe ym.; Myalgie, Muskelzucken; Hämaturie; Müdigk.; Herzgeräusch. Selten: Manie, Desorientierth., Libidostörg., Probl. b. d. Auge rosis, makul. Hautausschl., Erythem; Gynäkomastie. Sehr selten: Anämie, Leukopenie, Thrombozytopenie, thrombozytop. Purpu g. depr. Verstimmung, anorm. Denken, Apathie, stereotype Verh. Beweg., revers. ischäm. neurol. Defizit, NMS (schwach dokumen- tf.verschl.; periphere Kälte, Raynaud-Phänom.; gestört. Leberfunkt. tis, fix. AM-Exanthem; Muskelkrämpfe, plötzl. Herztod; erhöhte al- zahl, anorm. Zahl d. weißen Blutkörp. Nicht bekannt: Panzytopenie, provask. Erkrank. (einschl. Vaskulitis, Hirnblutungen, Schlaganf., Migräne, supra-ventrikul. Tachykardie, Bradykardie, ventrikul. Extra- beschw., Hyperpyrexie. Hinw.: Es wurden Fälle v. Abhängigk. u. Miss- sierungen. **Warnhinw.:** Enth. Sucrose. **Verschreibungspflichtig.** & Co. KG, 58638 Iserlohn. www.medikinet.de Stand: 05/2013

GEMEINSAM ADHS BEGEGNEN
MEDICE – DIE ERSTE WAHL

Moderate densities of TRH receptors have been detected in the frontal perirhinal and temporal cortex, the accessory olfactory bulb, the habenula, the medial septum, the bed nucleus of the stria terminalis, the nucleus accumbens, the preoptic area, the dorsomedial and ventromedial nucleus of the hypothalamus, the basomedial nucleus of the amygdala and in the posterior part of the pituitary.

Some data indicate that a second central nervous receptor exists (Itadani et al. 1998), which binds TRH and is different from the known TRH receptor. This receptor is labeled TRHR2. The TRHR2 receptor displays an overall homology of 50% with the common TRH1 receptor. The mRNA for TRHR2, however, appears to be restricted to, or at least predominately expressed in, the brain. Although both receptor subtypes are expressed in the rat brain, the pattern of TRHR2 mRNA expression is entirely different from that of the other TRH receptor. One striking example is the anterior pituitary gland, where there are large numbers of the common TRH receptor, but no expression of TRHR2. Conversely, TRHR2 is selectively expressed in the dorsal horn of the spinal cord.

4.32.5
Biological Effects

The primary physiological effect of TRH is to stimulate secretion of the thyroid-stimulating hormone (TSH). Secretion of TRH, however, is under control of the thyroid hormones T3 and thyroxin (T4). These hormones exert negative feedback control on thyreotropic cells of the pituitary.

TRH also stimulates the release of prolactin, but this is not regarded as the main stimulatory pathway for prolactin release. TRH release is also subject to control through neurotransmitters. Among these are norepinephrine (which stimulates TRH secretion) and dopamine (which inhibits the release of TRH).

TRH has been reported to reduce stress- and deprivation-induced eating, hypothetically by induction of satiation. Early work demonstrated thyroid extracts reduced alcohol intake; and recent research shows a TRH analog specifically inhibits alcohol preference. It seems likely that TRH is one of several functional elements in the integrative neuropeptide control of alcohol consumption via short-term satiation (Kulkowsky et al. 2000).

TRH levels seem to affect motor behavior. Peripherally or locally administered TRH stimulates motor activity, an effect which involves enhanced turnover of the mesolimbic dopaminergic system and of the noradrenergic system. TRH also affects spinal cord motor activity since TRH, as well as serotonin, facilitates the discharge of motoneurons. Tissue culture experiments show a direct involvement of TRH in the development and activity of spinal motoneurons. TRH induces trophic effects, stimulates axonal outgrowth and the activity of choline actyltransferases.

TRH is also a mediator of peripheral effects. Intravenously applied TRH can alter blood circulation. In the periphery of the body, TRH application induces vasoconstriction, whereas in the brain vasodilatatory effects predominate.

TRH also induces gastro-intestinal effects, since it stimulates secretion of pepsin. However, these effects are variable, depending on the species and the manner in which TRH is administered. Gastrointestinal effects of TRH include an involvement in the pathogenesis of ulcer. The formation of a gastric ulcer can be experimentally induced by intracisternal injection of TRH, which mediates an increase in peripheral vagal activity through an enhancement of discharge from vagal nuclei. Curiously, this effect is accompanied by a reduction of the secretion of gastric acids.

4.32.6
Neurological Disorders and Neurodegenerative Diseases

Several studies have focused on pathological modifications of TRH receptors or of TRH levels in the central nervous system. TRH is present, for example, in the ventral horn of the spinal cord. A decrease in TRH level and TRH receptor density has been found in amyotrophic lateral sclerosis, a degenerative disease that primarily afflicts primarily the motor neurons of the spinal cord. TRH was also found to enhance motor neuron firing, leading to increased muscle tone, contractibility and spinal reflexivity. Clinical studies have repeatedly reported beneficial effects of TRH administration in several motor neuron diseases. However, these improvements, for some unexplained reason, are of short duration (Brooke 1989).

TRH levels increase during epileptic convulsions. In spontaneously epileptic rats, the level of TRH as well as the level of its mRNA (coding for the precursor prepro-TRH) are enhanced in the hippocampus during the first 24 h after the onset of seizure activity.

Further Reading

Brooke, M. H. (1989): Thyrotropin-releasing hormone in ALS. Are the results of clinical studies inconsistent? *Ann. N.Y. Acad. Sci.* **553**: 422–430.

Cao, J., O'Donnell, D., Vu, H., Payza, K., Pou, C., Godbout, C., Jakob, A., Pelletier, M., Lembo, P., Ahmad, S., Walker, P. (1998): Cloning and characterization of a cDNA encoding a novel subtype of rat thyrotropin-releasing hormone receptor. *J. Biol. Chem.* **273**: 32281–32287.

Guillemin R. R. (2005): Hypothalamic hormones a.k.a. hypothalamic releasing factors. *J. Endocrinol.* **184**: 11–28.

Horita, A: (1998): An update on the CNS action of TRH and its analogs. *Life Sci.* **62**: 1443–1448.

Itadani, H., Nakamura, T., Itoh, J., Iwaasa, H., Kanatani, A., Borkowski, J., Ihara, M., Ohta, M. (1998): Cloning and characterization of a new subtype of thyrotropin-releasing hormone receptors. *Biochem. Biophys. Res. Commun.* **250**: 68–71.

Kerdelhue, B., Palkovits, M., Karteszi, M., Reinberg, A. (1981): Circadian variations in substance P, Luliberin (LH-RH) and thyroliberin (TRH) contents in hypothalamic and extra-hypothalamic brain nuclei of adult male rats. *Brain Res.* **206**: 405–413.

Kubek, M. J., Low, W. C., Sattin, A., Morzorati, S. L., Meyerhoff, J. L., Larsen, S. H. (1989): Role of TRH in seizure modulation. *Ann. N.Y. Acad. Sci.* **553**: 286–302.

Kulkosky, P. J., Allison, C. T., Mattson, B. J. (2000): Thyrotropin releasing hormone decreases alcohol intake and preference in rats. *Alcohol* **20**: 87–91.

Ladam, A., Bulant, M., Delfour, Montagne, J.J., Vaudry, H., Nocolas, P. (1994): Modulation of the biological activity of thyrotropin-releasing hormone by alternate processing of pro-TRH. *Biochimie* **76**: 320–328.

Lechan, R.M., Wu, P., Jackson, I.M.D., Wolf, H., Cooperman, S., Mandel, G., Goodman, R.H. (1986): Thyreotropin-releasing hormone precursor: Characterization in rat brain. *Science* **231**: 159–161.

Lee, S.L., Sevarino, K., Roos, B.A., Goodman, R.H. (1989): Characterization and expression of the gene-encoding rat thyrotropin-releasing hormone (TRH). *Ann. N.Y. Acad. Sci.* **53**: 14–28.

Ogawa, N., Yamawaki, Y., Kuroda, H., Ofuji, T., Itoga, E., Kito, S. (1981): Discrete regional distributions of thyrotropin releasing hormone (TRH) receptor binding in monkey central nervous system. *Brain Res.* **205**: 169–174.

Pazos, A., Cortes, R., Palacios, J.M. (1985): Thyreotropin-releasing hormone receptor binding sites: autoradiographic distribution in the rat and guinea pig brain. *J. Neurochem.* **45**: 1448–1463.

Perlman, J.H., Laakonen, L., Osman, R., Gershengorn, M.C. (1994): Thyrotropin-releasing hormone (TRH) receptor binding pocket. *J. Biol. Chem.* **23**: 23383–23385.

Pilotte, N.S., Sharif, N.A., Burt, D.R. (1984): Characterization and autoradiographic localization of TRH receptors in sections of rat brain. *Brain Res.* **293**: 372–376.

Schauder, B., Schomburg, L., Köhrle, J., Bauer, K. (1994): Cloning of cDNA encoding an entoenzyme that degrades thyrotropin-releasing hormone. *Proc. Natl Acad. Sci. USA* **91**: 9534–9538.

Tixier-Vidal, A., Faivre-Bauman, A. (1992): Ontogeny of TRH biosynthesis and release inhypothalamic neurons. *Trends Endocrinol. Metab.* **3**: 59–64.

Zabavnik, J., Arbuthnott, G., Eigne, K.A. (1993): Distribution of thyrotropin-releasing hormone receptor messenger RNA in rat pituitary and brain. *Neuroscience* **53**: 877–878.

4.33
The Tyr-MIF-1 Family

Four small peptides (Fig. 4.56) with almost identical primary structures belonging to the group of the "melanocyte-stimulating hormone release-inhibiting hormones" have been isolated from brain tissues. Alternatively, they are termed "melanostatins". The four peptides belong to the so-called "Tyr-MIF-1 family".

Two of the four peptides (Tyr-MIF-1 and Tyr-W-MIF-1) bind to specific Tyr-MIF-1-binding sites and in addition show binding capacity to the opioid μ receptor. Tyr-MIF-1 is a peptide with a molecular mass of 447.51 Da and has been isolated from bovine hypothalamus and human parietal cortex.

MIF-1	Pro-Leu-Gly-NH$_2$
Tyr-MIF-1	Tyr-Pro-Leu-Gly-NH$_2$
Tyr-W-MIF-1	Tyr-Pro-Trp-Gly-NH$_2$
Tyr-K-MIF-1	Tyr-Pro-Lys-Gly-NH$_2$

Fig. 4.56 Amino acid sequences of the four members of Tyr-MIF-1 family.

The third member of this family, MIF-1, does not bind to any of these receptors. It is, however, expressed in various structures of the brain, including the cortex, the striatum, the diencephalon, the midbrain and the pons.

The fourth member of the Tyr-MIF-1 family is Tyr-K-MIF-1 (Reed et al. 1994). This peptide recognizes a specific binding site in the brain and does not interact with μ receptors.

The Tyr-MIF-1 family interacts with functions mediated by the opioid system, by binding to $\mu 1$ and $\mu 2$ receptors. In this context, Tyr-W-MIF-1 functions as a mixed $\mu 2$-opioid receptor agonist and $\mu 1$-opioid receptor antagonist (Gergen et al. 1996). These findings lend support to the idea that Tyr-MIF-1 belongs to an endogenous central nervous antiopiate system. Binding sites for Tyr-MIF-1 are significantly decreased in animals chronically treated with morphine. By examining the effects of the Tyr-MIF-1 peptides in rats on morphine-induced analgesia in acute pain and on immobilization stress-induced antinociception, it has been shown that Tyr-MIF-1 peptides significantly decreased the analgesic effect of morphine.

In addition to its antiopiate effects, Tyr-MIF-1 interacts with non-opiod systems, including the dopaminergic and the GABAergic system.

Intravascular Tyr-MIF-1 (amino acid sequence: Tyr-Pro-Leu-Gly-NH$_2$) is metabolized to Tyr-Pro and to Leu-Gly-NH$_2$, indicating that the circulating Tyr-MIF-1 is not a precursor of MIF-1. However, Tyr-MIF-1 can be metabolized to MIF-1 in mitochondria of the brain.

The half-lives of MIF-1 and Tyr-MIF-1 in the blood of rodents (minutes) are very different from those of humans (days). A functional explanation for this difference is not yet available.

Tyr-MIF-1 and MIF-1 pass the blood–brain barrier in a differential manner. Tyr-MIF-1 and Tyr-W-MIF-1 are saturably transported from brain to blood. While Tyr-MIF-1 crosses the barrier from the parenchyma to the blood, MIF-1 is transported exclusively in the opposite direction. Tyr-MIF-1 and Tyr-W-MIF-1 bind to $\mu 1$ and $\mu 2$ receptors; and it is believed that they exhibit antagonistic or agonistic effects on these receptors. In conjunction with the opioid receptor interaction, Tyr-W-MIF-1 induces analgesia.

Destruction of the pituitary gland leads to a large increase in Tyr-MIF-1 concentration in the circulating blood. This increase can be prevented by chronic administration of corticosterone, but not by chronic administration of other hormones, e.g. thyroxin, estrogen, progesterone, or by short administration of corticosterone.

Further Reading

Bocheva, A., Dzambazova-Maximova, E. (2004): Antiopioid properties of the TYR-MIF-1 family. *Methods Find. Exp. Clin. Pharmacol.* **26**: 673–677.

Erchegyi, J., Kastin, A. J., Zadina, J. E. (1992): Isolation of a novel tetrapeptide with opiate and antiopiate activity from human brain cortex: Tyr-Pro-Gly-NH$_2$ (Tyr-W-MIF-1). *Peptides* **13**: 623–631.

Galina, Z. H., Kastin, A. J. (1986): Existance of antiopiate systems as illustrated by MIF-1/Tyr-MIF-1. *Life Sci.* **39**: 2153–2159.

Gergen, K. A., Zadina, J. E., Paul, D. (1996): Analgesic effects of Tyr-W-MIF-1: a mixed mu2-opioid receptor agonist/mu1-opioid receptor antagonist. *Eur. J. Pharmacol.* **316**: 33–38.

Hackler, L., Kastin, A. J., Zadina, J. E. (1994): Isolation of a novel peptide with a unique binding profile from human brain cortex: Tyr-K-MIF-1 (Tyr-Pro-Lys-Gly-NH$_2$). *Peptides* **15**: 945–950.

Kastin, A. J., Olson, R. D., Ehrensing, R. H., Berzas, M. C., Schally, A. V., Coy, D. H. (1979): MIF-1s differential actions as an opiate antagonist. *Pharmacol. Biochem. Behav.* **11**: 721–723.

Kastin, A. J., Stephens, E., Ehrensing, R. H., Fischman, A. J. (1984): Tyr-MIF-1 acts as an opiate antagonist in the tail-flick test. *Pharmacol. Biochem. Behav.* **21**: 937–941.

Kastin, A. J. Hahn, K., Zadina, J. E., Banks, W. A., Hackler, L. (1995): Melanocyte-stimulating hormone release-inhibiting factor-1 (MIF-1) can be formed from Tyr-MIF-1 in brain mitochondria but not in brain homogenate. *J. Neurochem.* **64**: 1855–1859.

Kastin, A. J., Hahn, K., Banks, W. A., Zadina, J. E. (1998): Regional differences in the metabolism of Tyr-MIF-1 and Tyr-W-MIF-1 by rat brain mitochondria. *Biochem. Pharmacol.* **55**: 33–36.

Kastin, A. J., Pan, W., Maness, L. M., Banks, W. A. (1999): Peptides crossing the blood–brain barrier: some unusual observations. *Brain Res.* **848**: 96–100.

Reed, G. W., Olson, G. A., Olson, R. D. (1994): The Tyr-MIF-1 family of peptides. *Neurosci. Biobehav. Rev.* **18**: 519–525.

Zadina, J. E., Kastin, A. J., Ge, L. J., Gulden, H., Bungart, K. J. (1989): Chronic, but not acute administration of morphine alters antiopiate (Tyr-MIF-1) binding sites in rat brain. *Life Sci.* **44**: 555–561.

Zadina, J. E., Kastin, A. J., Kersh, D., Wyatt, A. (1992): Tyr-MIF-1 and hemorphin can act as opiate agonists as well as antagonists in the guinea pig ileum. *Life Sci.* **51**: 869–885.

4.34
Vasoactive Intestinal Polypeptide

4.34.1
General Aspects and History

In the early 1970s (Said and Mutt 1970), vasoactive intestinal polypeptide (VIP) was isolated from intestinal extracts. Pharmaco-physiological studies indicated that VIP exhibits potent vasodilatatory activity (Smitherman et al. 1988). Because of this strong vasoactive effect, the peptide was termed vasoactive intestinal polypeptide.

Originally, the peptide was recognized as a gastrointestinal hormone, but later it was discovered that VIP is a neuropeptide with neuromodulatory properties. It is now generally accepted that VIP represents a widely distributed neuropeptide in the central and peripheral nervous system. VIP is a peptide of 28 amino acids with a molecular mass of 3326 Da.

It belongs to a family of structurally and functionally closely related proteins, the other members of which are: glucagon, secretin, corticotropin-releasing hormone (CRH), growth hormone-releasing hormone (GHRH) and pituitary adenylate cyclase-activating peptide (PACAP). The primary structure of VIP shows a high degree of conservation between human, rat, mouse, pig, sheep, dog and goat. The guinea pig VIP differs slightly (in three amino acids) from that of other mammals.

4.34.2
Localization Within the Central Nervous System

VIP-carrying neurons have been identified by immunocytochemistry and by *in situ* hybridization in various areas of the central nervous system (Fig. 4.57). Within the brain, the cerebral cortex exhibits the highest concentration of VIP. VIP-like immunoreactivity and its corresponding mRNA have been found in interneurons throughout the neocortical neurons. VIP is also present in the hippocampus, where it is confined to single interneurons, both in Ammon's horn and in the dentate gyrus.

VIP-positive neurons are found in several amygdaloid nuclei, in the thalamus where it occurs in the ventromedial, ventrolateral and ventroposterior nucleus, as well as in the reticular thalamic nucleus and in the geniculate nucleus. The hypothalamic suprachiasmatic nucleus shows a high concentration of VIP-positive neurons, while infrequent numbers are present in the supramamillary nucleus. In addition, VIP-containing neurons have been identified in the periaqueductal gray and the dorsal raphe.

VIP often colocalizes with additional neuromodulators and neurotransmitters, as is also the case for several other neuropeptides. Colocalization has been described with acetylcholine in cholinergic neurons and with GABAergic interneurons and CCK-containing neurons of the hippocampus. In the ventral thalamus, colocalization occurs with cholecystokinin, and in the reticular thalamic nucleus with GABA. Similarly, GABA coexists with VIP in hypothalamic neurons, some of which also contain galanin.

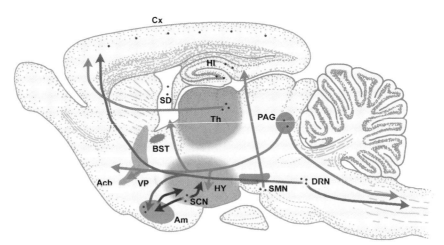

Fig. 4.57 Schematic overview of the distribution of VIP-synthesizing neurons. *Abbreviations*: Acb = *nucleus accumbens*; BST = bed nucleus of the *stria terminalis*; Cx = cortex; DRN = dorsal raphe nucleus; HI = hippocampus; HY = hypothalamus; PAG = periaqueductal gray; SCN = suprachiasmatic nucleus; SD = septum, dorsal; SMN = supramamillary nucleus; Th = thalamus; VP = ventral pallidum.

4.34.3
Biosynthesis and Degradation

The precusor gene of VIP has been cloned and its putative primary structure deduced. Sequence homology is about 80% between mice and human.

The human VIP gene is located on chromosome 6 in position p24. The gene contains seven exons, separated by corresponding introns. Each of these exons encodes a functional domain of the protein precursor. The initial product is the prepro-VIP (20 kDa). This 70-amino-acid peptide not only provides the precursor for VIP, but also for PHI (peptide with amino-terminal histidine and carboxy-terminal isoleucine amide) and PHM (peptide with amino-terminal histidine and carboxy-terminal methionine amide).

The precursor is processed through proteolytic cleavage to form VIP (Fig. 4.58) and PHI; and it shows tissue-dependent variability which gives rise to variable amounts of VIP and PHI.

The exact steps in the degradation of VIP have not been elucidated. Since VIP contains several basic amino acids in its sequence, it is considered to constitute a feasible target for unspecific peptidases in the central nervous system and in the periphery.

4.34.4
Receptors and Signal Transduction

Specific VIP-binding sites have been detected in diverse tissues, including the alimentary tract, the liver and the central nervous system. Two classes of VIP receptors can be differentiated. The first consists of the VIP receptors (VPAC1 and VPAC2) and the second consists of the PACAP receptor (PAC1), which binds pituitary adenylate cyclase-activating polypeptide (PACAP) and VIP.

This nomenclature is now generally accepted. For a better understanding of the literature, some of the older terms and their modern counterparts (in italics) to the current are listed below:
- The *VPAC1* receptor is also the VIP1, PVR2 or PACAP type 2 receptor.
- The *VPAC2* receptor is also the VIP$_2$, PVR3 or PACAP type 3 receptor.
- The *PAC1* receptor is also the PVR1 or PACAP type 1 receptor.

Both VPAC receptors have been cloned and their genes characterized. The human VPAC 1 gene is about 22 kb in length and consists of 13 exons. The human gene locus is 3p22. The putative primary sequence consists of 459 amino acids in rats and 457 amino acids in humans.

His-Ser-Asp-Ala-Val-Phe-Thr-Asp-Asn-Tyr-Thr-Arg-Leu-Arg-Lys-Gln-Met-Ala-Val-Lys-Lys-Tyr-Leu-Asn-Ser-Ile-Leu-Asn-NH$_2$

Fig. 4.58 Amino acid sequence of intestinal vasoactive polypeptide (VIP).

The VPAC1 receptor has a molecular mass of 52 kDa and belongs to the family of adenylate cyclase-stimulating G_s protein-coupled receptors, containing seven transmembrane domains. The VPAC1 receptor is structurally related to the secretin and calcitonin hormone receptors. The VPAC2 receptor gene consists of 13 exons and spans at least 40 kb. In humans, the gene locus is 7q36.3. The VPAC2 receptor also forms seven transmembrane domains.

Both receptors are coupled to G proteins; however the type of G protein (G_s, G_i or G_o) which is activated may be dependent upon species and cell type (Shreeve et al. 2000).

The biological activity of VIP requires its complete sequence. Fragments of VIP and some analogs can act as specific and competitive antagonists. Among these analogs are [4-Cl-D-Phe^6,Leu^{17}]-VIP, VIP(10–28), and an analog of the growth hormone-releasing factor, [N-Ac-Tyr^1, D-Phe^2]-growth hormone-releasing factor (1–29)-NH$_2$.

4.34.5
Biological Effects

A wide spectrum of biological effects has been attributed to VIP. Some of these effects are summarized in Table 4.9. Although the full range of VIP functions is still unclear, strong evidence indicates that VIP mediates several basic physiological effects including:
- relaxation of smooth muscle cells
- enhancement of electrolyte and diuresis
- regulation of neuroendocrine and endocrine functions.

Vasoactive intestinal peptide stimulates salt secretion by the mammalian intestine. VIP seems to modulate renal tubular reabsorption, by increasing urine volume, fractional excretion of sodium, chloride and potassium, as well as by osmolar clearance (Rosa et al. 1985).

The effect of VIP in relaxing blood vessels and smooth muscle cells is assumed to be mediated partially by a VIP-induced release of nitric oxide (NO).

Table 4.9 Some actions of VIP in different organs.

Place of action	Effect
Cardiovascular system	Hypotension, vasodilatation
Liver	Stimulation of water secretion
Pancreas	Release of insulin
Peripheral nervous system	Modulation of the cholinergic transmission
Pituitary, hypothalamus	Release of GH, LH and prolactin
Respiratory system	Bronchodilation, vasodilatation, increased ventilation
Stomach	Suppression of acid production

The suprachiasmatic nucleus (SCN) contains the predominant circadian pacemaker in mammals. Considerable evidence indicates that VPAC(2) and PAC(1), receptors for VIP and PACAP, play critical roles in maintaining and entraining circadian rhythms. Since VIP is expressed diurnally – with high levels during the dark periods – in the ventrolateral portion of the suprachiasmatic nucleus, a brain region responsible for the regulation of circadian rhythms, VIP is considered to play a substantial role in the control of diurnal functions; and VIP signaling between suprachiasmatic neurons is supposed to provide a paracrine reinforcing signal that is essential for sustained rhythm generation. It is thought that the daily VIP rhythm is first generated in the early developed clock-controlled rostral suprachiasmatic neurons and is later regulated by light-dependent neurons of the ventrolateral portion of the suprachiasmatic nucleus (Ban et al. 1997).

Several effects of VIP on sleep have also been described. Central administration of VIP enhances periods of rapid eye movement (REM) and increases the duration of REM sleep. The effects on sleep seem to involve prolactin secretion.

VIP is a well known secretagogue for pituitary prolactin, acting both at the level of both the hypothalamus and the pituitary gland. VIP-mediated prolongation of REM sleep can be prevented by co-administration of antiserum to prolactin.

A wide range of events fundamental to brain development (including cell proliferation, differentiation, neurite outgrowth and neuronal survival) is influenced by VIP. Additionally, VIP is upregulated after neural injury and has potent neuroprotective properties.

VIP also exerts some effects on glia cells. Since VIP expression is confined to neurons and has not been detected in glia, neuroglial communication via VIP seems to be unidirectional.

One prominent glial function influenced by VIP is energy metabolism in astrocytes. VIP (as well as norepinephrine) helps to control local energy homeostasis through the stimulation of glycogenolysis, an effect that occurs within minutes after VIP application in astrocytic cultures (Sorg and Magistretti 1991). Following glycogenolysis, VIP stimulates the resynthesis of glycogen, which peaks at levels which are 5- to 10-fold greater than before treatment. VIP is a neuropeptide that is expressed late in development. In prenatal rat brains, VIP secretion is almost absent, but increases rapidly postnatally, with a maximum about 2 weeks after parturition. In contrast to the delayed appearance of VIP in the brain, VIP-binding sites appear very early in development and are primarily, if not exclusively, localized in the central nervous system.

The presence of abundant VIP-binding sites in early embryonic brains, long before the onset of neuronal VIP expression, indicates that the brain is supplied from peripheral VIP sources during early neurogenesis.

During neurogenesis, VIP stimulates the proliferation and differentiation of brain neurons. The addition of VIP to embryonic mouse spinal cord cultures, for example, was found to increase neuronal survival and activity. Moreover, *in vitro* studies provide evidence that VIP induces neuronal mitosis and neuronal outgrowth. Blockade of VIP in early postnatal stages can induce neuronal malformation, indicative of some neurogenetic effects of VIP during critical periods of brain development. Interestingly, VIP has neuroprotective effects against excitotoxic lesions (induced by intracerebral administration of ibotenate) of the developing mouse brain.

4.34.6
Neurological Disorders and Neurodegenerative Diseases

The profound vasodilatory action of VIP on cerebral blood vessels and its localization in structures involved in pain pathways is taken as an indication that VIP may be involved in some kinds of headache.

Since VIP seems able to prevent the cell death of cortical neurons induced by beta amyloid, a neuroprotective effect in Alzheimer's disease has also been proposed.

Further Reading

Acsady, L., Arabadzisz, D., Freund, T. F. (1996): Correlated morphological and neurochemical features identify different subsets of vasoactive intestinal polypeptide-immnureactive interneurons in rat hippocampus. *Neuroscience* **73**: 299–315.

Ban, Y., Shigeyoshi, Y., Okamura, H. (1997): Development of vasoactive intestinal peptide mRNA rhythm in the rat suprachiasmatic nucleus. *J. Neurosci.* **17**: 3920–3931.

Di Cicco-Bloom, E. (1996): Region-specific regulation of neurogenesis by VIP and PACAP: direct and indirect modes of action. *Ann. N.Y. Acad. Sci.* **805**: 244–256.

Fahrenkrug, J. (1989): Transmitter role of vasoactive intestinal peptide. *Pharmacol. Toxicol.* **72**: 354–363.

Gressens, P. (1999): VIP neuroprotection against excitotoxic lesions of the developing mouse brain. *Ann. N.Y. Acad. Sci.* **897**: 109–124.

Harmar, A. J. (2003): An essential role for peptidergic signalling in the control of circadian rhythms in the suprachiasmatic nuclei. *J. Neuroendocrinol.* **15**: 335–358.

Harmar, T., Lutz, E. (1994): Multiple receptors for PACAP and VIP. *Trends Pharmacol. Sci.* **15**: 97–99.

Ishihara, T., Shigemoto, R., Mori, K., Takahashi, K., Nagata, S. (1992): Functional expression and tissue distribution of a novel receptor for vasoactive intestinal polypeptide. *Neuron* **8**: 811–819.

Kalamatianos, T., Kallo, I., Piggins, H. D., Coen, C. W. (2004): Expression of VIP and/or PACAP receptor mRNA in peptide synthesizing cells within the suprachiasmatic nucleus of the rat and in its efferent target sites. *J. Comp. Neurol.* **475**: 19–35.

Klimaschewski, L. (1997): VIP – a 'very important peptide' in the sympathetic nervous system? *Anat. Embryol. Berl.* **196**: 269–277.

Kohler, C. (1982): Distribution and morphology of vasoactive intestinal polypeptide-like immunreactive neurons in regio superior of the rat hippocampal formation. *Neurosci. Lett.* **33**: 265–270.

Kozicz, T., Vigh, S., Arimura, A. (1998): Immunohistochemical evidence for PACAP and VIP interaction with meta-enkephalin and CRF containing neurons in the bed nucleus of the stria terminalis. *Ann. N.Y. Acad. Sci.* **865**: 523–528.

Maggi, C. A., Giachetti, A., Dey, R. D., Said, S. I. (1995): Neuropeptides as regulators of airway functions: with special reference to VIP and the tachykinins. *Physiol. Rev.* **75**: 277–322.

Magistretti, P. J., Morrison, J. H. (1988): Noradrenaline- and vasoactive intestinal peptide-containing neuronal systems in neocortex: function convergence with contrasting morphology. *Neuroscience* **24**: 367–378.

Moody, T. W., Hill, J. M., Jensen, R. T. (2003): VIP as a trophic factor in the CNS and cancer cells. *Peptides* **24**:163–177.

Rosa, R. M., Silva, P., Stoff, J. S., Epstein, F. H. (1985): Effect of vasoactive intestinal peptide on isolated perfused rat kidney. *Am. J. Physiol.* **249**: E494–E497.

Sorg, O., Magistretti, P. J. (1991): Characterization of the glycogenolysis elicited by vasoactive intestinal peptide, noradrenaline and adenosine in primary cultures of mouse cerebral cortical astrocytes. *Brain Res.* **563**: 227–233.

Rostène, W. (1984): Neurobiological and neuroendocrine functions of VIP. *Prog. Neurobiol.* **22**: 103–129.

Said, S. I. (1995): Vasoactive intestinal peptide. In: *Airway Smooth Muscule: Peptide Receptors, Ion Channels and Signal Transduction.* ed. Raeburn, D., Giembycz, M. A., Birkhäuser, Basel, pp 87–113.

Said, S. I., Mutt, V. (1970): Polypeptide with broad biological activity: isolation from small intestine. *Science* **169**: 1217–1218.

Shreeve, S. M., Sreedharan, S. P., Hacker, M. P., Gannon, D. E., Morgan, M. J. (2000): VIP activates G(s) and G(i3) in rat alveolar macrophages and G(s) in HEK293 cells transfected with the human VPAC(1) receptor. *Biochem. Biophys. Res. Commun.* **272**: 922–928.

Smitherman, T. C., Dehmer, G. J., Said, S. I. (1988): Vasoactive intestinal peptide as a coronary vasodilator. *Ann. N. Y. Acad. Sci.* **527**: 421–430.

Sreedharan, S. P., Huang, J.-H., Cheung, M.-C., Goetzel, E. J. (1995): Structure, expression and chromosomal localization of the type 1 human vasoactive intestinal peptide receptor gene. *Proc. Natl. Acad. Sci. USA* **92**: 2929–2943.

4.35
Vasopressin and Oxytocin

4.35.1
General Aspects and History

Vasopressin (VP) and oxytocin (Fig. 4.59) represent the principal hormones of the posterior pituitary. Both substances are nonapeptides synthesized in the supraoptic and paraventricular nuclei of the hypothalamus, from which they are released into the posterior pituitary via axon terminals of the supraoptic–hypophyseal tract. The peptides are structurally closely related, differing in only two amino acids positions.

Oxytocin and vasopressin (also known as arginine vasopressin, AVP) are found exclusively in mammals, probably evolving from the ancestral peptide arginine–vasotocin, from which oxytocin and vasopressin differ in only a single amino acid. The traditional view of vasopressin and oxytocin being exclusive hormones with targets in peripheral organs has been revised. Both peptides serve potent neuromodulatory functions with major effects in the central nervous system.

Beside the known endocrine effects, vasopressin and oxytocin play a role in the regulation of body temperature and cardiovascular as well as behavioral mechanisms.

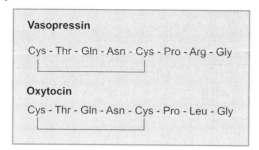

Fig. 4.59 Primary structure of vasopressin and oxytocin. Both neuropeptides share a great similarity in their amino acid sequence, since they differ in only two amino acids.

Vasopressin's peripheral target centers on the regulation of water homeostasis, since vasopressin is the main regulator of body fluid osmolality serving a function as the most potent antidiuretic hormone which has led to the synonym antidiuretic hormone (ADH) for this peptide.

4.35.2
Localization Within the Central Nervous System

The main central sources of vasopressin and oxytocin are the magnocellular neurons of the supraoptic nucleus and the paraventricular nucleus of the hypothalamus (Figs. 4.60, 4.61). The magnocellular neurons project to the pituitary, where they form vasopressinergic and oxytocinergic terminals, which converge to capillaries. Upon an adequate stimulus, the peptides are released into the blood circulation.

Vasopressinergic and oxytocinergic neurons are common among mammalians in both nuclei; however, the relative amount of vasopressin- and oxytocin-containing neurons varies considerably.

Although vasopressin and oxytocin colocalize in the same brain nuclei, they are not coexpressed in single neurons; and both nuclei exhibit variable amounts of hormone-carrying neurons. For instance, in the human supraoptic nucleus, vasopressin-containing neurons dominate over oxytocinergic ones. A further group of vasopressinergic neurons reside in the bed nucleus of the stria terminalis. These neurons project into the direction of the septum, the habenula and the diagonal band of Broca, as well as to the anterior cortical nucleus of the amygdala and to the olfactory bulb.

The latter vasopressinergic projection appears to be sexually dimorphic, since the extrahypothalamic vasopressin cells are androgen-dependent and markedly more abundant in males.

In addition, vasopressin-containing neurons of the medial nucleus of the amygdala project to the septum. Vasopressin-containing neurons have also been found in the locus coeruleus; however, their projections have yet not been defined.

In the central nervous system, vasopressin is differently distributed in males and females. Sexually dimorphic vasopressinergic nuclei are found in the lateral

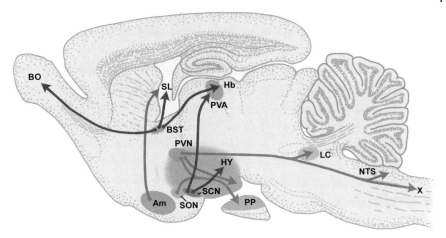

Fig. 4.60 Schematic representation of the distribution and main projections of vasopressin-containing neurons. *Abbreviations*: Am=amygdala; BST=bed nucleus of the *stria terminalis*; Hb=*habenula*; HY=hypothalamus; LC=locus coeruleus; NTS=nucleus of the *tractus solitarius*; PP=posterior pituitary; PVA=paraventricular nucleus of the thalamus; PVN=paraventricular nucleus of the hypothalamus; SCN=suprachiasmatic nucleus; SL=septum, lateral; SON=supraoptic nucleus; X=dorsal nucleus of the vagal nerve.

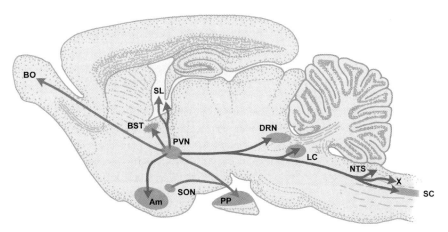

Fig. 4.61 Schematic representation of the main sources of neuronal oxytocin and the main projections of oxytocin containing neurons. *Abbreviations*: Am=amygdala; BO=*bulbus olfactorius*; BST=bed nucleus of the *stria terminalis*; DRN=dorsal raphe nucleus; LC=*locus coeruleus*; NTS=nucleus of the *tractus solitarius*; PP=posterior pituitary; PVN=paraventricular nucleus of the hypothalamus; SC=spinal cord; SL=lateral septum; SON=supraoptic nucleus; X=dorsal nucleus of the vagal nerve.

septum, medial amygdaloid nucleus, hypothalamus and bed nucleus of the stria terminalis, where mammalian males exhibit a greater number of cell and fibers than females. In the bed nucleus of the stria terminalis and in the amygdala, male rats have two to three times as many vasopressin-containing neurons as females.

All oxytocinergic projections to the parabrachial nucleus, to the raphe nuclei and to the dorsal nucleus of the vagal nerve and caudally to the spinal cord originate from the paraventricular nucleus. Oxytocinergic projections also occur in the limbic system which target to the lateral septum and the bed nucleus of the stria terminalis.

4.35.3
Biosynthesis and Degradation

Both genes encoding oxytocin and vasopressin have been cloned and sequenced (Land et al. 1982; Ivell and Richter 1984). The human genes are localized on chromosome 20p13, where they exhibit antiparallel orientation with a long interspersed repetitive DNA element (so-called LINE).

LINEs are endogenous mobile genetic elements accumulated in the human genome. Most of them are non-functional and truncated "dead on arrival" (DOA) copies, but some are in an active form (~6 kb long) transcribing an RNA-binding protein and an integrase–replicase complex, showing both endonuclease and reverse transcriptase activities. The genes for both peptides consist of three exons and two introns, with little changes between species from mouse to human.

Since oxytocin and vasopressin are not co-expressed in single cells, it is likely that activation of one gene exerts a suppressive effect on the other. Vasopressin is generated from a 166-amino-acid precursor (propressorphysin). This precursor also leads to the formation of neurophysin II. Oxytocin is also processed from a precursor which in addition serves as a precursor for neurophysin I.

Both vasopressin and oxytocin are bound to protein carriers called neurophysins. Neurophysins are hormone-specific binding proteins, which form dimeric complexes either with vasopressin or with oxytocin. Neurophysins are derived through proteolytic cleavage from the same larger precursor, as do their associated hormones, and they bind vasopressin and oxytocin through an identical 90-amino-acid sequence. The neurophysin–neuropeptide dimers are packed in granules in the neuronal soma. They are delivered by axonal transport to the nerve terminals.

Oxytocin and vasopressin seem to be largely resistant to degradation in the central nervous system and are mainly degraded in peripheral tissues by proteolysis (vasopressinase–oxytocinase).

4.35.4
Receptors and Signal Transduction

Four genes encode for the vasopressin and oxytocin receptors. The receptors are termed as OTR, V1a, V1b (also known as V3) and V2. All receptors have been cloned and sequenced. The genes are approximately equivalent in size, with 40–60% homology in their putative amino acid sequences. Each encodes for a receptor that belongs to the family of G protein-coupled receptors which possess seven transmembrane-spanning domains.

Large regions, particularly within the transmembrane domains and the second and third extracellular loops, are conserved across the four different receptor types. In addition to ligand binding, coupling to G proteins differs among this group. The OTR, V1a and V1b receptors use $G_q/11$ to induce phosphoinositol hydrolysis and increase intracellular calcium. The V2 receptor is coupled to G_s to stimulate adenylate cyclase.

Both oxytocin and vasopressin receptors are developmentally regulated and expressed to higher amounts in the immature brain.

The idea that vasopressin and oxytocin act as neuromodulators in the central nervous system originates form the fact that both substances are expressed in hypothalamic as well as extrahypothalamic areas with an associated specific receptor complement. With respect to vasopressin, only one of the three vasopressin-specific receptors has been identified in the brain.

Vasopressin receptors

The receptors V1a and V1b differ in their affinity for most of the peptidergic antagonists of vasopressin. So far, V1b receptors have exclusively been identified in the anterior hypophysis; and it is believed that they are involved in vasopressin-mediated effects on the secretion of ACTH from corticotropes.

V1a receptors are more widely distributed and they seem to mediate most of the common vasopressin-mediated effects. The V1a receptor gene has been cloned in several species, including humans. The human V1a receptor gene encodes a putative 394-amino-acid protein. The human V1a receptor gene is located on chromosome 12q14-q15. To date, the V2 receptor (human chromosomal gene locus Xq28) has not been localized in the brain.

Vasopressin receptors have been identified in the entorhinal cortex, the olfactory nuclei, the septum, the subiculum, the amygdala, the bed nucleus of the stria terminalis, the paraventricular nucleus of the thalamus, the ventromedial nucleus of the hypothalamus and in the nucleus of the vagal nerves.

Oxytocin receptor

The human OTR gene is roughly 17 kb in length, with four exons and three introns. The human OTR gene is localized to chromosome 3 locus p25-3p26. The OTR gene, which encodes a putative 388 amino acid protein, shows relatively little variance across species.

Oxytocin receptors have been identified in several areas of the central nervous system, including the hippocampus, the septum, the olfactory nuclei, the subfornical organ, the amygdala, the bed nucleus of the stria terminalis, the nucleus accumbens, the substantia nigra, the colliculus, the periaqueductal gray, the dorsal raphe nucleus, the dorsal tegmental nucleus, the nucleus of the tractus solitarius and the inferior olive. The highest density of the receptor in the hypothalamus has been detected in the ventro-medial hypothalamus.

4.35.5
Biological Effects

In general, vasopressin and oxytocin contribute to different kinds of physiological functions including:
- neuromodulatory effects in the central nervous system
- endocrine activity in the anterior hypophysis
- hormonal potencies in the periphery of the body, where they provide constrictive effects on smooth muscles (oxytocin) and antidiuresis on the collecting tubules of the kidney (vasopressin).

As hypophyseal factors, vasopressin is involved in the control of ACTH and oxytocin is involved in the control of prolactin. Conceptually, based on their projection pattern and the distribution of their receptors, both neuropeptides interact with three different central nervous systems:
- the autonomic system of the brain stem and the spinal cord
- neuroendocrine regulation in the hypothalamus
- control of behavior, especially of emotions and memory in the limbic system.

Both peptides seem also be involved in neurogensis. Vasopressin is found around embryonic day 16 in the developing rat brain and increases to high levels prior to birth. Oxytocin is also found in fetal brain, although it is not processed to its mature, amidated form until postnatal stages in rat. Chronic prenatal administration of both vasopressin and oxytocin has been found to be associated with altered brain weight in adults, with the most significant and persistent changes observed in the cerebellum.

Vasopressin
For a long time, it was believed that vasopressin is the primary factor regulating the release of ACTH. After the discovery of the corticotropin-releasing factor (CRF), it became apparent that CRF and not vasopressin constitutes the main regulatory factor for ACTH secretion. Nevertheless, vasopressin seems to contribute to ACTH secretion to some degree, since it is able to potentiate the effects of CRF. Vasopressin is also known as antidiuretic hormone (ADH), because it serves regulatory properties for body fluid osmolality. The secretion of vasopressin relies on hypothalamic osmoreceptors and osmoreceptors located in circumventricular organs (as e.g. the organum vasculosum of the lamina termi-

nalis; OVLT), which sense water concentration and stimulate vasopressin secretion when the plasma osmolality increases. Circulating vasopressin – after binding to its receptor sites in the collecting tublues – increases the re-absorbtion of water in the kidney, thereby yielding the excretion of concentrated urine with the consequence of a drop in body osmolality. An increase in plasma vasopressin levels also induces thirst.

In addition, vasopressin has central regulatory effects on body temperature; and during fever vasopressin is released to counter-regulate increasing body temperature. Since vasopressin modulates the thermoregulatory response during fever, it referred to as an endogenous antipyretic. Evidence suggests that there may be a synergistic relationship between vasopressin receptors and the cyclo-oxygenase enzyme during antipyresis; and the presence of vasopressin may enhance the efficacy of nonsteroidal antipyretic drugs (Richmond 2003).

Several studies provide evidence that vasopressin and oxytocin modulate learning and memory processes. Evidently, vasopressin facilitates learning of both active and passive avoidance behaviors in a dose-dependent manner. The peptide appears to influence both the consolidation and retrieval phases of learning, possibly with different metabolites affecting consolidation and retrieval, independently. Oxytocin effects on learning are also dose-dependent, but generally opposite to that of vasopressin, with an inhibition of extinction in both passive and active avoidance paradigms. Some evidence for a neuromodulatory function of vasopressin in the limbic system of rats derives from the fact that vasopressin is synthesized in the medial amygdaloid nucleus in the presence of steroids and is transported to other limbic structures such as the hippocampus and septum, and is secreted from there by a calcium-dependent process.

In the hippocampus, vasopressin acts on cerebral microvessels and local circuit interneurons. Its excitatory action on the inhibitory interneurons produces near-total shutdown of electrical activity of the efferent fibers of pyramidal cells.

Stimulation of the medial amygdala and release of endogenous vasopressin duplicates these effects, which can be blocked by the intraventricular application of vasopressin antagonists. Stimulation of vasopressinergic neurons produces alterations in sexual behavior, in a pattern consistent with the idea that the medial amygdala organizes the appetitive phase of recognition of an appropriate partner and sexual arousal.

A link between aggression and low levels of serotonin is a generally accepted paradigm; however, high levels of vasopressin may also play a role in aggression. Several animal studies have gained evidence in favor of this idea. In golden hamsters, for instance, injection of vasopressin into the hypothalamus increases the number of biting attacks on intruders. Administration of a drug that blocks vasopressin from binding to receptors in the brain, conversely, inhibits offensive aggression.

Oxytocin

Oxytocin is released together with ACTH or prolactin and inhibits the release of gonadotrope hormones: luteinizing hormone (LH) and follicle-stimulating hormone (FSH).

The possible interaction of oxytocin and prolactin is still the subject of controversy. Some studies have indicated the stimulatory effects of oxytocin on the secretion of prolactin; however, opposing arguments indicate that this action is most likely to be an indirect effect, since the main regulator of prolactin secretion is supposed to be the vasoactive intestinal polypeptide (VIP).

Oxytocin is secreted in a diurnal rhythm, whereby the secretion reaches its maximum during the daytime and its nadir at night. Oxytocin is also involved in the regulation of water and salt homeostasis.

Osmotic stress induces the secretion of oxytocin (as well as that of vasopressin). Oxytocin plays a major role in sexual and social behavior. For instance, it exerts strong effects on sexual behavior in both females and in males and, in addition, induces negative effects on maternal behavior.

Successful reproduction in mammals requires that mothers become attached to and nourish their offspring immediately after birth. Conversely, it is important that nonlactating females do not manifest such nurturing behavior. The same event that causes the peripheral effects of myometrial contraction and lactation at the time of parturition also affects the brain significantly. During delivery, the increase in central concentrations of oxytocin plays a major role in establishing the above maternal behavior. Evidence in support of this dual role of oxytocin came form two types of experiments. First, infusion of oxytocin into the ventricles of the brain of virgin rats or nonpregnant sheep rapidly induced maternal behavior. Second, administration of oxytocin-neutralizing antibodies or oxytocin antagonists into the brain prevented mother rats from accepting their pups.

Recent studies, however, have cast some doubt on the necessity of oxytocin in regulating delivery functions and maternal behavior. Mice that are unable to secrete oxytocin due to targeted disruption of the oxytocin gene will mate, deliver their pups without apparent difficulty and display normal maternal behavior. However, they do show deficits in milk ejection and have subtle deficits in social behavior. According to these data, it seems more appropriate to consider oxytocin as a major facilitator of parturition and maternal behavior rather than an obligatory factor.

Males synthesize oxytocin in the same regions of the hypothalamus as do females. In addition, it is also found in the testes. Pulses of oxytocin can be detected during ejaculation in seminal fluid, where it is suggested to be involved in facilitating sperm transport within the male reproductive tract and also in the female genital tract. Some evidence indicates that it may also be involved in male sexual behavior.

A number of factors can influence oxytocin release; and a most prominent inhibitory effect is delivered by acute stress. For example, oxytocin neurons are repressed by catecholamines, which are released from the adrenal gland in response to many types of stress reactions, including fright.

Both the production of oxytocin and the response to oxytocin are also modulated by circulating levels of sex steroids. The maternal burst of oxytocin at birth seems to be triggered in part by uterine and vaginal stimulation through the fetus, an effect which is supported by abruptly declining concentrations of progesterone. Another well studied phenomenon of steroid hormones is the marked increase in uterine oxytocin receptors late in gestation in response to the increasing concentrations of circulating estrogen.

4.35.6
Neurological Disorders and Neurodegenerative Diseases

Vasopressin plays an important role in a rare form of diabetes, known as *diabetes insipidus*. This disease derives from a deficiency in secretion of vasopressin from the posterior pituitary. Causes of this disease include head trauma and infections or tumors involving the hypothalamus. The most striking symptom of *diabetes insipidus* is the inability of the patients to reabsorb water in the collecting tubules of the kidney, which results in massive production of urine which can exceed $10 \, l \, day^{-1}$.

Oxytocin is released from the pituitary gland in response to a variety of stressful stimuli, including noxious stimuli, conditioned fear and exposure to novel environments. In response to various stressors, oxytocin is released not only from neurohypophysial terminals into the blood, but also within distinct brain regions, as e.g. the supraoptic and paraventricular nuclei of the hypothalamus, the septum and the amygdala, depending on the quality and intensity of the stressor. Thus, oxytocin secretory activity may accompany the response of the hypothalamic–pituitary–adrenal (HPA) axis to a given stressor (Neumann 2002; Onaka 2004).

Further Reading
Bourque, C. W., Renaud, L. P. (1990): Electrophysiology of mammalian magnocellular vasopressin and oxytocin neurons. *Front. Neuroendocrinol.* **11**: 183–212.
Buijs, R. M. (1983): Vasopressin and oxytocin – their role in neurotransmission. *Pharm. Ther.* **22**: 127–141.
Caffe, A. R., van Leeuwen, F. W. (1983): Vasopressin immunoreactive cells in the dorsomedial hypothalamic region, medial amygdala and locus coeruleus of the rat. *Cell Tiss. Res.* **233**: 23–33.
de Wied, D., Diamant, M., Fodor, M. (1993): Central nervous system effects of the neurohypophyseal hormones and related peptides. *Front. Neuroendocrinol.* **14**: 251–302.
Gainer, H., Wray, S. (1992): Oxytocin and vasopressin: from genes to peptides. *Ann. N.Y. Acad. Sci.* **652**: 14–28.
Hara, Y., Battey, J., Gainer, H. (1990): Structure of mouse vasopressin and oxytocin genes. *Mol. Brain Res.* **8**: 319–324.
Hawthorn, J., Ang, V. T. Y., Jenkins, J. S. (1984): Comparison of the distribution of oxytocin and vasopressin in the rat brain. *Brain Res.* **307**: 289–294.
Insel, T. R. (1992): Oxytocin: a neuropeptide for affiliation – evidence from behavioral, receptor autoradiographic, and comparative studies. *Psychoneuroendocrinology* **17**: 3–33.
Ivell, R., Richter, D. (1984): Structure and comparison of the oxytocin and vasopressin genes from rat. *Proc. Natl Acad. Sci. USA* **81**: 2006–2010.

Jenkins, J. S., Ang, V. T. Y., Hawthorn, J., Rossor, M. N., Iversen, L. L. (1980): Vasopressin, oxytocin and neurophysin in the human brain and spinal chord. *Brain Res.* **291**: 111–117.

Land, H., Schutz, G., Schmale, H., Richter, D. (1982): Nucleotide sequence of cloned cDNA encoding bovine ariginine vasopressin–neurophysin II precursor. *Nature* **295**: 299–303.

McCann, S. M., Antunes-Rodrigues, J., Jankowski, M., Gutkowska, J. (2002): Oxytocin, vasopressin and atrial natriuretic peptide control body fluid homeostasis by action on their receptors in brain, cardiovascular system and kidney. *Prog. Brain Res.* **139**: 309–328.

McKinley, M. J., Mathai, M. L., McAllen, R. M., McClear, R. C., Miselis, R. R., Pennington, G. L., Vivas, L., Wade, J. D., Oldfield, B. J. (2004): Vasopressin secretion: osmotic and hormonal regulation by the lamina terminalis. *J. Neuroendocrinol.* **16**: 340–347.

Mühlethaler, M., Dreifuss, J. J., Gahwiler, B. H. (1982): Vasopressin causes excitation of hippocampal neurons. *Nature* **296**: 749–751.

Neumann, I. D. (2002): Involvement of the brain oxytocin system in stress coping: interactions with the hypothalamo-pituitary–adrenal axis. *Prog. Brain Res.* **139**: 147–162.

Olpe, H.-R., Balzer, V. (1981): Vasopressin activates noradrenergic neurons in the rat locus coeruleus: a microiontophoretic investigation. *Eur. J. Pharmacol.* **73**: 377–378.

Onaka, T. (2004): Neural pathways controlling central and peripheral oxytocin release during stress. *J. Neuroendocrinol.* **16**: 308–312.

Poulain, D. A., Wakerley, J. B. (1982): Electrophysiology of hypothalamic magnocellular neurons secreting oxytocin and vasopressin. *Neuroscience* **7**: 773–808.

Rhodes, M. E., Rubin, R. T. (1999): Functional sex differences ('sexual diergism') of central nervous system cholinergic systems, vasopressin, and hypothalamic–pituitary–adrenal axis activity in mammals: a selective review. *Brain Res. Rev.* **30**: 135–152.

Richard, P., Moos, F., Freund-Mercier, M. J. (1991): Central effects of oxytocin. *Physiol. Rev.* **71**: 331–370.

Richmond, C. A. (2003): The role of arginine vasopressin in thermoregulation during fever. *J. Neurosci. Nurs.* **35**: 281–286.

Thomas, R., Insel, T. R., O'Brien, D. J., Leckman, J. F. (1999): Oxytocin, vasopressin, and autism: is there a connection? *Biol. Psychiatr.* **45**: 145–157.

4.36
Deorphanized Neuropeptides

This final section will deal with a collection of neuropeptides which have been identified mostly through the deorphanization strategies of G protein-coupled receptors. As already indicated in Section 1.2, the deorphanization of GPCRs (also referred as "reverse pharmacology") is a powerful tool to identify unkown neurotransmitters or neuromodulators. The basic approach is to use the orphan exogenously expressed GPCR as a bait to extract potential ligands from tissues extracts that contain a putative transmitter.

The first successful attempt was the discovery of the neuropeptide nociceptin/orphanin FQ as the transmitter of the orphan GPCR ORL-1. This first event was followed by a series of further successful endeavors, among which the hypocretins (orexins) and ghrelin identification have already been detailed in foregoing sections. Here, we will briefly describe some of the deorphanized neuropeptides with potential pharmacological relevance.

4.36.1
Apelin

The apelin peptide family has been recently identified with "reverse pharmacology" as the cognate ligands of the APJ gene. In 1993, O'Dowd and coworkers identified a gene revealing high homology to the angiotensin receptor (AT-1, see Section 4.3). This gene was termed APJ and, in spite of its high homology to the AT-1 receptor, angiotensin-II did not activate cells expressing APJ. Tatemoto and coworkers identified a selective ligand of 36-amino peptides from bovine stomach for this receptor. Concurrently, cDNA encoding a 77-amino-acid prepropeptide of the new ligand (which was called apelin according to its binding to the APJ receptor) was identified from human and bovine tissues. Carboxy-terminal fragments of varying sizes have further been shown to activate the receptor.

Messenger RNA encoding both APJ and apelin as well as the carboxy-terminal peptides are highly expressed in the central nervous system, indicating an important role for the apelin system in regulating central pathways. In the brain, APJ expression was discovered in a variety of areas, including the cerebral cortex, hypothalamus, hippocampus and pituitary gland. The mRNA was most abundant in neurons of the paraventricular and supraoptic nucleus of the hypothalamus. Apelin immunoreactivity was found in brain nuclei of the thalamus and hypothamalus, and APJ-like immunoreactivity in neurons of the rat cerebellar and hypothalamus. Apelin mRNA was also found in a number of peripheral tissues including heart, liver, kidney, testis, ovary and adipose tissues, achieving highest levels in the lung and mammary glands.

A number of studies indicate that the APJ/apelin system plays an essential role in cardiovascular regulation. The major effect consists of a significant drop of systolic and diastolic blood pressure after intravenous infusion. Intraventricular administration of apelin-13 (a carboxy-terminal fragment of the apelin-36 peptide) did not change mean arterial blood pressure, while intravenous administration resulted in a significant drop associated with an increase in heart rate. Pharmacological ganglionic block abolished the compensatory increase in heart rate, indicating that activation of the sympathetic nervous system may be responsible for the positive chronotropic effect.

The abundance of the APJ receptor and apelin protein in the paraventricular and supraoptic nucleus is suggestive for a function of this system in the regulation of fluid homeostasis. Functional evidence supporting this concept derives from a study showing that intraventricular infusion of apelin-13 significantly dercreased circulating apelin. In mice deprived of water, intraventicular administration of apelin-13 significantly reduced the water intake in the initial 30 min after reexposure to drinking water. Although increased water intake was also reported after intraventricular apelin-13 administration, a general effect of the APJ/apelin system on the regulation of fluid homeostasis is the most likely centrally related activity of this system.

A further interesting aspect is that APJ seems to be one of the GPCRs proposed as alternative HIV coreceptors; and it has been shown to support the entry of a number of HIV-1 and simian immunodeficiency viruses in CD4-positive cell lines. The expression of the APJ receptor in the brain is suggestive of its function as an HIV entry site *in vivo*.

4.36.2
Kisspeptin/Metastin

A further class of peptides discovered by the deorphanization of GPCRs is the so-called kisspeptin/metastin family and its cognate receptor GPR54 (G protein-coupled receptor 54). In a library screening, a clone was isolated that encodes for a GPCR receptor of 396 amino acids which shares significant identities with rat galanin receptors. *In situ* hybridization indicated that GPR54 is expressed in brain regions of the pons, midbrain, thalamus, hypothalamus, hippocampus, amygdala, frontal cortex and striatum. However, expression of GPR54 in COS cell did not reveal specific binding for galanin or galanin-like peptides. Later, high-affinity ligand-binding assays led to the discovery of a peptide derived by proteolytic processing from the product ot the KiSS-1 gene. KiSS-1 encodes a 145-amino-acid peptide with displays typical features of secreted neuropeptides: a signal sequence, several potential dibasic cleavage site and a cleavage/amidation site. The predicted cleavage processing gives rise to a 54-amino-acid secreted product termed metastin or kisspeptin-54. Additional naturally occurring cleavage fragments of 13 and 14 amino acids also convey biological activity. Kisspeptin/metastin was first defined on its putative role in cancer metastasis. KiSS-1 was originally defined as a human melanoma suppressor gene. The recent discovery of loss-of-function mutations in the gene encoding GPR54 has highlighted a previously unrecognized pathway in the physiologic regulation of puberty and reproduction. Patients with idiopathic hypogonadotropic hypogonadism (IHH), a condition characterized by the absence of spontaneous pubertal development, low levels of sex steroids and inappropriately low gonadotropins, indicates that this receptor is essentially required for the activation of puberty and implies a major role of this pathway in the neuroendocrine control of reproduction. This concept is supported by studies in rodents and primate models that have demonstrated:
- localization of KiSS-1 mRNA in the hypothalamus;
- colocalization of GPR54 in GnRH neurons;
- GnRH-dependent activation of LH and FSH release by intracerebroventricular administration of kisspeptin;
- increased hypothalamic KiSS-1 and GPR54 mRNA levels at the onset of puberty.

Many questions remain about the fine-tuning of this system with respect to its integration into the GnRH-releasing cascade, but the discovery of the system itself is a further example of the powerful potential of the deorphanizing approach.

4.36.3
Opiod-modulating Peptides (NPFF and NPAF)

Neuropeptide FF and the octadecapeptide neuropeptide AF were first isolated from bovine brain and were initially defined as anti-opiod peptides. They both derive from a prepropeptide by proteolytic cleavage. The gene product is primarily expressed in the CNS, where the posterior pituitary and dorsal spinal cord reveal the highest concentrations. Further biochemical and immunochytochemical studies indicated that all NPFF-like peptides may not derive from the original NPFF gene, but that other genes may exist which encode related peptides. Intraventricular injections of NPFF elicits antiopiod effects, while intrathecal administration potentiates the analgesic effects of morphine. The bivalent action of the antiopiod peptides indicates that they act as modulators of opiod function participating in a homeostatic system that tends to balance the endogeous opiod system. Knowledge of the molecular properties of their specific receptor complement will certainly futher our understanding of the control of opiod function.

From the above, it becomes evident that deorphanization of GPCRs comprises a powerful technique that allows the identification of unknown neurotransmitters and neuromodulators. One has to expect some more novel neuroactive candidates in the near future, although the climax of the deorphanization success seems to have passed; and more direct searches will certainly celebrate a resurrection.

Further Reading
Cesselin, F. (1995): Opiod and anti-opiod peptides. *Fundam. Clin. Pharmacol.* **9**: 409–433.
Civelli, O. (2005): GPCR deorphanization: the novel, the known and the unexpected transmitters. *Trends Pharmacol Sci.* **26**: 15–19.
College, W.H. (2004): GPR54 and puberty. *Trends Endocrinol. Metab.* **15**: 448–458.
Davenport, A.P. (2003): Peptide and trace amine orphan receptors: prospects for new therapeutic targets. *Curr. Opin. Pharmacol.* **3**: 127–134.
Hinuma, S., Onda, H., Fujino, M. (1999): The quest for novel bioactive peptides utilizing orphan seven-transmembrane-domain receptors. *J. Mol. Med.* **77**: 495–504.
Kaiser, U.B., Kuohung, W. (2005): KiSS-1 and GPR54 as new players in gonadotropin regulation and puberty. *Endocrine* **26**: 277–284.
Katugampola, S., Davenport, A. (2003): Emerging roles for orphan G protein-coupled receptors in the cardiovascular system. *Trends Pharacol. Sci.* **24**: 30–35.
Kleinz, M.J., Davenport, A.P. (2005): Emerging roles for apelin in biology and medicine. *Pharamacol. Ther.* **107**: 198–211.
Maguire, J.J. (2003): Discovering orphan receptor function using human *in vitro* pharmacology. *Curr. Opin. Pharmacol.* **3**: 135–139.
Masri, B., Knibiehler, B., Audgier, Y. (2005): Apelin signalling: a promising pathway from cloning to pharmacology. *Cell Signal.* **17**: 415–426.
Mollereau, C., Roumy, M., Zajac, J.M. (2005): Opiod-modulating peptides: mechanisms of action. *Curr. Top. Med. Chem.* **5**: 341–355.
Panula, P., Kalso, E., Nieminen, M., Kontinen, V.K., Brandt, A., Pertovaara, A. (1999): Neuropeptide FF and modulation of pain. *Brain Res.* **848**: 191–196.
Philip, A.E., Poupaert, J.H., McCurdy, C.R. (2005): Opiod receptor-like 1 (ORL1) molecular "road map" to understanding ligand interaction and selectivity. *Curr. Top. Med. Chem.* **5**: 325–340.

Rothman, R. B. (1992): A review of the role of anti-opiod peptides in morphine tolerance and dependence. *Synapse* **1992**: 129–138.

Roumy, M., Zajac, J. M. (1998): Neuropeptide FF, pain and analgesia. *Eur. J. Pharmacol.* **345**: 1–11.

A
Appendix

A1
Amino Acids

Name	Abbreviation	Structure
Alanine	Ala (A)	
Arginine	Arg (R)	
Asparagine	Asn (N)	
Aspartic acid	Asp (D)	
Cysteine	Cys (C)	
Glutamine	Gln (E)	

A Appendix

A1
Amino Acids

Name	Abbreviation	Structure
Glutamic acid	Glu (Q)	
Glycine	Gly (G)	
Histidine	His (H)	
Isoleucine	Ile (I)	
Leucine	Leu (L)	
Lysine	Lys (K)	
Methionine	Met (M)	
Phenylalanine	Phe (F)	
Proline	Pro (P)	

A1
Amino Acids

Name	Abbreviation	Structure
Serine	Ser (S)	HO–CH(NH$_2$)–COOH
Threonine	Thr (T)	HO–CH(CH$_3$)–CH(NH$_2$)–COOH
Tryptophan	Trp (W)	(indole)–CH$_2$–CH(NH$_2$)–COOH
Tyrosine	Tyr (Y)	HO–C$_6$H$_4$–CH$_2$–CH(NH$_2$)–COOH
Valine	Val (V)	(CH$_3$)$_2$CH–CH(NH$_2$)–COOH

A2
Nucleotides

Name	Abbreviation	Structure
Adenine	A	
Cytosine	C	
Guanine	G	
Thymine	T	
Uracil	U	

A3
Abbreviations for Neurotransmitters and Neuromodulators

5-HT	5-Hydroxytryptamine (serotonin)
ACh	Acetylcholine
ACTH	Adrenocorticotropic hormone
ADH	Antidiuretic hormone = Vasopressin
ADP	Adenosine diphosphate
AMP	Adenosine monophosphate
ANF	Atrial naturetic factor
ATP	Adenosine triphosphate
AMPA	a-Amino-3-hydroxy-5-methyl-4-isoxazolepropionate
AVP	Arginine–vasopressin

BDNF	Brain-derived neurotrophic factor
BNP	Brain natriuretic peptide
CACA	Cis-aminocrotonic acid
CCK	Cholecystokinin
CGRP	Calcitonin gene-related peptide
CLIP	Corticotropin-like intermediary peptide
CNP	C-type natriuretic peptide
CO	Carbon monoxide
CPON	C-Flanking peptide of neuropeptide Y
CRF	Corticotropin-releasing factor
CRH	Corticotropin-releasing hormone
CT	Calcitonin
DA	Dopamine
FSH	Follicle-stimulating hormone
GABA	Gamma-amino butyric acid
GALP	Galanin-like peptide
GAP	GnRH-associated peptide
GH	Growth hormone
GHRH	Growth hormone-releasing hormone
GMAP	Galanin message-associated peptide
GnRH	Gonadotropin-releasing hormone
GRP	Gastrin-releasing peptide
IFN	Interferon
IL	Interleukin
JP	Joining peptide
LH	luteinizing hormone
LHRH	Luteinizing hormone-releasing hormone
LPH	Lipotropin
MCH	Melanin-concentrating hormone
MIF	Melanocyte-stimulating hormone-inhibiting factor
MSH	Melanocyte-stimulating hormone
NA	Noradrenaline (norepinephrine)
NE	Norepinephrine (noradrenaline)
NEI	Neuropeptide glutamate-isoleucine
NG	Neuropeptide glycine-glutamate
NGF	Nerve growth factor
NKA	Neurokinin A
NKB	Neurokinin B
NmB	Neuromedin B
NMDA	N-Methyl-D-aspartate
NmN	Neuromedin N
NO	Nitric oxide
NPK	Neuropeptide K
NPY	Neuropeptide Y (Y = Tyrosine)
NPγ	Neuropeptide γ

NT	Neurotrophin
OT	Oxytocin
PACAP	Pituitary adenylate cyclase-activating peptide
PGMP	Pregalanin message-associated peptide
PHI	Peptide histidin-isoleucine
PHM	Peptide histidine-methionine amide
POMC	Proopiomelanocortin
PPCE	Post-proline-cleavage enzyme
PRP	PACAP-related protein
PYY	Peptide YY
SRIF	Somatotropin releasing-inhibiting factor (Somatostatin)
SP	Substance P
SS	Somatostatin
TRH	Thyrotropin-releasing hormone
TSH	Thyroid-stimulating hormone
TNF	Tumor necrosis factor
Tyr-MIF-1	Tyr-Pro-Leu-Gly-NH_2
Tyr-K-MIF-1	Tyr-Pro-Lys-Gly-NH_2
Tyr-W-MIF-1	Tyr-Pro-Trp-Gly-NH_2
VIP	Vasoactive intestinal polypeptide
VP	Vasopressin

A4
Miscellaneous Abbreviations (Enzymes and Transporters)

AADC	Amino acid decarboxylase
ACE	Angiotensin-converting enzyme
AChE	Acetylcholine esterase
ChAT	Choline acetyltransferase
COMT	Catechol-O-methyltransferase
DAT	Dopamine transporter
DBH	Dopamine-β-hydroxylase
DDC	DOPA-decarboxylase
DOPA	Dihydroxyphenylalanine
DOPAC	Dihydroxyphenylacetic acid
EAAT	Excitatory amino acid transporter
GABA-T	GABA transaminase
GAD	Glutamic acid decarboxylase
GAT	GABA transporter
GPCRs	G protein-coupled receptors
HD (or HDC)	L-Histidine-decarboxylase
HO	Heme oxygenase
HVA	Homovanillic acid
LOX	Lipoxygenase

MAO	Monoamine oxidase
NET	Norepinephrine transporter
NOS	Nitric oxide synthetase
PAMase	Peptidyl-glycine a-amidating monoxygenase
PC	Proconvertases
PE	Proline endopeptidase
PGA	Pyroglutamyl aminopeptidase
PNMT	Phenylethylamine-N-methyl-transferase
Sert	Serotonin transporter
TH	Tyrosine hydroxylase

Subject Index

a

acetylcholine 21, 47–49, 51–53, 55–58, 111, 119, 146, 153, 214, 231, 233–234, 245, 276, 295, 297, 320, 325, 332, 335, 350
acetylcholine esterase 49, 57–58
acetylCoA 48
acetylsalicylic acid 208
acromegaly 247
ACTH 120, 139, 144–148, 165, 173, 192, 194, 196, 233, 239, 258, 266–269, 271, 276, 315, 318–319, 359–360, 362, 370
adenohypophysis 246, 319
adenosine 12, 33, 100, 320–321, 325, 337
adenylate cyclase 10, 12, 53–54, 67–69, 102, 119, 129, 138, 145, 152, 155, 181, 192, 195, 202, 208, 212, 231, 244, 246, 268, 276, 311, 321–322, 329, 337, 349, 351–352, 359, 371
ADP 320–322, 370
adrenal cortex 146, 197, 268
adrenal gland 125, 144, 148, 218, 250, 269, 319, 362
adrenoceptor 127, 129
adrenocorticotropic hormone 144, 192, 258, 315, 318
adrenomedullin 181
aggression 264, 269, 361
agonist 9, 13, 57, 68, 82, 85, 99–101, 105–106, 112, 118–119, 121, 128–129, 139–140, 195, 197, 212–214, 216, 270, 335, 348
alcohol 17, 214, 345
aldosteron 173
allosteric 2, 10, 13, 53, 57, 80, 84, 87–88, 97, 105, 107, 111–112, 119
alytesin 175
amino acid decarboxylase 62, 134
aminopeptidase 160, 162–163, 219, 242, 263, 307, 344, 372
aminopeptidase N 160, 307

amnesia 276
AMP 140, 320–321, 370
AMPA 93, 98–102, 106, 227, 370
amperometry 25
amphibians 177–178, 218, 235, 266, 311–312
amygdala 47, 55, 57, 68, 73, 85, 91, 100–101, 112, 124, 131, 133, 136–138, 145, 159, 166, 169, 172, 175, 177, 185, 188, 193, 195, 200–202, 212–213, 217, 229, 232–233, 240, 242, 267–269, 273, 276, 280, 287–288, 294, 306, 308–309, 312, 316, 322, 328, 335, 342, 344–345, 356, 358–361, 363
amyotrophic lateral sclerosis 86, 92, 209, 303, 339, 346
analgesia 139, 156, 182, 204, 211, 213–214, 216, 279, 308, 348
anandamide 150–152, 155–156, 205, 207
angiotensin 31, 144, 158–160, 162–163, 165–167, 172, 194, 196, 258, 336
angiotensin-converting enzyme 166
angiotensinogen 159–160, 166–167
antagonist 13, 30, 51, 53, 69, 80, 82, 84–85, 87, 99, 105, 111–112, 120–121, 128–129, 137, 139, 155, 166–167, 182, 197, 208, 214, 227, 231, 233, 238, 247, 252, 256, 260–261, 270, 279, 282, 309, 339, 348
anxiety 43, 69, 73, 85–87, 107, 131, 139–140, 147, 156, 173, 189, 193, 197, 234, 238, 264, 276, 308–309
anxiolytic 80, 87–88, 140, 156, 173, 189, 276, 309
apelin-12 365
apelin-13 365
apelin-36 365
APJ gene 365
Aplysia 208

apolipoprotein E 258
arachidonic acid 143–144, 150–152, 205–208, 337
arborization 48
area postrema 17, 137, 159, 165, 169, 171, 181, 188, 217, 280, 342
arginine 100, 355
arginine–vasopressin 192, 238
aspartate 90, 92–93, 96, 105–106, 113
astrocytes 2, 14, 19, 72, 92, 109, 155, 163, 209, 222, 225, 255–258, 303, 339, 353
astrogliosis 258–259
ATP 10, 13, 320–325, 370
ATPase 208, 321
atrial naturetic factor 169, 370
atropine 50, 52
autoradiography 21, 30–31, 33, 138–139, 187, 212, 231, 280, 297
autoreceptor 8, 68, 119
avoidance 42, 189, 264, 309, 361

b

baclofen 84–85
band of Broca 47, 240, 249, 282, 342, 356
basal ganglia 15, 68–70, 98, 133, 136, 187, 212, 222
BDNF 140, 226, 286–294, 370
benzodiazepine 79, 85, 87
β adrenoceptors 129
bicuculline 82, 84–85
bilirubin 296, 298
Bioassays 22
biogenic amines 4, 45, 133
bipolar disorders 278
birds 152, 235, 312
blood pressure 37, 120, 147, 158–159, 162, 166, 173, 181, 197, 208, 277, 309, 338
blood vessels 165, 179, 181–183, 295, 352, 354
blood–brain barrier 16–18, 72, 108, 115, 118, 133, 158, 165, 167, 189, 236, 247, 251, 254, 258, 348
bombesin 175–177
Bombina bombina 175
bovine 230, 268, 279, 321, 347
bradykinin 295
brain natriuretic peptide 169, 173
brain stem 15, 52, 55, 68, 109, 111–114, 116, 130, 137, 162, 170, 175, 183, 193, 200, 229, 232, 274, 288, 297, 313, 335, 360
brain-derived neurotrophic factor 140, 286, 288

bungarotoxin 50–51
butyrylcholine esterase 49

c

C-type natriuretic peptide 169, 173, 370
Caenorhabditis elegans 152, 221
caerulein 184
calcitonin 178–183, 246, 352
calcitonin receptor-stimulating peptide 183
calmodulin 11, 98, 104, 208, 242, 299
calretinin 226
cAMP 8–12, 40, 53, 55, 69, 81, 130, 138, 152, 180, 195, 202, 232, 246–247, 263, 290, 321, 329–330
cannabinoid 149–150, 152, 155–156, 268
Cannabis sativa 149
carbachol 50, 52
carbamyl ester 49
carbon monoxide 7, 45, 143, 295–297
carboxypeptidase 160, 162, 263
cardiotrophin-1 256
carrier-mediated mechanisms 17
caspase-1 256
catalepsy 155
catatonia 211, 214
catechol-O-methyltransferase 64–65
catecholamine 59, 62, 125, 127, 313
cathepsin 160, 329
caudate-putamen 55, 100–103, 105, 118, 153, 180, 200, 212–213, 217–218, 232, 263, 267, 280, 287, 297, 327
cell death 14, 72, 106, 227, 259, 302, 303, 314, 354
cerebellum 69, 75, 82, 100–102, 105, 109, 112, 115, 138, 146, 152–153, 171, 180, 193, 195, 206, 208, 221–223, 233, 255–256, 269, 273, 283, 288, 297, 301, 312, 322–323, 360
cerebrospinal fluid 16, 24, 38, 190, 197, 251, 284
cGMP 8, 10, 12, 172, 295–296, 300–303
CGRP 178–183, 335, 370
chick 321
cholecystokinin 173, 180, 184, 186–187, 189–190, 281, 350
choline 46, 48, 57, 146, 345
choline acetyltransferase 48
chromatography 21, 26–28
chromosome 68–69, 114, 126, 136, 138, 170–171, 176, 180, 186, 194, 223, 231–232, 237, 245–246, 250–251, 274, 282, 307, 321, 330–331, 344, 351, 358–359
ciliary neurotrophic factor 256

circadian rhythm   130, 148, 180, 208, 254, 264, 276, 353
circumventricular organ   17, 159, 165, 171, 360
cis-aminocrotonic acid   84, 370
citrulline   298
classic conditioning   41–42
Clostridium tetani   113
clozapine   69
cocaine   64, 126
cognition   69, 166, 213, 234, 325
collaterals   47, 63, 115, 124
colliculus   47, 52, 55, 75, 82, 91, 138, 195, 287, 297, 336, 360
corpus callosum   222, 287
cortex   47, 52, 55–57, 61, 68–69, 73, 75, 82, 91, 98, 100–104, 112–113, 115, 118–120, 124, 128, 130, 133, 136–138, 141, 144, 146, 153, 172, 174, 180, 186–188, 190, 195, 197, 208, 217, 221–223, 227, 229, 232–233, 236, 245, 249, 255–257, 262–263, 267–269, 274, 278, 283, 287, 293, 297, 301–303, 306, 308, 313–314, 322, 327–328, 332, 335–336, 338–339, 341–342, 344–345, 347–348, 350, 359
corticoliberin   192
corticotropin   130, 139, 144–145, 165, 173, 192–193, 196–197, 244, 258, 276, 281, 302, 309, 311, 315, 318–319, 332, 349, 360
corticotropin-like intermediary peptide   315
corticotropin-releasing factor   130, 173, 192–193, 196–197, 244, 276, 309, 318, 332, 360
corticotropin-releasing hormone   145, 165, 192, 258, 302, 311, 319, 349
cortisol   146–148, 172–173
coulometry   25
CPON   274, 370
CRH   144, 192, 194, 238–239, 258, 281, 311, 318–319, 349, 370
curare   50
cyklooxygenase   206, 208
cytochrome P_{450} 207, 299
cytokine   13, 140, 254–256, 258, 272, 324
cytokine receptors   12
cytoplasm   10–11, 17, 48–49, 98, 206, 290
cytosolic receptors   8

d

D-aspartate   93, 105, 109, 112, 371
D-serine   105, 107
de Wied   5

dentate gyrus   55, 100–103, 175–176, 196, 200, 202, 226, 228, 238, 255–256, 287, 297, 309, 332–333, 350
depression   73, 104, 112, 131, 140, 147, 155, 173, 193, 197, 204, 211–212, 214, 216, 220, 234, 271, 277, 293, 301, 308, 332, 340
desensitization   13, 68, 100, 118, 128–129, 136, 323, 337
development   5, 34, 41, 68, 84, 88, 100, 104–105, 111, 120, 162, 190, 204, 213, 216, 221, 225, 231, 247, 264, 277, 283, 292–294, 301, 309, 313, 318–319, 337, 344–345, 353–354
diacylglycerol   9–11, 53, 128, 162, 242
diazepam   79, 87
dihydroxyphenylacetic acid   64
dihydroxyphenylalanine   62, 71, 134
diuresis   169, 172–173, 309, 352
dog   41, 275, 321–322, 349
DOPA decarboxylase   62
dopamine   12, 19, 52, 59–60, 62–66, 69–73, 115–116, 119, 123, 125–126, 134, 141, 186, 188, 190, 211, 219–220, 225, 233, 245, 264, 270, 281, 283–284, 303, 310, 318–319, 325, 332, 336, 339, 342, 345
dopamine-β-hydroxylase   60, 123, 125–126
dopamine transporter   63–64
down-regulation   13, 131
Drosophila melanogaster   152
dynorphin   196, 199–204, 214, 305–306

e

Ehrlich   16
eicosanoids   152, 205–208
Electrochemical Detection   24
electron microscopy   21, 35
eledoisin   334
embryogenesis   105, 225
eminentia mediana   145–146, 213, 240, 306
endocannabinoid   151–153, 156
endocytosis   13
endopeptidase   159, 263, 281, 307, 336, 344, 372
endoplasmic reticulum   11, 170, 288, 299
endorphin   70, 196–197, 211–212, 214, 219, 315, 318
endosome   13
endothelium   17
enkephalin   18, 194, 200–202, 216–220, 245, 281
enkephalinase   202, 219, 336
entopeduncular nucleus   75, 153

epilepsy   77, 86–87, 106, 113, 131, 204, 259–260, 277, 293, 332
epinephrine   59, 62, 115, 123, 125–126, 129, 144, 196–197, 281, 313
epoxygenase   207
EPSP   45, 93, 293
Escherichia coli   178
estrogen   230, 348, 363
ethanol   112, 214, 283
excitatory amino acid transporters   92
excitotoxity   14
exocytosis   11, 17, 49, 295, 299, 301, 330
extracellular recording   37–38

f

Falck-Hillarp method   60, 123
fatty acids   7, 205
fear   41, 73, 147, 177, 197, 276, 309, 363
fear conditioning   41
fever   120, 190, 208, 254, 271, 302, 361
fibroblast growth factor   221
fibroblasts   257
fishes   178, 235, 312, 328
follicle-stimulating hormone   5, 240, 243, 362, 370
formatio reticularis   123–124
free radicals   12, 106, 207, 302
fura-2   296

g

G protein-coupled receptors   10
G proteins   9–10, 13, 52, 54–55, 67, 69, 93, 96, 102, 119, 127, 129, 137, 152, 162, 202, 208, 232, 275, 292, 329, 352, 359
G substrate   12
GABA   3, 14, 47, 49, 52, 73, 75–80, 82, 84–88, 90, 92, 108–109, 112, 186, 196, 245, 273, 281, 297, 309, 319, 325, 328, 335, 342, 350, 370, 372
GABA transaminase   77, 372
GABA transporters   77, 86
GABA$_A$ receptors   27, 78–80, 84–87, 112
GABA$_B$ receptor   81
GABA$_C$ receptor   84
galanin   229–234, 243, 245, 350
galanin message-associated peptide   229
galanin-like peptide   231
gap junction   1, 19, 35, 155
gaseous molecules   45, 143
gastrin   175, 184, 187–188, 332
gastrin-releasing peptide   175
gastrointestinal tract   85, 152, 175, 179, 212–213, 250, 326, 335, 341

gene transcription   8
gephyrin   111–112
ghrelin   12, 235–239
GHRH   144, 192, 233, 244–248, 311–312, 349, 370
glia   2, 77–78, 92, 100, 102, 118, 129, 137, 206, 259, 271, 336, 353
glioma   188
gliosis   260
Gln14-ghrelin   236
globus pallidus   75, 136, 153, 201, 217–218
glucagon   192, 197, 244–245, 269, 311, 331, 349
glucocorticoid   146–147, 198, 268
glutamate   3, 14, 52, 73, 76–77, 85, 90–93, 96, 98–106, 108, 112–113, 130, 153, 227, 259, 263, 301, 303, 309, 319, 324–325, 371
glutamate decarboxylase   76
glutamine   92, 100, 343
gluthathione   109
glycemia   197
glycine   15, 19, 26, 50, 97, 100, 105–114, 219, 263, 343, 371–372
glycogenolysis   130, 353
glycoprotein   50, 64, 110, 118, 256, 344
glycosylation   10, 98, 232, 344
GnRH   18, 196, 233, 240–243, 313, 370–371
GnRH-associated peptide   240–241
Golgi apparatus   1, 13, 144, 329
Golgi technique   34
gonadoliberin   240
gonadotropin-releasing hormone   18, 196, 209, 233, 241
gonadotropins   139, 243, 276, 313
gp130   256–257
GPCrs   367
GPR54   366
growth hormone-releasing hormone   144, 192, 311, 349
GTP-binding subunit   10
guanosine   12, 172, 295
guanosine monophosphate   12, 172, 295
guanylate cyclase   11–12, 171–172, 299, 301
guinea-pig   321

h

H1 receptor   118, 120–121
H2 receptor   118, 120
H3 receptor   119, 121
5-HPETE   206
5-HT receptors   135–136, 138

5-HT1 receptors   135–137
5-HT2 receptor   135, 137, 139, 206
5-HT3 receptors   137
5-HT4 receptors   138
5-HT5A receptor   138
5-HT6 receptor   138
5-HT7 receptor   138
5-hydroxyindolic acid   135
5-hydroxytryptophan   134, 139
12-HETE   206
12-HPETE   206, 208
15-HETE   207
habenula   47, 101, 138, 171, 176, 249, 336, 338, 345, 356
half-life   7, 49, 115, 117, 131, 144, 171, 298, 301
headache   140, 183, 354
heart   56, 130, 152, 158, 173, 179, 197, 268, 309, 313
heme oxygenase   296–298, 302–303
heparan sulfate   223–224
heroin   216
heteroreceptors   9
hindbrain   68, 137, 169, 176, 222, 245, 322
hippocampus   47, 52, 55–57, 68–69, 75, 85, 91, 98, 100–105, 112, 115, 118–119, 124, 131, 133, 136, 138, 146, 152, 171–172, 177, 180–181, 185–188, 190, 195–196, 201, 203–204, 206, 208, 212–213, 217, 221–223, 225–228, 231–234, 238, 240, 242, 255–259, 264, 267–269, 273, 276–277, 281, 283, 287–288, 293, 297, 301–303, 306, 308, 312–313, 322–324, 328, 332, 336, 338, 342, 344, 346, 350, 360–361
histamine   45, 114–117, 119–121, 233, 277, 282, 339, 342
histidine   116, 120, 351, 371
homeostasis   2, 5, 146, 158–159, 167, 172–173, 178, 182, 267, 270, 314, 353, 356, 362
homocysteate   96
homocysteic acid   90
homovanillic acid   64
horseradish peroxidase   32, 35
HPLC   25, 27–28
human   11, 67–69, 72, 111, 114, 119, 128, 131, 136–139, 153–154, 170–171, 176, 180, 186, 188, 190, 194, 221–223, 231–232, 236–239, 241, 245–246, 250–252, 255, 262, 268–269, 274–275, 277, 282–283, 293, 306–307, 312, 316, 321–322, 330–331, 337, 339, 343, 347, 349, 351, 356, 358–359

hyperekplexia   113
hyperglycemia   269–270, 283
hyperlocomotion   220
hyperphagia   270
hypertension   120, 147, 182
hypocretin   12, 122, 248
hypoglycemia   106, 271
hypotension   173, 182, 211, 214, 266, 283
hypothalamus   47, 52, 60, 68, 70, 91, 98, 115–116, 119–120, 124, 131, 136, 138–139, 145–146, 148, 159, 162, 166, 169, 171–172, 179–180, 185, 187–189, 193–195, 200–202, 209, 213, 217–218, 221–222, 229, 231–233, 235–236, 238, 240, 243–245, 247, 249, 251–252, 255–258, 262–263, 267–269, 273, 276–277, 279–282, 287–288, 297, 306, 308–309, 311–312, 316, 319, 327–328, 332, 336, 342–343, 345, 347, 353, 355–356, 358–363
hypothermia   120, 155, 190, 279, 283
hypoxia   320, 325

i

IL-1ra   256
immunohistochemistry   21, 31–35, 75, 145, 222, 236, 249, 280, 297, 306
inositol-trisphosphat   10, 11, 276
in situ hybridization   15, 31, 33–35, 75, 101, 138–139, 175, 223, 231–232, 263, 280, 297, 306, 308, 350
instrumental conditioning   42
insulin   12, 134, 163, 269, 313, 331
interferon   156, 255–256, 259
interferon-gamma-inducing factor   255
interleukin   12, 156, 171, 209, 254–256, 258, 260
internalization   13, 128, 172, 202, 212
intracellular recording   38–39
in vivo recording   36
ion channel   2, 9–10, 45, 49–51, 84, 88, 93, 96–99, 101, 110, 112, 137, 208, 212, 276, 322–323
ionotropic receptors   9
ischemia   106, 166–167, 173, 259, 293, 302–303, 314, 320, 325
isocitrate   109
isocitrate lyase   109
isoform   68, 98, 111, 119, 206, 209, 245, 256, 290–291, 297–298, 336
isoleucine   134, 263, 351, 371
isoreceptors   8

j

JAK kinase 257
joining peptide 315

k

kainate 93, 98, 101–102, 208, 227–228, 332
kassinin 334
kindling 277, 332
kisspeptin/metastin 366
Krebs cycle 77, 92

l

L-histidine-decarboxylase 116, 372
learning 1, 41–42, 57, 69, 85, 103–105, 119, 139, 147, 159, 166, 190, 196, 201, 203, 208, 221, 226, 233, 259, 264, 266, 269, 276, 293, 296, 303, 309, 324, 361
Leu-enkephalin 18, 201–202, 216, 218–219
leucine 134, 201
leukocytes 119, 257
leukotriene 209
levocabastine 282
Lewy bodies 71, 303
LHRH 240, 276, 371
lipopolysaccharides 255, 258
litorin 175
locus coeruleus 15, 47, 57, 123–124, 129–130, 169, 193, 195–196, 213, 221, 229, 232, 249, 274, 280, 287–288, 290, 308, 316, 327, 336, 356
long-acting natriuretic peptide 171
long-term depression 104, 153, 301
long-term potentiation 3, 85, 103–104, 119, 166, 177, 203, 207, 226
luteinizing hormone 196, 209, 240, 243, 258, 264, 362, 371
luteinizing hormone-releasing hormone 240
lymphocytes 156, 180, 198, 214, 242, 256, 258, 313, 339
lysergic acid diethylamide 121

m

macrophages 149, 155–156, 256, 258, 295, 323–324, 339
mast cell 115, 119, 155, 323, 339
MCAO 167, 227
median eminence 17, 60, 159, 169, 192, 222, 231, 241, 245, 280, 312, 316, 327–328
medulla oblongata 109, 146, 159, 166, 188, 233, 238, 335–336, 342
melanin-concentrating hormone 261, 263
melanocortin 266, 268–271
melanocyte-stimulating hormone 145, 238, 266, 315, 347
melanocyte-stimulating hormone release-inhibiting hormones 347
melanophores 261
melanostatins 347
Memantine 107
membrane potential 9, 38–39, 84, 113, 302
memory 1, 41, 43, 57, 85, 103–105, 107, 119, 139, 155, 159, 166, 177, 190, 196, 201, 208, 221, 226, 233, 259, 264, 266, 269, 276, 293, 296, 301, 309, 324, 360–361
mesencephalon 68, 115, 181–182, 212, 225, 263
mesocortical 62, 70
mesolimbic 62, 70, 73, 190, 264, 284, 345
Met-enkephalin 18, 216, 218–219
metabotropic receptors 10
metalloendopeptidase 202, 281
metastin 12
metencephalon 212, 225
methylene blue 16
microdialysis 24–26, 156
microglia 2–3, 209, 257–258, 322, 324, 339
midbrain 71, 182, 222, 227, 348
MIF-1 18, 347–348, 371
mineralocorticoid 146
mitochondria 1, 48, 77, 109, 348
mitogen-activated protein kinase 129
monoamine oxidase 63–65, 72, 117, 126, 131
morphine 189, 199, 211–212, 214, 216, 219, 348
Morris water maze 43, 233, 309
motilin 236–237
MPTP 71–72, 227, 310
MSH 17, 70, 145–146, 245, 261, 263–264, 266–271, 315, 318–319, 371
multi-unit recording 36
muscarinic 47, 49, 51–57
muscimol 79, 85
Myastenia gravis 58
myelencephalon 225
myelin 2, 292

n

N-acetylaspartylglutamate 90
N-arachidonyl-phosphatidyl-ethanolamine 150, 207
N-methyltransferase 117, 125–126
N-palmitoyl-ethanolamine 152

NADPH-diaphorase 296
NAPE 150, 207
narcolepsy 249, 252
natriuresis 169, 172–173
neocortex 100–101, 172, 193, 200, 212–213, 283, 287–288
neonatal 111, 162
neostriatum 61, 182, 195, 288, 301, 342
nerve growth factor 62, 286
neuroblastoma 37, 165
neurogenesis 146, 169, 182, 221, 225, 270, 301, 353
neurokinin 334–336, 338–339
neuromedin B 175
neuromedin K 334
neuromedin L 334
neuromedin N 281, 283
neuropeptide B 12
neuropeptide FF 189, 367
neuropeptide γ 334–335, 338
neuropeptide glutamate-isoleucine 263
neuropeptide glycine-glutamate 263
neuropeptide K 334–335, 338
neuropeptide S 12
neuropeptide W 12
neuropeptide Y 238, 272, 274, 319, 370
neurophysin 358
neuroprotective 57, 106, 167, 174, 227–228, 234, 301, 310, 314, 320, 325, 353–354
neurosteroids 80, 88
neurotensin 194, 279–284
neurotrophin 286–287, 289–293
neurotrophin-3 286–287
NGF 62, 126, 286–293, 336, 371
nicotinic 47, 49–52, 55, 57–58, 98–99
Niemann–Pick 89
nigro-striatal 62, 64, 70, 86
Nissl-staining 34
nitric oxide 7, 12, 45, 143–144, 166, 208, 295–303, 321, 352
nitric oxide synthase 295–296, 298, 301, 303
nitroblue tetrazolium 297
NMDA 85, 93, 96–100, 102, 104–107, 109, 112–113, 117, 119, 204, 206, 208, 227, 296, 298, 300, 302–303, 309, 338, 371
nociceptin 12, 231, 305–310
nociception 139, 156, 188–189, 203, 212, 219, 284, 296, 308, 324
Nogo 292
norepinephrine 15, 52, 59–60, 62, 65, 72–73, 115, 119, 123–127, 129–131, 140, 144, 165, 186, 196–197, 219, 233, 264, 276–277, 281, 325, 345, 353, 371
norepinephrine transporter 126, 264
NPY 124, 129, 180, 238, 243, 245, 270, 272–278, 297, 319, 328, 342, 371
nucleotide 12, 33, 108, 176, 186, 293, 321–322
nucleus accumbens 52, 55, 62, 69, 73, 91, 100, 103, 115, 138, 153, 180, 185, 187–189, 195, 201, 212–213, 263, 269, 280–281, 309, 322, 327, 336, 345, 360
nucleus accumbens 91
nucleus basalis of Meynert 47, 57, 221

O

6-OHDA 71–72, 310
olfaction 213, 296, 302
olfactory bulb 47, 55, 60, 98, 102–103, 109, 138, 153, 169, 171, 175, 188, 222, 232, 240, 249, 274, 283, 297, 302, 336, 342, 345, 356
olfactory tract 146, 287–288
olfactory tubercle 55, 62, 69, 103, 138, 185, 187, 262–263, 322
oligodendrocytes 2, 209, 222, 322
operant conditioning 41–42
orexin 122, 238, 248–253
organum vasculosum of the lamina terminalis 17, 171, 240, 361
orphanin FQ 12, 305
osteoblasts 257
osteoclasts 182
osteoporosis 182
overexcitation 14
oxytocin 70, 196, 355–356, 358–363

P

PACAP 192, 244, 311–314, 349, 351, 353, 371
PACAP-related protein 312, 371
pain 107, 140, 156, 183, 204, 208–209, 212, 214, 216, 219, 233, 309, 324–325, 338, 348, 354
pancreas 158, 181, 188, 238, 240, 250, 267, 273–274, 313, 327, 329, 331, 341
pancreatic polypeptide 272, 275
pancreozymin 184
Papaver somniferum 216
parabrachial nucleus 57, 175, 193, 213, 217, 232, 250, 358
paracrine messenger 295
parathormone 182
parenchyma 16–18, 165, 348

patch-clamp 39–40
pentagastrin 184, 187
pepsin 332, 346
peptide transport system 18
peptide YY 272
periaqueductal gray 60, 145, 156, 186, 200, 203, 217, 268, 280, 283, 306, 308, 327, 335, 342, 350, 360
periglomerular cells 60
peroxynitrite 298, 303
pertussis toxin 53, 128–129, 155, 232, 344
PHA-L 34
phenylalanine 62, 134, 187, 219
phenylalanine hydroxylase 62
phenylethylamine-N-methyl-transferase 125
PHI 351, 371
PHM 351, 371
phosphatidyl-4,5-bisphosphate 11
phospholipase A 54, 206, 337
phospholipase C 10, 53, 102, 118, 128, 137, 153, 162, 188, 231–232, 237, 242, 276, 282, 321, 337, 344
phospholipase D 54, 150, 207, 324
phosphorylation 9–10, 12–13, 98, 111, 125, 129, 195, 225, 232, 257, 268, 288, 301, 337
physalaemin 334
picrotoxin 84–85
pineal gland 17
pituitary 5, 60, 70, 84, 104, 120, 139–140, 144–145, 147–148, 158, 170, 172, 180–181, 185, 192, 195–197, 200–201, 203, 208–209, 211, 213–214, 218, 222, 231, 233, 235, 238–240, 242–244, 246–247, 256, 261, 267, 269, 276, 280, 297, 311–313, 315–316, 318–319, 326–327, 330–331, 341, 343–345, 348–349, 351, 353, 355–356, 363
placenta 158, 240, 268, 341
plasminogen activator inhibitor-1 163
polymorphism 167, 190, 260, 293
pons 47, 91, 109, 146, 180, 208, 222, 238, 250, 297, 312, 335, 348
post-proline-cleavage enzyme 242, 371
postsynaptic 1–2, 4–5, 8–9, 11, 15, 37–38, 45, 49, 57, 64, 78, 81–82, 84, 91, 93, 102–104, 111–112, 115, 126, 129, 131, 152, 188, 284, 301, 309
preoptic area 146, 180, 189, 193, 196, 208, 213, 221, 240, 264, 267, 280–281, 316, 328, 345
prepro-GHRH 245
preproorexin 250–251
presynaptic 1, 4–5, 8–9, 15, 45, 52, 62–66, 81–82, 92–93, 102, 104, 110, 115, 119, 129, 131, 152, 188, 295, 301, 303, 322, 325
pro-enkephalin A 218
pro-enkephalin B 200–201, 218
proconvertase 263
prodynorphin 200–202
profilin 112
progesterone 80, 348, 363
prolactin 12, 70, 120, 139, 211, 214, 233, 238, 240, 343, 345, 353, 360, 362
proline 162, 242, 290, 343–344, 371
proline endopeptidase 344
proNGF 292
proopiomelanocortin 144–145, 196, 211, 266–267, 306, 315–316
prostacyclin 205–206
prostaglandine 206, 208
protein kinase 9–13, 54, 98, 102, 129, 172, 207–208, 237, 246, 268, 290, 299–300, 303
protein kinase C 320
proteomics 28
protirelin 341
Ps4 343
Ps5 343
pseudogene 317
purines 320, 324
purinoceptors 321–324
Purkinje cell 89, 102, 118, 256, 283, 323
pyridoxal phosphate 76
pyroglutamate aminopeptidase 242
pyruvate 48

q
quisqualate 93, 99, 102, 227

r
rabbit 275, 322
radioimmunoassay 23, 192
radioisotope assays 23
ranatensin 175
raphe 47, 133, 136, 139, 175, 193, 195, 229, 238, 249–250, 268, 283, 287, 306, 308, 336, 342, 350, 358, 360
rat 41–42, 59, 68, 111, 115, 119, 124, 138, 153–154, 162, 171, 176–177, 188–189, 195–196, 222–223, 226, 231–232, 241, 245, 249, 255, 257, 264, 269, 275, 282–283, 288, 293, 298, 307–308, 321–322, 337, 343–345, 349, 353, 360
re-uptake 4, 14, 45–46, 48, 64, 92, 126, 131, 135, 140, 143, 146, 150–151, 208, 295, 298, 339

receptor activity modifying proteins   181
redistribution   13
REM sleep   252, 353
renin   158–160, 165, 167, 173, 258
renin–angiotensin system   158–159, 165, 167, 258
Renshaw cell   47, 339
respiration   203, 213, 216, 338
retina   19, 60, 84, 103, 109, 111, 158, 329
retrograde messenger   207–208, 296, 301, 303
retrograde signaling   152
RNA editing   100–101

s

salt appetite   139, 165
satiety   189
Scatchard diagram   23
Schaffer collateral   104
schizophrenia   60, 70, 73, 106, 121, 190, 220, 278, 284
scopolamine   50, 52, 57
SDHACU   48
second messengers   8
secretin   244–246, 311, 349, 352
septum   85, 91, 100, 102–103, 124, 138, 146, 185, 194–195, 221, 229, 267–269, 280, 335–336, 345, 356, 358–361, 363
serine   10, 102, 105, 107, 109, 128, 162, 187, 237–238, 245, 290
serine hydroxymethyltransferase   109
serine protease   187
serotonin   12, 52, 60, 73, 115, 119, 132–135, 137, 139–141, 165, 186, 196, 233, 319, 325, 332, 335, 339, 342, 345, 361, 370
serotonin transporter   135
sexual behavior   139, 147, 159, 165, 361–362
sexual dimorphism   139, 147, 230
sheep   194, 269, 322, 349, 362
signal propagation   9
signal transduction   8
silver-staining   34
single unit recording   36
skeleton   182
skin   147, 175, 261, 267, 269
sleep   87, 120–121, 139, 197, 247, 249, 252, 264, 325, 353
slice preparation   37–38, 40
somatoliberin   244
somatostatin   180, 186, 196, 212, 231, 238, 245–247, 263, 273, 326–333, 335

somatotropin releasing-inhibiting factor   326
spinal cord   15, 47, 55, 68, 85–86, 92, 108–109, 112–114, 124, 133, 137, 169–170, 175, 179–180, 185, 187–189, 200, 202–203, 208, 213, 217, 222, 230, 232–233, 267, 269, 274, 280, 282, 288, 301, 308, 313, 316, 323–325, 327–328, 335, 337–339, 341–342, 345–346, 353, 358, 360
splicing   68, 79, 98–99, 111, 119, 138, 180, 245, 257, 288, 307, 313, 317, 344
startle disease   113
steroid   88, 144, 208, 363
stomach   115, 188, 235–238, 268–269, 313, 329
stress   73, 85, 106–107, 120, 126, 130–131, 147, 159, 173, 182, 192, 195–197, 204, 214, 219, 263, 265, 278, 283, 293, 302–303, 309, 319, 345, 348, 362
striatum   19, 47, 52, 55, 57, 60, 62, 67–72, 75, 100, 105, 131, 136, 138, 190, 202, 220–222, 255, 257, 269, 273, 284, 287, 308, 322, 325, 328, 335–337, 339, 348
strychnine   105, 107, 110–113
Strychnos nux vomica   113
subcommissural organ   17
subfornical organ   17, 159, 165, 169, 171, 360
substance K   334
substance P   166, 180, 182, 186, 245, 334–336, 338–339, 342
substantia gelatinosa   133, 188
substantia innominata   47, 159, 249
substantia nigra   47, 60–61, 69–71, 75, 91, 130, 141, 153, 167, 175, 180, 185–186, 188, 190, 201, 217, 219–223, 238, 250, 263, 269, 280–284, 287–288, 290, 293, 303, 308–310, 316, 335, 339, 342, 360
subventricular zone   145, 225
succinic acid semialdehyde   77
superoxide   298
supraoptic nucleus   102, 172, 181, 200, 218, 222, 283, 297, 356
synaptic cleft   1, 4, 9, 19, 45, 48–49, 64, 70, 109, 126, 344
synaptic transmission   19, 45–46, 75, 99, 103, 108, 112, 143, 264, 339
synaptogenesis   97, 337
syncytium   19
synenkephalin   219

t

tachykinin   334, 336–339
taurine   112

tectum opticum 133
tertiary topology 8
testis 267–268, 313
tetanus toxin 49, 86, 113
tetrahydrobiopterin 62, 125
thalamus 47, 52, 55, 57, 68–69, 75, 82, 101–103, 119–120, 124, 131, 136, 138, 145–146, 159, 185, 212–213, 221–222, 232, 249, 263, 268, 280, 288, 297, 308, 312–313, 316, 322, 341–342, 350, 359
THC 149–150, 156
thermoregulation 270
thioperamide 121
thirst 172, 265, 338, 361
thromboxane 206, 209
thyreoliberin 341
thyroid 140, 178–179, 183, 238, 341, 345
thyroid-stimulating hormone 341
thyrotropin 341
thyrotropin-releasing hormone 341
thyroxin 341, 345, 348
tight junction 17
TNF-α 156, 254, 259, 339
TNF-β 171, 254
Torpedo mamorata 50
tracing techniques 34
tractus solitarius 60, 123, 169, 188, 195, 212–213, 229, 269, 274, 280, 287, 306, 316, 328, 336, 360
trancytosis 17
transferrin 17
transforming growth factor 171
transporter 14, 17, 48, 63–64, 77, 92, 109, 134–135, 227, 372
tremor 70, 73, 156
trigeminal ganglia 179, 323
trisomy 21 260
tryptophan 133–134
tryptophan hydroxylase 134
tuberomammillary nucleus 115–116, 120–121
tumor 156, 180, 239, 247, 254, 289, 332
tumor necrosis factors 254
Tyr-MIF-1 18, 347–348, 371
Tyr-W-MIF-1 347–348, 371

tyrosine 9–12, 62, 123–126, 130, 162, 196, 219, 224, 257, 272, 289–290, 330, 344, 371
tyrosine hydroxylase 62, 123, 125–126, 130, 196, 290
tyrosine kinase 9, 11–12, 162, 224, 257, 289–290

u

up-regulation 13, 86, 126, 190, 204, 242, 258, 264
urocortin 194–195, 197

v

valine 134
vasoactive intestinal peptide 126, 144
vasoconstriction 129–130, 132, 165–166, 277, 345
vasodilatation 130, 165, 173, 182–183, 283, 338
vasopressin 70, 144, 165–166, 172–173, 186, 192, 196, 276, 302, 355–356, 358–363, 370
ventral tegmental area 61, 175, 185–186, 195, 221, 238, 240, 250, 263, 268, 280, 282–283, 287, 335, 339
vesicle 11, 64
VIP 62, 126, 130, 144, 192, 194, 244–246, 311–313, 349–354, 362, 372
Voltammetry 25
volume transmission 18–19, 202

w

wakefulness 120–121, 249, 296
Werman 4
wiring transmission 18–19, 202

x

xanthine 322
Xenopus 98–99, 221, 316

z

zinc 112
zona incerta 60, 70, 249, 262, 265, 280